Third Edition

Pile Foundation

Design and Construction

Including Well Foundation

a useful book for

- Students
- Architects
- Practicing Engineers and
- Consultants

Dr. Satyendra Mittal

BE (Civil), ME (Soil Dyn.) PhD, MIGS, FIE, MISET, MIWRS, MISSMFE, MISTE

Pro Vice-Chancellor
University of Patanjali, Haridwar

Ex Professor
Department of Civil Engineering
Indian Institute of Technology
Roorkee 247667
India

CBSPD

CBS Publishers & Distributors Pvt Ltd

New Delhi • Bengaluru • Chennai • Kochi • Kolkata • Lucknow • Mumbai
Hyderabad • Jharkhand • Nagpur • Patna • Pune • Uttarakhand

Disclaimer

Science and technology are constantly changing fields. New research and experience broaden the scope of information and knowledge. The author has tried his best in giving information available to him while preparing the material for this book. Although, all efforts have been made to ensure optimum accuracy of the material, yet it is quite possible some errors might have been left uncorrected. The publisher, printer and the author will not be held responsible for any inadvertent errors or inaccuracies.

ISBN: 978-93-5466-722-0

Third Edition: 2025

First Edition: 1988

Reprint: 1993, 1997, 1999, 2008, 2009, 2010, 2012, 2013, 2014, 2016

Second Edition: 2017

Reprint: 2019

Published by **Satish Kumar Jain** and produced by **Varun Jain** for

CBS Publishers & Distributors Pvt Ltd

4819/XI Prahlad Street, 24 Ansari Road, Daryaganj, New Delhi 110 002, India.

Ph: 011-23266838, 23289259

Website: www.cbspd.com

e-mail: delhi@cbspd.com

Corporate Office: 204 FIE, Industrial Area, Patparganj, Delhi 110 092

Ph: 011-4934 4934 Fax: 011-4934 4935

e-mail: publishing@cbspd.com; publicity@cbspd.com

Branches

- **Bengaluru:** Seema House 2975, 17th Cross, K.R. Road, Banasankari 2nd Stage, Bengaluru 560 070, Karnataka, India.
  Ph: +91-80-26771678/79 Fax: +91-80-26771680 e-mail: bangalore@cbspd.com
- **Chennai:** 18/8B, Subbarayan Street, Shenoy Nagar, Chennai 600 030, Tamil Nadu, India
  Ph: +91-044-42032115, 26681266 e-mail: chennai@cbspd.com
- **Kochi:** 42/1325, 1326, Power House Road, Opp. KSEB Power House, Ernakulam 682 018, Kochi, Kerala, India.
  Ph: +91-484-4059061-65,67 Fax: +91-484-4059065 e-mail: kochi@cbspd.com
- **Kolkata:** 147, Hind Ceramics Compound, 1st Floor, Nilgunj Road, Belghoria, Kolkata 700056, West Bengal, India
  Ph: +033-25633055, 25633056 e-mail: kolkata@cbspd.com
- **Lucknow:** Basement, Khushnuma Complex, 7 Meerabai Marg (Behind Jawahar Bhawan), Lucknow-226001, UP, India
  Ph: +0522-4000032 e-mail: tiwari.lucknow@cbspd.com
- **Mumbai:** PWD Shed, Gala no 25/26, Ramchandra Bhatt Marg, Next to JJ Hospital Gate no. 2, Opp. Union Bank of India, Noorbaug, Mumbai-400009, Maharashtra, India
  Ph: 022-66661880/89 e-mail: mumbai@cbspd.com

Representatives

- **Hyderabad** 0-9885175004
- **Patna** 0-9334159340
- **Jharkhand** 0-9811541605
- **Pune** 0-9664372571
- **Nagpur** 0-8692091830
- **Uttarakhand** 0-9716462459

Printed at: Glorious Printers, Daryaganj, Delhi, India

Third Edition

Pile Foundation

Design and Construction

Including Well Foundation

a useful book for

- Students
- Architects
- Practicing Engineers and
- Consultants

to

Rohan, Rishi, Nandita and Kartikey
with fond memories of
my divine parents

Preface to the Third Edition

The Second edition of this book was published in 2017 which was reprinted in 2019 after an unprecedented response by reader and users. The last edition of this book had been highly beneficial to not only students but also to professional engineers (as per time to time feedback from various people to author).

The present edition removes grammatical errors (which had been left due to oversight in previous edition), besides adding some latest knowledge on pile load test. I am sure, this edition shall also attract new readers and benefit everybody who come across this book. I am thankful to Ms. Nehul Tyagi and Ms. Hemlata Negi for preparing the new additions for this third edition.

Satyendra Mittal

Preface to the First Edition

Soil has been used as a foundation and construction material since the earliest days of recorded history. By the time the scientific method becomes generally recognized as a fruitful approach to the solution of engineering problems, monumental buildings, bridges, dams, etc. have not only been built but some had served their useful purpose for many centuries. It was inevitable, therefore, that earthwork and foundation engineering developed primarily as an art stepped in tradition and empirical practices based on earlier successful accomplishments.

Pile foundation is becoming popular day by day due to its usefulness and economy. It was felt that there is no design book available as on date for design and construction of all kinds of pile foundation. With this end view, the first edition of the book is presented to the readers. Nevertheless, this book attempts to develop a broad picture of the pile foundation which would be useful not only to soil engineering students but also at the interfaces of the specialists involved in relevant decision-making, namely, structural engineers, soil engineering specialists, foundation consultants and architects.

I take this opportunity to thank all my friends who have helped me to prepare this book. Thanks are due to Shri RK Bhatia for preparing all the drawings. Special thanks are due to CBS Publishers and Distributors for accepting my handwritten manuscript and printing the book in such a short period.

The author is grateful to his friends and son Sunny for extending fullest cooperation during writing of the book and assisting in every possible way in its successful completion. Any suggestion to improve the book shall be heartily welcomed.

Satyendra Mittal

Contents

Chapter 1

Introduction

When a good bearing stratum is not available near the ground surface or at relatively shallow depths, the loads of superstructures have to be 'transmitted' to firm strata capable of catering such loads though may be at much depth. Such foundations made are called **deep foundations**. Thus, deep foundations are those foundations where the soil support is found at appreciable depth below the main structure. The deep foundations may be pile foundations or wells and caissons. A pile is relatively small diameter shaft which is driven into the ground or otherwise introduced into the soil by suitable means so as to support all the loads. Wells and caissons are usually installed by excavation of sub-soil. Thus in case of wells and caissons, visual inspection of the firm stratum on which they rest, may be made.

The piles are of two types, i.e.

i. Friction pile
ii. End bearing pile

i. Friction pile

The pile which supports the structure load due to friction between the pile and the neighbouring soil is known as **friction pile** (Fig. 1.1).

ii. Point bearing or end bearing pile

The pile whose lower end rests on a hard strata is called **end bearing pile** (Fig. 1.2).

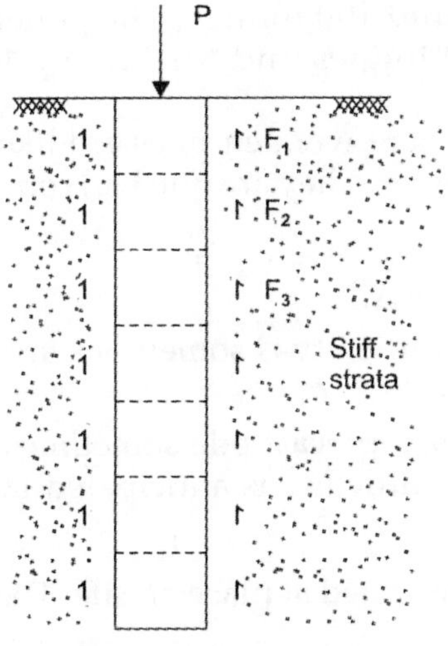

Fig. 1.1: Friction pile

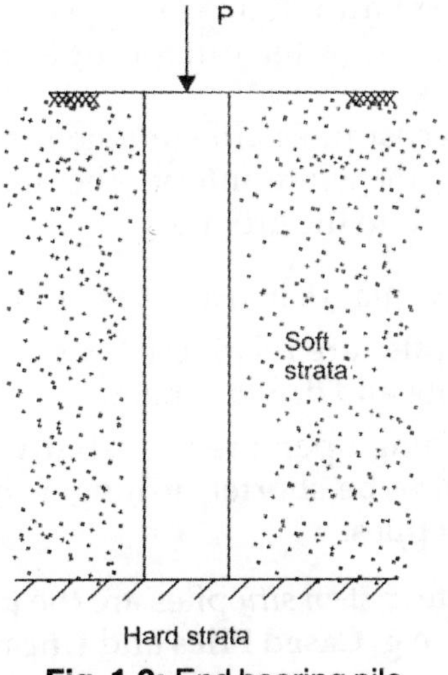

Fig. 1.2: End bearing pile

1.1 KINDS OF PILES

Depending upon the location, requirement, economy, etc. the piles may be of following kinds according to their material compositions:

a. Timber/bamboo piles

b. Concrete piles

c. Steel piles

The uses of different piles may be different. The characteristics and uses of various types of piles are summarized below:

1.1.1 Timber/Bamboo Piles

These are the piles made of tree trunks and logs. They are proved very cheap at places where timber is available in plenty. The lengths of such piles may vary from 2 to 8 m. However, more lengths can also be used with due care in bonds between two consecutive pieces. The precaution must be taken that the wood should be of good quality capable to carry loads without buckling.

The diameter of such piles may vary from 25 to 45 cm and at the tip around 12.5 to 20 cm.

The wood should be applied with a good wood preservative to withstand the alternate change in the climate or surrounding situations. Use of creosote @ 225–250 kg/m^3 of piles in fresh water and the same @ 350 kg/m^3 in saline water is recommended to be used. The piles fully submerged in water last long. The branches off shooting the main stem should be trimmed off before applying the load.

After driving to final depth, the pile head should be made square. Before concrete pouring for the pile cap, the head of the treated pile should be protected by zinc coat, lead paint or by wrapping the pile head with clothes.

The maximum design load per pile in such piles should not exceed 25 tonnes.

1.1.2 Concrete Piles

The concrete piles may be either driven or cast-*in situ* piles. The precast piles are made of uniform section with pointed tips. The square or octagonal sections are normally preferred owing to the ease in casting them. The necessary reinforcement is also provided in such piles.

The maximum load on a precast pile is around 100 tonnes. The general load capacity of smaller piles range from 25 to 70 tonnes and for large piles upto 200 tonnes.

The precast piles are suitable for the places where wooden piles are likely to be weaken by the white ants or other insects and termites. They are quick in construction and laying into the ground.

The disadvantages in using precast piles are as follows:

i. These piles are relatively heavy hence a great difficulty is sometimes involved in handling and driving them.

ii. Depending upon the availability of hard strata, a precast pile sometimes may be proved to be shorter or longer, hence a great difficulty is anticipated in case of precast piles.

Concrete cast-*in situ* piles are the piles which are casted at place on site. They are of two types, e.g. **Cased Piles** and **Uncased Piles.**

a. Cased Piles

Where a thin metal casing is driven first into the ground and subsequently the concrete is poured inside the casing they are called **cased piles**. The casing in such piles may be left in the ground permanently. They are of following types:

i. Raymond piles

In such type of piles, a corrugated thin steel sheet tapered shell is driven into the ground with the collapsible steel mandrel inside it to a desired depth. The shell diameter is around 20 cm at the tip and has the tapering of 3.35 cm/m length (Fig. 1.3). After driving into the ground, the mandrel is withdrawn and the shell is fied with concrete. The lengths vary between 6 to 15 m for such piles. Such pile is named after the patent licence of the pile with M/s Raymond Concrete Pile Co., USA.

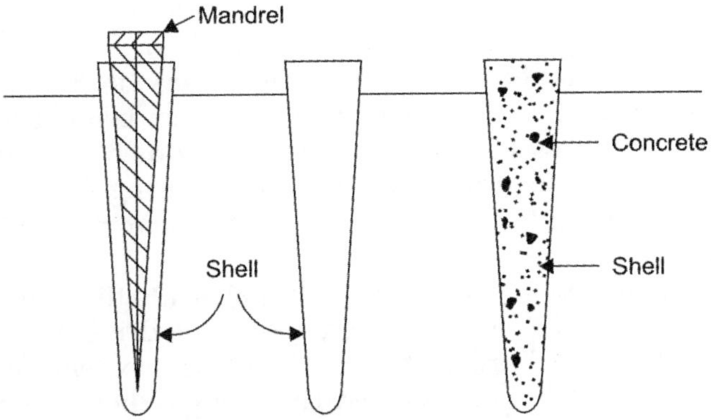

Fig. 1.3: Raymond piles

ii. Mac-arthur pile

In such piles heavy steel casing is driven into the ground (with a core inside it) to a desired depth. The core is then pulled out and a corrugated steel shell is inserted. Finally, cement concrete is poured and outer casing is withdrawn.

iii. Pipe or tube pile

In this type of piles, a tapered steel shell without any mandrel is used as casing. The dia of the pipe is 20 to 50 cm with thickness as 0.5 to 1.25 cm. The pipe is closed at the bottom. The pipe section is filled with concrete. If the pipe is driven with its bottom open, the soil inside the pipe is usually jetted out with air and water under pressure.

b. Uncased Piles

In such piles, a thin or relatively thick metal casing is driven first into the ground. The casing is then taken out gradually and concrete thus makes bond with the neighbouring soil.

The uncased cast-in-situ concrete piles are of the following types:

i. Simplex piles

In such piles, a hollow cylindrical steel casing with one pointed iron shoe attached to its bottom is driven into the ground to the desired depth. Concrete is then poured into the casing pipe for a depth of about one metre and then it is compacted. The casing pipe is then lifted up in steps. In this way the entire length of hole is filled up with compacted concrete which is a pile (Fig. 1.4).

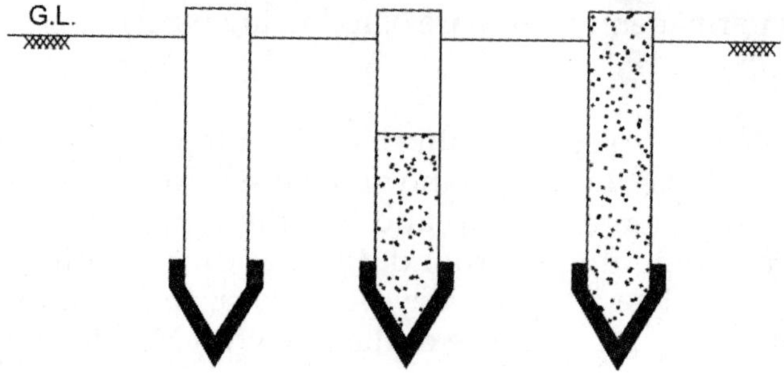

Fig. 1.4: Simplex piles

ii. Franki piles

In such piles, a steel casing of 50 cm diameter is driven into the ground for a small depth and cleaned of soil. Then the concrete (of low water-cement ratio) is placed and rammed. This concrete forms a solid plug at the end which grips the casing pipe so tightly so as to become an integral part of the casing during the driving process. The ram works inside the casing to fall from a height of 3 to 6 m on to the plug. Fig. 1.5 shows the mechanism of Franki piles.

The pipe due to impact of the falling ram penetrates into the ground. When the pile reaches its final elevation, the casing is anchored to the driving rig and the concrete plug is forced out of the pipe. Further the ramming of this expelled concrete is done to obtain a bulb of approximately 1 m diameter. The casing is then raised while successive charges of concrete are rammed in place to form as rough a surface of the finished pile as possible. Length of such piles varies up to 30 m and it is designed to cater the loads up to 100 tonnes.

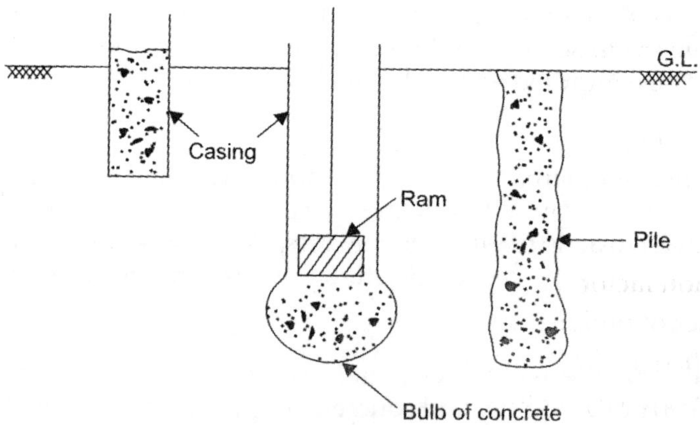

Fig. 1.5: Franki piles

iii. Vibro piles

A hollow steel tube with a cast iron shoe at its bottom is driven into the ground to the desired depth. Then the reinforcement is placed in position. The tube is then filled with concrete and extracting links are fitted to it. The extraction of the tube and formation of the concrete pile are affected by upward and downward movement of the hammer.

The shoe is left into the ground only. The section of the pile is enlarged at the base and at other location by suitably ramming the freshly laid concrete.

iv. Bored piles

They consist of holes made with auger filled with concrete. Such piles can be used in relatively firm ground. Under-reamed piles are the special type of bored piles. In such piles, a hole is bored by means of an auger. After digging the hole a under-reamer is lowered into the hole. It is then pressed down and rotated. By this, the blades get widen due to pressing and they cut the soil from the sides by rotations. A bucket is attached under the under-reamer. When this bucket is full, the under-reamer is pulled out and the soil is removed. Again the under-reamer is lowered and the soil is removed and thus a bulb is made at the required depth. Then the reinforcement is lowered into the hole and then concrete is poured in.

Such piles are very useful and economical in expansive soils like black-cotton soils. Such soil is mostly achieved in Madhya Pradesh in India. The usual size of these piles is 15 to 25 cm in diameter and 3 to 4 m long. The diameter of the bulb is 2 to 3 times the diameter of the piles.

v. Pressure piles

The pile which is made by pouring cement concrete inside a casing pipe in layers of 30 cm and each layer compacted by compressed air is called a **pressure pile.** While constructing this pile, the casing pipe is gradually lifted and the pile is thus completely casted. Such piles can resist shocks and vibrations to a greater extent.

vi. Under-reamed pile

Under-reamed piles are cast-in situ concrete piles which have bulb (called **under-ream**) on its periphery (Fig. 1.6). These piles are often used as load bearing and anchor piles and are suitable in all type of soils. For clayey soils or expansive soils, such piles are a boon. A single under-reamed pile is suitable for anchor pile, while double under-reamed piles are used to increase the load carrying capacity of the pile. The design of such piles is given in subsequent chapters of this book. The approx capacity of under-reamed pile may be computed by following equation also:

$$Q_u = \Sigma \propto C_{u_1} f_s + \frac{\pi d_0^2}{4} C_{u_2} N_c \qquad \text{... (1.1)}$$

where,

C_{u_1} = Undrained shear strength of clay along pile shaft

$\propto$ = Reduction factor

d_0 = Diameter of under-ream

C_{u_2} = Undrained shear strength of clay at pile tip

N_c = Bearing capacity factor.

The specialty of such piles is that for double under-reamed pile, the soil between the bulbs acts as part of the pile and the frictional resistance for the length (L_0) may be calculated for the diameter, d_0. While for rest of the pile length above the under-ream, the skin friction develops over the diameter (d).

Under-reamed piles are normally not recommended for dry cohesionless soils (or collapsible soils), owing to the doubts about formation of bulb in soils, because of no cohesion in such soils.

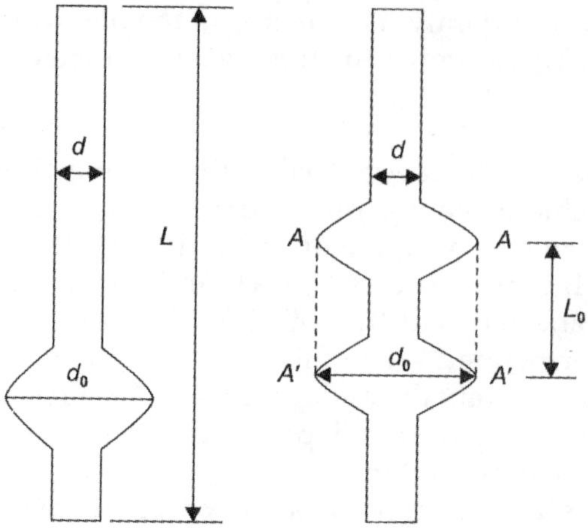

Fig. 1.6: Under-reamed pile

c. Steel Piles

Steel piles are usually rolled H-shapes or pipe piles. H-piles are proportioned to withstand large impact stresses during hard driving (Fig. 1.7). Pipe piles are either welded or seamless steel pipes which are driven either open ended or closed ended. Pipe piles are often filled with concrete after driving, although in some cases it may not be necessary. The load range on steel piles is 40–125 tonnes.

Fig. 1.7: H-Piles (Steel Piles)

1.2 BATTER PILES (INCLINED PILES)

For inclined loads or high lateral loads, piles can be provided at some angle (called **batter**). A maximum batter of 30° is possible for piles under dry ground conditions and around 10–15° for submerged grounds. Batter under-reamed piles (Piles with underground bulb) are used under transmission line tower foundations and as anchors under direct and lateral loads. The batter pile is illustrated in Fig. 1.8.

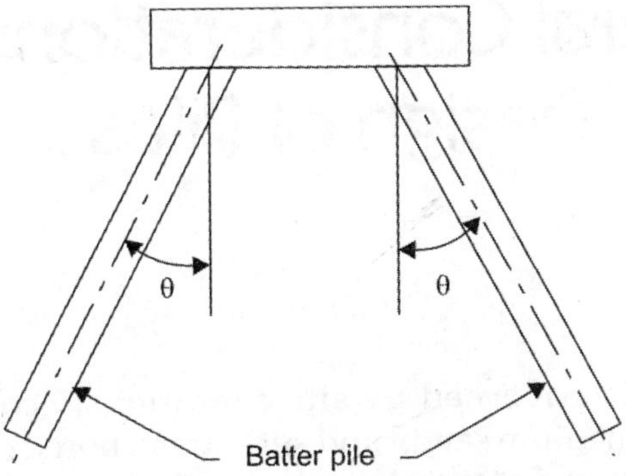

Fig. 1.8: Batter piles

1.3 TEST PILE

It is a pile which is selected for load testing over it. The test pile may become a working pile if it is subjected to routine load test with upto 1.5 times the safe load.

1.4 WORKING PILE

It is the pile which forms the part of the foundation of a structural system.

1.5 TRIAL PILE

The piles which are not working piles are installed initially to assess the load carrying capacity of a pile. They may be one or more in numbers. These piles are tested either to their ultimate load capacity or upto twice the estimated safe load. Such piles are called *Trial piles*.

Chapter

2

General Considerations for Design of Piles

2.1 INTRODUCTION

The choice of the pile is governed by site conditions, economics and time considerations. For loose to medium sandy and silty strata, bored compaction piles should be used since in such piles, the compaction process increases the load bearing capacity of piles. In case of expansive soils, e.g. black cotton soils or filled up soils, under-reamed piles with bulb(s) provide a good anchorage. It is found that provision of bulbs in the under-reamed piles increases the lateral load capacity of piles. For the multi-storeyed apartments, high chimneys or heavier structures, multiunder-reamed piles are generally used. In general, under-reamed piles can be used where the structures are subjected to various loading conditions including those due to wind and seismic forces. Such piles can be constructed at batter also depending upon the requirements (Fig. 2.1).

2.2 DATA FOR PILE DESIGN

Following are the data required for design and construction of a pile.

a. **Soil exploration data:** An extensive field test should be carried out before taking up the pile construction on site. There should be sufficient bore holes made on site to study the soil properties on sites. The laboratory tests should also be carried out to study the permeability, shear and cohesive properties of soil. There should be a few dynamic cone tests as well as standard penetration tests. The boreholes should be sufficiently deep. The extent of soil exploration depends on type of structure, possible loads, layout of structure, economics, etc.

b. **Depth of water table:** The location of water table should be carefully studied and possible variations in water table during rainy and dry seasons should be considered in designing the pile lengths.

c. **Chemical properties of soil:** The ingredients of soil should also be studied. For example there may be the factories waste, human wastes etc. in the soil. Hence chemical properties of such soil should be evaluated and studied so that there is no loss to piles after construction of them.

d. **Nature of soil:** The expansive nature of soil should be checked. This may be checked by the standard differential free swell test described below (Mittal and Shukla, 2013):

   Take two samples of the dried soil weighing approximately 50 g each and passing through 425 micron sieve. Put one sample in a 250 ml graduated glass cylinder containing kerosene oil whereas put other sample in a similar cylinder but containing distilled water. Allow both the samples to settle without disturbance for 24 h and then their volume is noted.

The differential free swell (DFS) is expressed as:

$$DFS = \left[\frac{\text{Soil volume in water} - \text{Soil volume in kerosene}}{\text{Soil volume in kerosene}}\right] \times 100 \qquad \ldots (2.1)$$

The degree of expansiveness of soil may be qualitatively understood from the following Table 2.1.

Table 2.1: Degree of expansiveness of soil

Degree of expansiveness	*DFS(%)*
Low	< 20
Moderate	20–37.5
High	37.5–50
Very high	> 50

For the soils indicating high or very high DFS, the pile foundation is used.

2.3 VARIOUS FACTORS FOR PILE FOUNDATION DESIGN

2.3.1 Spacing of Piles other than Under-reamed Piles

The centre to centre spacing of pile is considered from the following aspects:

1. Practical aspects of installing the piles.
2. The nature of the load transfer to the soil and possible reduction in the bearing capacity of a group of piles thereby. The choice spacing is normally made on semi-empirical approach. In case of foundations of tanks rested on ground, after evaluating the number of piles, the spacing should be such so as the piles are distributed uniformly beneath the tank structure. The piles should be staggered in triangle pattern.
3. In case of piles founded on a very hard stratum and deriving their capacity mainly from end bearing, the spacing will be governed by the competency of the end bearing strata. The minimum spacing in such cases shall be 2.5 times the diameter (circumscribing circle corresponding to the cross-section) of the shaft. In case of piles resting on rock, the spacing of two times the said diameter may be adopted.
4. Piles deriving their bearing capacity mainly from friction shall be sufficiently apart to ensure that the zones of soils from which the piles derive their support do not overlap to such an extent that their bearing values are reduced. Generally, the spacing in such cases shall not be less than three times the diameter of the shaft.
5. In case of loose sand or filling, spacing closer than in dense sand may be provided since displacement during the piling may be absorbed by vertical and horizontal compaction of the strata. Minimum spacing in such strata may be two times the diameter of the shaft.

2.3.2 Spacing of the Under-reamed Piles

The piles are provided in single as well as group. The minimum spacing of piles in a group is kept generally two times the bulb diameter. However, the spacing may be more also (if the site conditions demand so) but increasing spacing needs larger volume of pile cap which may become uneconomical. If the piles have to be provided closer than two times the bulb diameter, a reduction should be applied in the allowable loads. It is estimated that for a spacing of 1.5 times the bulb diameter, a 10 per cent reduction may be made. In case of piles of two different diameters kept adjacent, the spacing considerations are the same but the average bulb diameter may be considered for calculation.

In expansive soils, where grade beams are clear of the ground, the positioning of piles in relation to plinth plan is important. Piles are essentially provided on the corners of the buildings and wall junctions. For intermediate piles, they should be arranged in such a way that the doors and windows openings lie centrally as far as possible. For the structures with columns, e.g. multistoried complex, etc. the piles should first be laid for columns and then for walls. The maximum spacing between two piles in a beam and pile construction should not normally exceed 3.0 m.

2.3.3 Length of Piles

The depth or length of pile is governed by the soil condition and position of water table. By conducting the extensive soil test at various points on the site where the foundation work is to be carried out, the soil profile can be known. For example, by conducting standard penetration tests, the location of water table and type of soil strata can be known. The foundation should be rested only on firm strata and keeping this in view, the length of pile can be ascertained.

2.3.4 Length of Under-reamed Piles

The depth or length of under-reamed pile is governed by the soil data and depth of water table. For expansive soils, the zone of ground movement should be established carefully to select the pile length. In case of deep deposits of expansive soils, the minimum length of pile should not be less deposits of expansive soils, the minimum length of pile should not be less than 3.50 m. If the expansive soil deposits are shallow, piles of smaller lengths may be also be adopted. If the soil is of homogeneous nature throughout, the pile lengths may be same under the same structure. In sloppy ground, under-reamed piles may be of different depths; however, it is not advisable to provide the piles of different depths under the same pile cap.

Spacing of under-ream bulbs and their diameter, etc. are shown in Figs 2.1a and b. Johnson (1979) recommended the following minimum lengths of piles. A single under-reamed pile is shown in Fig. 2.2.

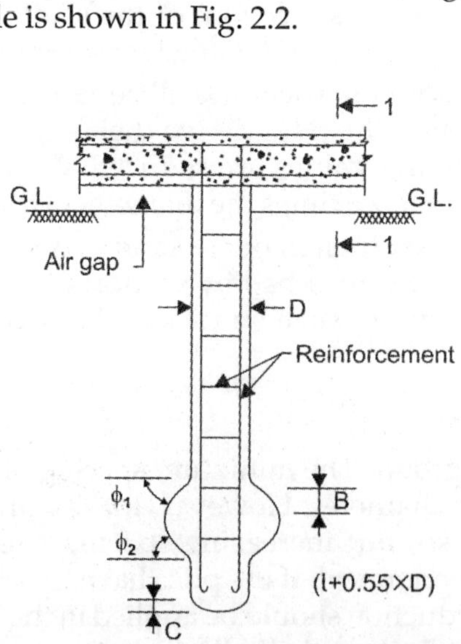

Legend

D	Diameter of pile
l	Bucket length
ϕ_1	Upper angle of bulb with horizontal
ϕ_2	Lower angle of bulb with horizontal
C	Cover
B	Edge width of bulb
D u	Under-ream bulb diameter
S	c/c spacing between two bulbs

Fig. 2.1(a): Single under-reamed pile

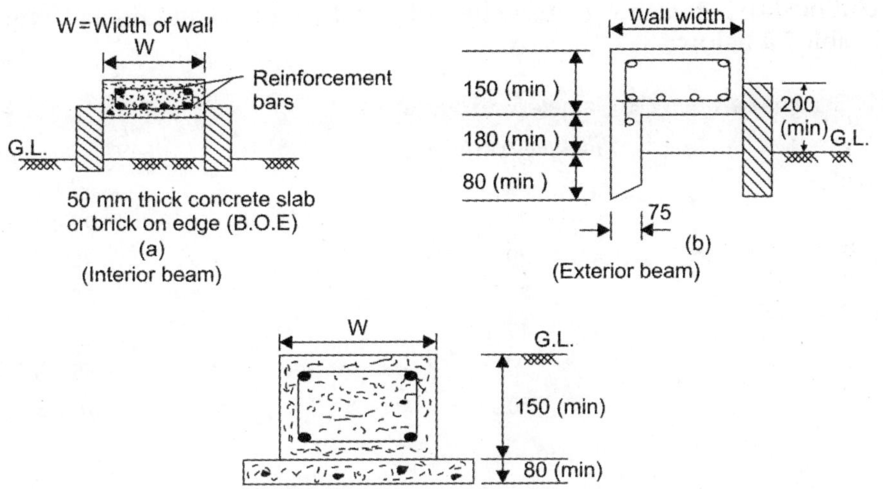

Fig. 2.1(b): Section 1.1

$$L = (2 \times Z') - \{1.42 \times (Du/D)^{2.5}\} \text{ while } D = 0.48 \text{ m} \quad \ldots (2.2)$$

$$L = (2 \times Z') - \{1.76 \times (Du/D)^{3.0}\} \text{ while } D = 0.76 \text{ m} \quad \ldots (2.3)$$

Where D = Pile shaft diameter

Du = Diameter of the enlarged base or under-ream diameter (m)

L = The minimum length of pile (m)

Z' = Depth of the active zone (m)

Equations 2.2 and 2.3 are the empirical relationships and are valid for shaft adhesion of equal to or less than 96 kN/m^2.

David and Komornik (1980) suggested that the depth of embedment of piles should be 1.5 times the depth where the swelling pressure is equal to the overburden.

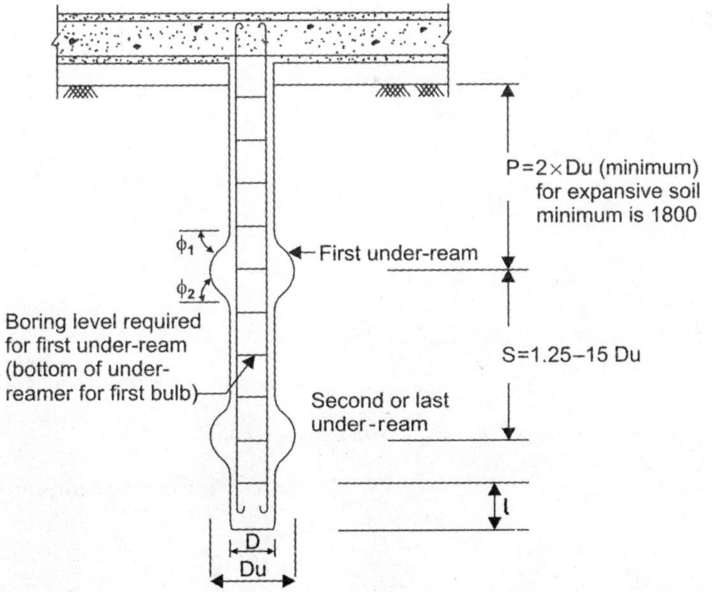

Fig. 2.2: Single under-reamed pile

The relationship between various values of pile diameter D and Bucket length *l* are given in Table 2.2 below:

Table 2.2: Relation between pile diameter and bucket length

Sl. no.	*Pile diameter (D) (cm)*	*Bucket length (l) (cm)*
1	20	40 ± 5
2	25	40 ± 5
3	30	45 ± 5
4	37.5	50 ± 5
5	40	55 ± 5
6	45	65 ± 5
7	50	70 ± 5

Chapter

3

Design of Pile Foundations

3.1 INTRODUCTION

The piles are designed for the structures like tanks, towers or buildings, etc. While designing a pile, an estimation of total load expected to be borne by the pile is done and accordingly the pile is designed. Therefore before designing, it is worthwhile to understand the terminology as below.

3.1.1 Allowable Load

This is the load which may be applied on to a pile after taking into account its ultimate load capacity, pile spacing, overall bearing capacity of the ground below the pile, allowable settlements, negative skin friction and the loading conditions including reversal of loads, etc.

3.1.2 Safe Load

It is the load (after taking into consideration the factor of safety) which a pile can safely withstand.

3.1.3 Ultimate Load Capacity

It is the maximum load which a pile can carry before failure of ground (when the soil fails by shear).

3.1.4 Working Load

It is the load which is assigned to a pile as per its design.

3.1.5 Cut-off Level

It is the level where the installed pile is cut-off to support the pile caps or beams or any other structural components at that level.

3.1.6 Total Elastic Displacement

This is the magnitude of the displacement of the pile due to rebound caused by the top after removal of a given test load. This comprises of two components, e.g.

i. Elastic displacement of the soil participating in the load transfer.
ii. Elastic displacement of the pile shaft.

3.1.7 Total Displacement/Gross Displacement

This is defined as the total movement of the pile top under a given load.

3.1.8 Set

This is defined as the net distance by which the pile penetrates in the ground due to a stated number of blows of the hammer.

3.1.9 Net Displacement

It is the net movement of the pile top after the pile has been subjected to a test load and subsequently released.

3.1.10 Drop or Stroke

The distance through which the driving weight is allowed to fall for driving the pile is called **Drop** or **Stroke.**

3.1.11 Factor of Safety

It is the ratio of the ultimate load capacity of a pile to the safe load of a pile.

3.2 PILES IN SAND

Pile foundation when constructed in sand serves the following purposes.

i. Provides adequate bearing if the top soil consists of soft clays.
ii. Pile driving compacts the sand and increases the bearing capacity.

The pile driven into sand derives most of its load carrying capacity by bearing. A typical load settlement curve for a pile embedded in sand is shown in Fig. 3.1. The settlement continues to increase with increasing loads. The limiting load is therefore based on settlement value rather than from considerations of shearing failure.

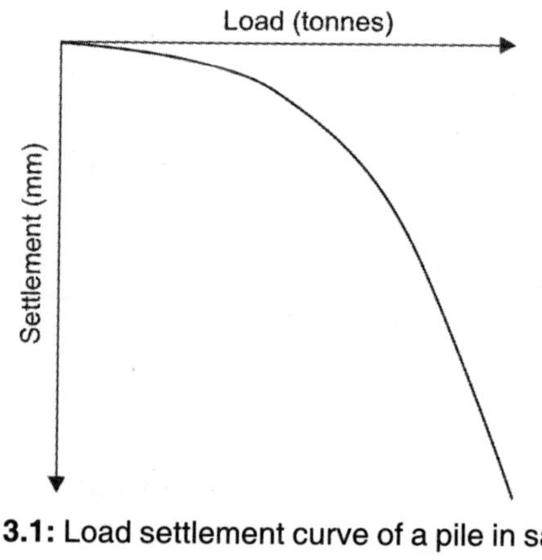

Fig. 3.1: Load settlement curve of a pile in sand

3.3 PILES IN CLAY

The allowable load taken by a single pile in clay can be determined by a static load test (described in latter parts). The load settlement curve is shown in Fig. 3.2. The failure of the soil occurs at any point along X. The allowable load Q_a is obtained by applying a factor of safety to the failure load Q_d.

Thus,

$$Q_a = Q_d/F.S \qquad \dots(3.1)$$

Pile load test (described later) is the only means of determining allowable load on a friction pile. The test should be conducted after minimum three days of driving the piles as the soil regains strength due to consolidation and thixotropy on being allowed to stand after remolding effect of pile driving.

A rough estimate of the load carrying capacity Q_d of a friction pile may be obtained from the following relationship:

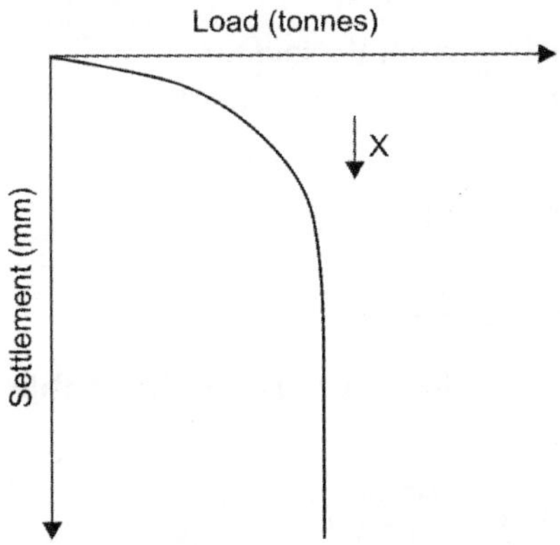

Fig. 3.2: Load settlement curve of a pile in clay

$$Q_d = \pi.D \times L.C \quad \dots (3.2)$$

where,

B = Diameter of the pile
L = Embedded length of pile
C = Unit cohesion of soil

For a single pile, C should be replaced by C_a

where $$C_a = m \times c \quad \dots (3.3)$$

The value of m ranges from 0.40 for stiff clays to 1.0 for soft clays.

3.4 DETERMINATION OF PILE CAPACITY

3.4.1 Pile Capacity by Static Formulae

The static bearing capacity of a pile is calculated from the consideration of the properties of the soil medium through which a pile passes. Thus, the bearing capacity is calculated as the sum of the total ultimate skin friction resistance and the total ultimate end bearing resistance. Thus,

$$Q_u = Q_p + Q_f \quad \dots (3.4a)$$

or $$Q_u = A_p q_p + A_s S_s \quad \dots (3.4b)$$

where,

Q_u = Ultimate bearing capacity of the pile
Q_p = Load carried by point bearing (for point resistance)
q_p = Unit point resistance or toe resistance

$$Q_p = A_p (CN_c + 0.4\gamma BN + qN_q) \quad \dots (3.5)$$

Q_f = Load carried by friction along perimeter of pile or shaft resistance
$= \Sigma (\Delta L) (A_s . S_s)$
A_p = Area of pile tip
ΔL = Increment of pile length
a_s = Area of pile in length ΔL in contact with soil in m^2/m, i.e. perimeter of the pile.
A_s = Area of surface of embedded length of pile
S_s = Unit shaft resistance

N_c, N_q and N_γ are the bearing capacity factors

γ = Unit weight of soil

q = Surcharge (γD)

B = Least dimension of pile tip

C = Cohesion of soil

D = Depth of embedment of pile

The value of N_c, and $N\gamma$ are taken from following Table 3.1 and value of N_q is taken from Fig. 3.3.

Table 3.1: Bearing capacity factors

ϕ'	N_c	N_q	N_γ	ϕ'	N_c	N_q	N_γ
0	5.14	1.00	0.00	26	22.25	11.85	12.54
1	5.38	1.09	0.07	27	23.94	13.20	14.47
2	5.63	1.20	0.15	28	25.80	14.72	16.72
3	5.90	1.31	0.24	29	27.86	16.44	19.34
4	6.19	1.43	0.34	30	30.14	18.40	22.40
5	6.49	1.57	0.45	31	32.67	20.63	25.99
6	6.81	1.72	0.57	32	35.49	23.18	30.22
7	7.16	1.88	0.71	33	38.64	26.09	35.19
8	7.53	2.06	0.86	34	42.16	29.44	41.06
9	7.92	2.25	1.03	35	46.12	33.30	48.03
10	8.35	2.47	1.22	36	50.59	37.75	56.31
11	8.80	2.71	1.44	37	55.63	42.92	66.19
12	9.28	2.97	1.69	38	61.35	48.93	78.03
13	9.81	3.26	1.97	39	67.87	55.96	92.41
14	10.37	3.59	2.29	40	75.31	64.20	109.41
15	10.98	3.94	2.65	41	83.86	73.90	130.22
16	11.63	4.34	3.06	42	93.71	85.38	155.55
17	12.34	4.77	3.53	43	105.11	99.02	186.54
18	13.10	5.26	4.07	44	118.37	115.31	224.64
19	13.93	5.80	4.68	45	133.88	134.88	271.76
20	14.83	6.40	5.39	46	152.10	158.51	330.35
21	15.82	7.07	6.20	47	173.64	187.21	403.67
22	16.88	7.82	7.13	48	199.26	222.31	496.01
23	18.05	8.66	8.20	49	229.93	265.51	613.16
24	19.32	9.60	9.44	50	266.89	319.07	762.89
25	20.72	10.66	10.88				

Source: (B.M. Das, 2013)

3.4.2 Pile Capacity by Dynamic Formulae

To determine the static capacity of a pile, dynamic pile formulae are also used. The assumptions made for this are:

i. The kinetic energy of the falling pile hammer is consumed partly in driving the pile into the soil and partly in the form of various friction and other losses.

ii. The soil resistance to the dynamic penetration of pile (dynamic soil resistance) is the same as the penetration of pile under static or sustained loading (i.e. static soil resistance).

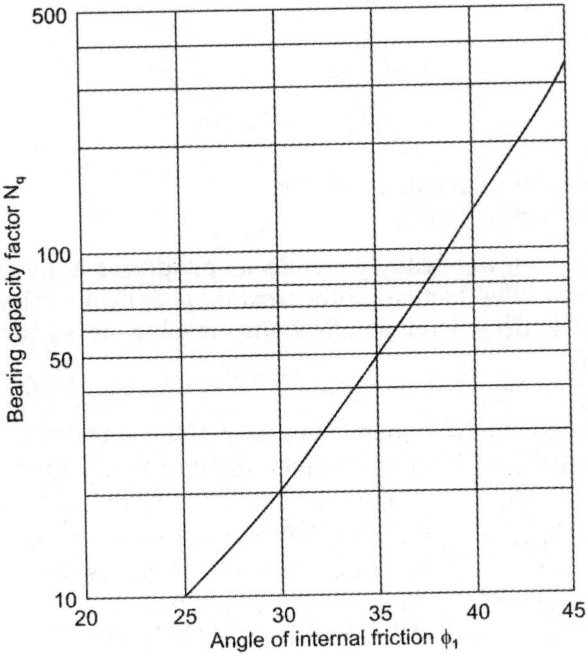

Fig. 3.3: Relationship between bearing capacity factor (N_q) and angle of internal friction

The dynamic pile formulae (also called the **rational pile driving formulae**) are based on impulse-momentum principles. The equation for determination of ultimate pile resistances as proposed by Hiley is given by

$$Q_u = \frac{E_h W_r h}{S + \frac{1}{2}(k_1 + k_2 + k_3)} \frac{W_r + n^2 W_p}{W_r + W_p} \quad \ldots (3.6)$$

(For drop hammers and single acting hammers)

where,

Q_u = Ultimate pile resistance

E_h = Efficiency of hammer

W_r = Weight of ram

h = Height of fall of ram

S = Amount of net point penetration cm, per blow

k_1 = Elastic compression of cap

k_2 = Elastic compression of piles

k_3 = Elastic compression of soil

n = Co-efficient of restitution

W_p = Weight of pile, including weight of pile caps, driving shoe and anvil for double acting hammers

Let $$c = \frac{1}{2}(k_1 + k_2 + k_3) \quad \ldots (3.7)$$

Here $$k_2 = \frac{Q_u L}{A.E} \qquad \text{... (3.8)}$$

where,

L = Length of pile
E = Young's modulus of elasticity of pile
A = Area of cross-section of pile

According to Chellis, the Hiley formula as proposed above underestimates the capacity of long-heavy piles. Hence another formula has been derived from equation 3.6. This formula, most widely used is called **engineering news formula**. According to this, the value of $\frac{1}{2}(k_1 + k_2 + k_3)$ in equation 3.6 is taken as 2.5 for drop hammers and 0.25 for single acting steam hammers. The co-efficient of restitution (n) is taken as 1.0, the efficiency also equal to 1.0 and the safety factor as 6. The formula is given as

$$Q_a = \frac{W_r h}{6(S + 2.5)} \text{ (for drop hammers)} \qquad \text{... (3.9)}$$

And $$Q_a = \frac{W_r h}{6(S + 0.25)} \text{ (for single acting steam hammers)} \qquad \text{... (3.10)}$$

3.4.3 Capacity of Bored Pile in Clay

For the piles situated in saturated clay, the ultimate bearing capacity of pile can be determined by the equation 3.4 a, b, i.e.

$$Q_n = Q_p + Q_f \qquad \text{... (3.4a)}$$
$$= A_p q_p + A_s S_s \qquad \text{... (3.4b)}$$

In case of saturated clays,

$$Q_p = c.N_c \qquad \text{... (3.11)}$$

where c = Cohesion of the soil

The value of N_c may be taken as 9.0 for deep foundations (After Skempton, 1953) and that for shallow footings as 5.14. Meyerhof (1976) suggested that in saturated homogeneous clay under drained conditions, the value of N_c below the critical depth varies with the sensitivity and deformation conditions of clay from 5 (for very sensitive brittle normally consolidated clay) to about 10 (for insensitive stiff over consolidated clays). However, a value of 9 is generally used for bearing capacity computations of driven and bored piles. The average value of S_s is also given by:

$$S_s = m.c. \qquad \text{... (3.3)}$$

where m = Empirical cohesion factor for reduction of average undrained shear strength C_u of undisturbed clay with embedded length of pile

Thus,

$$Q_u = cN_c A_p + \text{mc}A_s \qquad \text{... (3.12)}$$

The value of unit shaft resistance S_s is given by

$$S_s = c + K_s P_o \tan\phi \qquad \text{...(3.13)}$$

where,

c = Average unit cohesion
K_s = Average co-efficient of earth pressure on shaft

P_0 = Average effective overburden pressure along shaft

ϕ = Angle of internal friction

Analyzing the field data, Meyerhof (1976) stated that the ultimate skin friction of piles in saturated clay can approximately be estimated from the undrained shear strength of remoulded soil for which the cohesion may usually be taken as zero. On this basis, equation 3.13 can be rewritten as

$$S_s = \beta P_o \quad \text{... (3.14)}$$

where,

β = Skin friction factor

The value of β may be calculated as

$$\beta = K_s \tan\phi \quad \text{... (3.15)}$$

For bored piles and for piles driven into saturated soft clay, the ultimate co-efficient K_s of earth pressure on the shaft may be assumed to be close to earth pressure on at rest (K_o) conditions as for loose sand. Thus, the value of K_s may be taken from the values of K_o as

$$K_o = 1 - \sin\phi \quad \text{... (3.16)}$$

And, the skin factor β may be taken as

$$\beta = \tan\phi\ (1 - \sin\phi) \quad \text{... (3.17)}$$

The value of β ranges from 0.2 to 0.3 for typical range of ϕ for clay (Meyerhof, 1976). Analyzing the data on driven piles in soft and medium clays, β decreases with length of piles from a range of about 0.25 to 0.5 for short piles to about 0.1 to 2.25 for long piles.

3.4.3.1 Piles in Cohesive Soils

The ultimate load capacity (Q_u) of piles, in kN, in cohesive soils is given by the following formula:

$$Q_u = A_p N_c c_p + \Sigma_{i=1}^{n} \alpha_i c_i A_{si} \quad \text{... (3.18)}$$

The first term gives the end-bearing resistance and the second term gives the skin friction resistance.

where

A_p = cross-sectional area of pile tip, in m^2;

N_c = bearing capacity factor, may be taken as 9;

c_p = average cohesion at pile tip, in kN/m^2;

$\Sigma_{i=1}^{n}$ = summation for layers 1 to n in which the pile is installed and which contribute to positive skin friction;

α_i = adhesion factor for the ith layer depending on the consistency of soil, (*see* Note);

c_i = average cohsion for the ith layer, in kN/m^2; and

A_{si} = surface area of pile shaft in the ith layer, in m^2.

Note for (α): The value of adhesion factor, α_i depends on the undrained shear strength of the clay and may be obtained from Fig. 3.5.

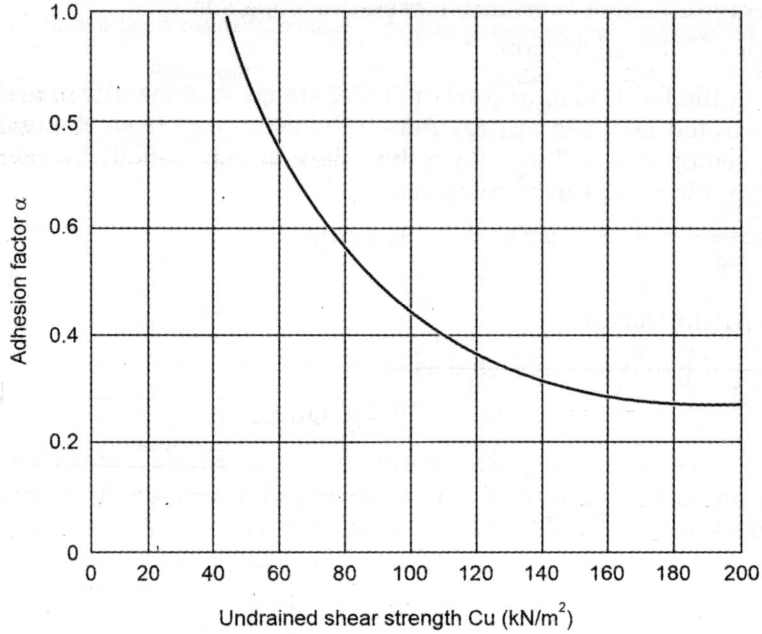

(For Cu < 40 kN/m² TAKE α = 1)

Fig. 3.4: Fig. 3.5: Variation of α with C_u

3.4.4 Piles in Granular Soils (As per BIS code)

The ultimate load capacity (Q_u) of piles, in kN in granular soils is given by the following formula.

$$Q_u = A_p(\tfrac{1}{2}D\gamma N_t + P_D N_q) + \Sigma_{i=1}^{n} K_i P_{Di} \tan\delta_i A_a \qquad \ldots (1)$$

The first term gives end-bearing resistance and the second term gives skin friction resistance.

where,

A_p = cross-sectional area of pile tip, in m²;

D = diameter of pile shaft, in m;

γ = effective unit weight of the soil at pile tip, in kN/m²;

N_t = bearing capacity factors depending upon

and N_q the angle of internal friction, ϕ at pile tip;

P_D = effective overburden pressure at pile tip, in kN/m² (*see* Note 5);

$\Sigma_{i=1}^{n}$ = summation for layers 1 to n in which pile is installed and which contribute to positive skin friction;

K_i = coefficient of earth pressure applicable for the ith layer (*see* Note 3);

P_{Di} = effective overburden pressure for the ith layer, in kN/m²;

δ_1 = angle of wall friction between pile and soil for the ith layer; and

A_{si} = surface area of pile shaft in the ith layer, in m²;

Note:

1. $N\gamma$ factor can be taken for general shear failure according to IS 6403.
2. N_q factor will depend on the nature of soil, type of pile, L/B ratio and its method of construction. The values applicable for bored piles are given in Fig. 3.4.

3. K_t the earth pressure coeffiecient depends on the nature of soil strata, type of pile, spacing of pile and its method of construction. For driven piles in loose to dense and F with ϕ varying between 30° and 40°. K_t values in the range of 1 to 1.5 may be used.
4. δ, the angle of wall friction may be taken equal to the friction angle of the soil around the pile stem.
5. In working out pile capacity by static formula, the maximum effective overburden at the pile tip should correspond to the critical depth, which may be taken as 15 times the diameter of the pile shaft for $\phi \leq 30°$ and increasing to 20 times for $\phi \geq 40°$.
6. For piles passing through cohesive strata and terminating in a granular stratum, a penetration of at least twice the diameter of the pile shaft should be given into the granular stratum.
7. For refusal, *if*$(N)_{SPT} \geq 30$, $\phi = 36°$ (or more) for that $Nq = 35$ (or more)

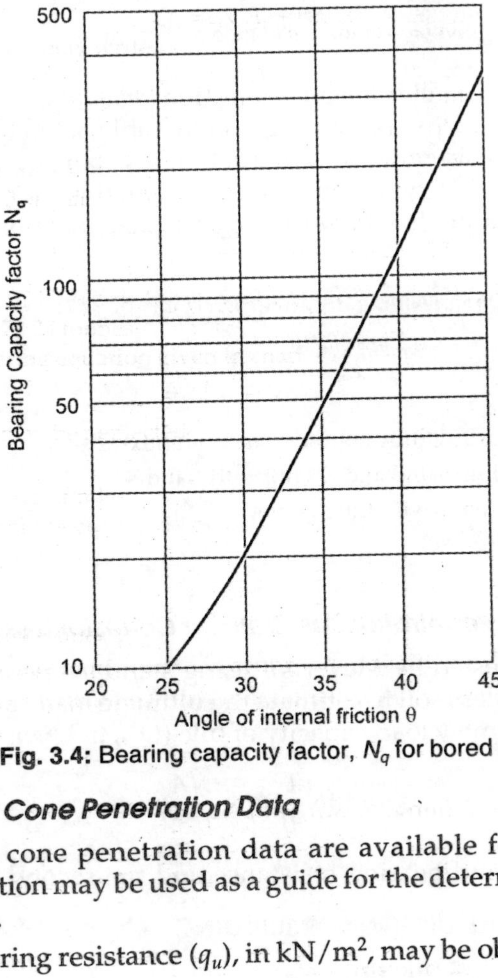

Fig. 3.4: Bearing capacity factor, N_q for bored piles

3.4.4.1 Use of Static Cone Penetration Data

A. When full static cone penetration data are available for the entire depth, the following correlation may be used as a guide for the determination of ultimate load capacity of a pile.

B. Ultimate end bearing resistance (q_u), in kN/m², may be obtained as:

$$q_u = \frac{\frac{q_{c0} + q_{c1}}{2} + q_{c2}}{2}$$

where

q_{c0} = average static cone resistance over a depth of $2D$ below the pile tip, in kN/m²;

q_{c1} = minimum static cone resistance over the same $2D$ below the pile tip, in kN/m²;

q_{c2} = average of the envelope of minimum static cone resistance values over the length of pile of $8D$ above the pile tip, in kN/m²; and

D = diameter of pile shaft.

C. Ultimate skin friction resistance can be approximated to local side friction (f_s), in kN/m², obtained from static cone resistance as given in Table 3.2.

Table 3.2: Side friction for different types of Soil

Sl No. (1)	*Type of Soil* (2)	*Local side friction, f_s kN/m²* (3)
(i)	q_c less than 1000 kN/m²	$q_c/30 < f_s < q_c/10$
(ii)	Clay	$q_c/25 < f_s < 2q_c/25$
(iii)	Silty clay and silty sand	$q_c/100 < f_s < q_c/25$
(iv)	Sand	$q_c/100 < f_s < q_c/50$
(v)	Coarse sand and gravel q_c = cone resistance, in kN/m²	$q_c/100 < f_s < q_c/150$

D. The correlation between standard penetration resistance, N (blows/30 cm) and static cone resistance, q_c, in kN/m² as given in Table 3.3 may be used for working out the end-bearing resistance and skin friction resistance of piles. This correlation should only be taken as a guide and should preferably be established for a given site as they can substantially vary with the grain size, Atterberg limits, water table, etc.

Table 3.3: Co-relation Between N and q_c for Different Types of Soil

Sl No. (1)	*Type of Soil* (2)	*q_c/N* (3)
(i)	Clay	150–200
(ii)	Silts, sandy silts and slightly cohesive silt-sand mixtures	200–250
(iii)	Clean fine to medium sand and slightly silty sand	300–400
(iv)	Coarse sand and sands with little gravel	500–600
(v)	Sandy gravel and gravel	800–1000

3.4.4.2 Use of Standard Penetration Test Data for Cohesionless Soil

A. The correlation suggested by Meyerhof using standard penetration resistance, N in saturated cohesionless soil to estimate the ultimate load capacity of bored pile is given below. The ultimate load capacity of pile (Q_u), in kN, is given as:

$$Q_u = 13N\frac{L}{B}A_p + \frac{\bar{N}A_s}{0.50} \qquad ...(3)$$

The first term gives end-bearing resistance and the second term gives frictional resistance.

where

N = average N value at pile tip;

L = length of penetration of pile in the bearing strata, in m;

B = diameter or minimum width of pile in m;

A_p = cross-sectional area of pile tip, in m²;

$\bar{N}$ = average N along the pile shaft; and

A_s = surface area of pile shaft, in m².

Note: The end-bearing resistance should not exceed $130\,NA_p$.

B. For non-plastic silt or very fine sand the equation has been modified as:

$$Q_u = 10N\frac{L}{B}A_p + \frac{\bar{N}A_s}{0.60} \qquad ...(4)$$

The meaning of all terms is same as for equation 3.

3.4.4.3 Factor of Safety

The minimum factor of safety for arriving at the safe pile capacity from the ultimate load capacity obtained by using static formulae shall be 2.5.

3.4.4.4 Piles in Stratified Soil

In stratified soil/C-ϕ soil, the ultimate load capacity of piles should be determined by calculating the skin friction and end-bearing in different strata by using appropriate expressions given in **B-1** and **B-2**.

3.4.4.5 Piles in Hard Rock

When the crushing strength of the rock is more than characteristic strength of pile concrete, the rock should be deemed as hard rock. Piles resting directly on hard rock may be loaded to their safe structural capacity.

3.4.4.6 Piles in Weathered/Soft Rock

For pile founded in weathered/soft rock different empirical approaches are used to arrive at the socket length necessary for utilizing the full structural capacity of the pile.

Since it is difficult to collect cores in weathered/soft rocks, the method suggested by Cole and Stroud using 'N' values is more widely used. The allowable load on the pile, Q_a, in kN, by this approach, is given by:

$$Q_a = c_{u1}N_c.\frac{\pi B^2}{4F_s} + \alpha c_{u2}.\frac{\pi BL}{F_s}$$

where

c_{u1} = shear strength of rock below the base of the pile, in kN/m^2 (*see* Fig. 3.6);
N_c = bearing capacity factor taken as 9;
F_s = factor of safety usually taken as 3;
α = 0.9 (recommended value);
c_{u2} = average shear strength of rock in the socketed length of pile, in kN/m^2 (*see* Fig. 3.6);
B = minimum width of pile shaft (diameter in case of circular piles), in m; and
L = socket length of pile, in m.

Note: For $N \geq 60$, the stratum is to be treated as weathered rock rather than soil.

3.4.5 Capacity of Under-reamed Pile Foundation in Clays

For clayey soils, the ultimate load carrying capacity of an under-reamed pile is worked out from following equation:

$$Q_u = A_p.N_c.C_p + A_a.N_c.C'_a + C'_a.A'_s + \alpha.C_a.A_s \qquad ... (3.18)$$

where,

Q_u = Ultimate bearing capacity (kg)
A_p = Cross-sectional area of pile stem at two levels (cm^2)
N_c = Bearing capacity factor, usually taken as 9.
C_p = Cohesion of the soil at toe level (kg/cm^2)
A_a = Difference in area between the under-ream and the pile stem (cm^2)

Shear Strength kN/m²	Approx IV Value	Strength/ Consistency	Grade	Breakability	Penetration	Strength
40000		Strong	A	Difficult to break against splid object with hammer		Can not be scratched with knife
20000						
	600	Moderately strong	B	Broken against solid object with hammer		Can just be scratched with knife
10000						
8000						
6000						
	400		C	Broken in hand by hitting with hammer		Scratched with knife. Can just be scratched with thumb-nail
4000		Moderately Weak				
2000	200	Weak	D	Broken by leaning on sample with hammer	No penetration with knife	Scratched with thumb-nail
1000			E	Broken by hand	Penetration to about 2 mm with knife	
800	100					
600	80	Hard or Very Weak	F	Easily Broken by hand	Penetration to about 5 mm with knife	
400	60					
200	40	Very Stiff			Penetrated by thumb-nail and to about 1.5 mm with knife	
100	20	Stiff			Indented by thumb	
80						
60	10	Firm			Penetrated by thumb with effort	
40	8	Soft			Easily Penetrated by thumb	
	6					
20	4					
10	2	Very Soft				

NOTE—Standard penetration test may not be practicable for N values greater than 200. In such cases, design may be done on the basis of shear strength of rock.

Fig. 3.6: Consistency and shear strength of weathered rock

C' = Average cohesion of soil around the under-reamed bulbs (kg/cm²)

α = Reduction factor (usually taken as 0.5)

A_s' = Surface area of the cylinder circumscribing the under-reamed bulb (cm²)

C_a = Average cohesion of the soil along the pile stem (kg/cm²)

A_s = Surface area of the stem.

Equation 3.18 holds good for the usual spacing of under-reams on pile stem (i.e. not more than one and a half diameters apart). For single under-reamed piles, the third term does not apply.

3.4.6 Capacity of Under-reamed Pile Foundation in Sandy Soils

For sandy soils, the following expression may be used:

$$Q_u = A_p\,(\tfrac{1}{2}D\gamma\,N\gamma + \gamma\,d_f N_q) + A_a\left(\tfrac{1}{2}D_u n\gamma\,N\gamma + N_q \sum_{r=1}^{n} d\gamma\right)$$

$$+ \tfrac{1}{2}\pi D\lambda.\text{k}.\ \tan\delta \times \left(d_1^2 + d_f^2 - d_n^2\right) \quad \dots (3.19)$$

where,

$A_p = \frac{\pi}{4}D^2\,(\text{cm}^2)$, D being the stem diameter

$A_a = \frac{\pi}{4}\left(D_u^2 - D^2\right)(cm^2)$ D_u being the bulb diameter

n = No. of under-reamed bulbs

γ = Average unit weight of soil (take submerged unit weight in strata below W.T.) (kg/cm^3)

N_q and N_γ = Bearing capacity factors depending on the value of ϕ, i.e. the angle of internal friction (Ref. Fig. 3.3 for value of N_q only. The value of $N\gamma$ is taken from table given in para 3.4.1 (a)

d_r = Depth of the centre of different under-reamed bulbs below ground level (cm)

d_f = Total depth of pile below ground level (cm)

K = Earth pressure co-efficient (usually = 1.75 for sandy soils)

δ = Angle of wall friction (may be taken equal to the value of ϕ)

d_1 = Depth of the centre of first under-reamed bulb (cm)

d_n = Depth of the centre of the last under-reamed bulb

The values of N_q as obtained from Fig. 3.3 are for driven piles but for bored piles, including under-reamed piles, these are reduced to half while estimating the ultimate bearing capacity. A factor of safety of 2.5 is used for determining the safe load.

In case of single under-reamed pile the ultimate load from equation 3.19 shall be determined by taking $n = 1$.

3.4.7 Uplift Load Capacity of Under-reamed Pile

For working out the uplift loads from equations 3.18 and 3.19, there will be no contribution by the pile toe as the direction of load application is reversed. Therefore, the first term in the equations shall not appear. For determination of safe loads in uplift, a factor of safety of 3 is recommended.

In field, purely sand or clay strata for the entire length are very rare and uncommon. Therefore, depending upon the predominant fraction of sand or clay, only one equation, i.e. whether 3.18 or 3.19 is used. **For piles resting on rock, the bearing should be obtained by multiplying the bearing capacity of rock with bearing area of pile stem plus the bearing provided by the bulb portion. However, the ultimate capacity considered should not exceed the structural strength of the pile stem.** If c and ϕ values are not available, they may be approximated by the standard penetration test data, as below Tables 3.4 and 3.5.

Table 3.4: Shear parameters for cohesionless soil

Penetration resistance value ('N' blows)	*Approximate ϕ (degrees)*	*Density index (%)*	*Description (i.e. compactness)*	*Approximate moist density (t/m³)*
0–4	25–30	0	Very loose	112.160
4–10	27–32	15	Loose	144184
10–30	30–35	35	Medium	176.208
30–50	35–40	65	Dense	176.224
>50	38.43	85	Very dense	208.240

Source: Mittal and Shukla, 2014.

Table 3.5: Shear parameters for cohesive soil

Penetration resistance value ('N' blows)	*Unconfined compressive strength (t/m²)*	*Saturated density (t/m²)*	*Consistency*
0–2	<2.5	–	Very soft
2–4	25–50	1.60–1.92	Soft
4–8	5–10	1.76–2.08	Medium
8–15	10–20	–	Stiff
15–30	20.40	1.92–2.24	Very stiff
>30	>40	–	Hard

3.4.8 Other Methods to Evaluate Pile Capacity

Penetration test can also be used to estimate the pile capacity. The value of angle of internal friction (ϕ) can be evaluated through standard penetration tests. Cone penetration tests can also be conducted to estimate the pile capacity. The ultimate end bearing resistance of the pile is taken as the resistance of the cone. The cone resistance usually varies hence to overcome it, Vander Veen (1957) suggested to evaluate the cone resistance over a depth equal to 3 times the diameter of pile above the pile point level and one pile diameter below pile point level. Meyerhof (1956) suggested the following relationship between skin friction on the pile shaft and average cone resistance q_c.

For displacement piles:

$$\text{Ultimate unit skin friction} = \frac{q_c}{2} \qquad \ldots (3.20)$$

where q_c = Average cone resistance in kg/cm² over the length of pile shaft under consideration.

Meyerhof stated that for straight sided displacement piles, the ultimate unit skin friction has a maximum value of 1 kg/cm² and for H beam piles, it is to a maximum of 5 kg/cm².

Example 3.1

A 20 cm side precast concrete pile of square cross-section is driven through a deposit of clay of medium consistency (unconfined compressive strength $q_u = 0.65$ kg/cm²). The length of pile is 12 m. Calculate the ultimate bearing capacity of pile.

Solution:

$$c = \frac{q_u}{2} = \frac{0.65}{2} = 0.325 \text{ kg/cm}^2 = 3.25 \text{ t/m}^2 \qquad \text{(From eqn. 3.3)}$$

Using equation 3.3, C_a = m.c.

$C_a = 0.9 \times 3.25 = 2.925$ (Taking an average value of m = 0.9)

A_s = Area of surface of embedded length of pile

$= \pi \times 0.20 \times 12 = 7.5 \text{ m}^2$

Substituting the values in equation 3.12, i.e.

$$Q_u = cN_c.N_cA_p + \text{m.c. As} \quad \text{(Eqn 3.12)}$$
$$= 3.25 \times 9.0 \times 0.786\,(0.2 \times 0.2) + 2.925 \times 7.5$$
$$= 44.9 \text{ tonnes}$$

Thus, the ultimate bearing capacity of pile is 44.9 tonnes. **Ans.**

(For computation of safe load, the value of 44.9 t may be divided by suitable F.O.S.)

3.5 STRUCTURAL DESIGN OF UNDER-REAMED PILES (AFTER CBRI HANDBOOK)

The pile should be safe against settlement and shear. It should have adequate strength to sustain load. By ultimate load theory, the ultimate capacity (P_u) in compression may be determined from the following equation:

$$P_u = 0.4\,\sigma_{cu}\,(A_p - A_s) + \sigma_{xy} \,.\, A_s \quad \ldots (3.21)$$

where, A_p = Area of cross-section of pile stem (cm^2)

A_s = Cross-sectional area of longitudinal steel in pile (cm^2)

σ_{xy} = Yield stress of steel reinforcement (kg/cm^2)

σ_{cu} = Ultimate cube strength in compression (kg/cm^2)

It is worth mentioning here that equations 3.18 and 3.19 are used for determination of ultimate capacity of pile from soil properties obtained from laboratory and field tests.

Sometimes, for the compressive loads, the cross-sectional area of concrete of pile stem is sufficient and there is no need to provide longitudinal steel. However, a minimum of 0.4% of cross-sectional area of pile stem of steel may be provided all through the pile as longitudinal reinforcement. The transverse reinforcement of circular rings should also be provided at a spacing of less than the pile stem diameter but not more than 30 cm.

For piles subjected to uplift loads, adequate steel should be provided to take care of uplift and it should be same all through the pile length. If some moments are also to be acting on the pile, it should be duly checked to be safe against moment and required amount of steel should be provided.

3.6 DETERMINATION OF SAFE LOADS ON UNDER-REAMED PILES

The safe loads on under-reamed piles can be determined from static formula using the soil properties while applying a suitable factor of safety. The another approach for determination of safe loads on pile is by performing field load tests on piles (Fig. 3.7) and determine safe loads from load-deflection curve (like Fig. 3.1 or Fig. 3.2, as the case may be). Field test on pile is illustrated in Fig. 3.7 below. A typical example of field load test is also given in Chapter 5, for interpretation of field load test data for computation of load capacity of pile.

Based on the load-deflection behaviour observed during several field tests conducted by Central Building Research Institute Roorkee, a table has been prepared for ready reference and approximate design. In Table 3.6, the safe loads have been given for almost all kinds of soils which are generally available in abundance. The table is self-explanatory.

Fig. 3.7: A view of pile load test in field
(Here load is imparted through gunny bags, alternatively load can be applied through anchors also)

3.7 GRANULAR PILE FOUNDATIONS

Granular piles are the piles which require no cement. These piles are now a days being used most extensively particularly for industrial sites where cohesive soils are available. The piles are constructed by making a bore with an auger of required size and then filling the bore by sand and gravels duly computed. Each layer of sand and gravel is compacted by a heavy hammer. These piles have a very good bearing capacity.

Such piles are being used these days for big tanks for oil industries, sugar industries etc. for storage of liquid, molasses etc. Where bearing capacity is very less. First author of this book has designed many such piles for the purpose of ground improvement even for housing projects. A skirt wall is also sometimes provided circumscribing all the piles to provide a positive confinement.

The granular piles are like stone columns and they also facilitate the underground water to pass through them.

For the design of such piles, a suitable percentage of total load is assumed to be borne by piles and balance by the surrounding soil, the load capacity of each pile may be calculated by the pressure metre theory as proposed by Gibson and Anderson (1961). Knowing the load capacity of each pile, the no. of piles are determined by dividing the total load to be supported by the piles by the load capacity of individual pile. Thus, based on the theory proposed by Gibson and Anderson,

$$Q_d = K_p \times (8\,C_u + O_e + O_v) \times A_p \qquad \ldots (3.22)$$

where,

C_u = Undrained shear strength of soil

O_e = Effective average stress at the critical pile depth (kg/cm^2)

O_v = Effective normal pressure (kg/cm^2)

A_p = Area of pile stem

K_p = Passive pressure co-efficient

Table 3.6: Safe load for vertical under-reamed piles in sandy and clayey soils including black cotton soils
[As per IS : 2911 (Part III) 1980]

Size		*Length*		*Reinforcement*			*Safe Loads*									
							Bearing resistance				*Uplift resistance*				*Lateral thrust*	
Diameter of pile (cm)	*Under-reamed diameter (cm)*	*Single under-reamed (m)*	*Double under-reamed (m)*	*Longitudinal Reinforcement*		*Rings spacing of 6 mm dia bars (cm)*	*Single under-reamed (t)*	*Double under-reamed (t)*	*Increase per 30 cm length (t)*	*Decrease per 30 cm length (t)*	*Single under-reamed (t)*	*Double under-reamed (t)*	*Increase per 30 cm length (t)*	*Decrease per 30 cm length (t)*	*Single under-reamed (m)*	*Double under-reamed (m)*
				No.	*Dia (mm)*											
(1)	(2)	(3)	(4)	(5)	(6)	(7)	(8)	(9)	(10)	(11)	(12)	(13)	(14)	(15)	(16)	(17)
20	50	3.5	3.5	3	10	18	8	12	0.9	0.7	4	6	0.65	0.55	1.0	1.2
25	62.5	3,5	3.5	4	10	22	12	18	1.15	0.9	6	9	0.85	0.70	1.5	1.8
30	75	3.5	3.5	4	12	25	16	24	1.4	1.1	8	12	1.05	0.85	2.0	2.4
37.5	94	3.5	3.75	5	12	30	24	36	1.8	1.4	12	18	1.35	1.10	3.0	3.6
40	100	3.5	4.0	6	12	30	28	42	1.9	1.5	14	21	1.45	1.15	3.4	4.0
45	112.5	3.5	4.5	7	12	30	35	52.5	2.15	1.7	17.5	25.75	1.60	1.30	4.0	4.8
50	125	3.5	5.0	9	12	30	42	63	2.4	1.9	21	31.5	1.80	1.45	4.5	5.4

Note:

1. *The values given in above table are Indicative only. Readers/ Users are advised NOT to use these values given in this table as the final recommended values of load carrying capacity of pile, as the every pile is a unique pile, and its capacity depends on the type of strata present at site, position of water table, soil layers at site, etc.*
2. *Values given in Col. (16) and (17) for lateral thrusts may not be reduced for changes in pile lengths. The higher values may be adopted after conducting lateral load tests on single or group of piles. The table values should be appropriately increased for wind and earthquake forces, etc. For broken wire conditions in the design of tower foundations, these values are to be increased by 50%.*
3. *The table values are to be reduced by 10% if the piles are provided at a spacing of one and a half times the bulb diameter.*

Therefore, $Q_{safe} = \dfrac{Q}{F.S}$

The various terms in equation 3. 22 are expressed as below:

$$\text{Passive pressure co-efficient} = \tan^2 (45 + \theta/2) \quad \dots (3.23)$$

$$O_e = d \times \gamma_{sat}, d \text{ being the pile depth} \quad \dots (3.24)$$

$$O_v = K_{B.q} \quad \dots (3.25)$$

where,

$$K_B = 1 - 1 - \left\{ \frac{1}{1 + \left(\frac{a}{Z}\right)} \right\}^{3/2} \quad \dots (3.26)$$

a = Radius of circular load (if load is circular, else higher dimension of rectangular load)

Z = Pile depth at failure = 4 to 5 times the pile diameter

q = The load intensity = $\dfrac{\text{Total load}}{\text{Area}}$

$C_u = 2 \times O_e$

A solved example has been given as the complete design of granular pile foundation in chapter 6 of this book. Fig. 3.8 below shows the general view of construction of granular pile (also called as **stone column**).

Example 3.2: What will be the penetration of square R.C. pile per blow which must be obtained in driving the pile with a 5 tonnes drop hammer falling through 1.2 metre. Allowable load is 30 tonnes.

Solution: The Engineering news formula for drop hammer is given by

$$Q_a = \frac{W_r h}{6(S + 2.5)} \quad \dots (3.27)$$

here Q_a = 30 tonnes

W_r = 5 tonnes

H = 1.2 m = 120 cm

Thus, substituting the values

or $S = 0.83$ cm **Ans.**

Example 3.3: A pile of 0.25 m × 0.25 m cross-sectional area is penetrated into soft soil having $c = 0.75$ kg/cm² for a length of 15 m and finally it rests on hard soil. Determining its load carrying capacity by skin friction.

Solution: $Q = \pi d \times L \times C_a$ (for circular piles) From eqn (3.2a)

and $Q = 4d \times L \times C_a$ (for square piles) From eqn (3.2b)

$C_a = m \times c$ From eqn (3.3)

Taking the value of m = 0.4 for stiff clay, and

m = 1.0 for soft clay

Thus, adopting an average value of 0.9

$$Q = 4 \times d \times L \times C_a$$

$$= 4 \times 0.25 \times 15 \times 7.5 \times 0.90$$

Thus, Q = 101.25 tonnes **Ans.**

Example 3.4: Determine the safe load capacity of pile in bearing (compression) and uplift, by referring the Table 3.1 for the pile of following data.

Stem diameter = 30 cm (single bulb)

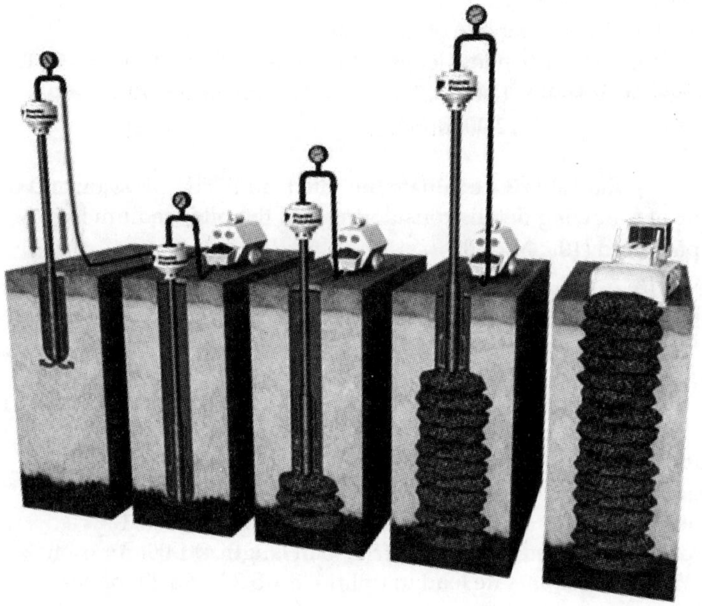

(a)

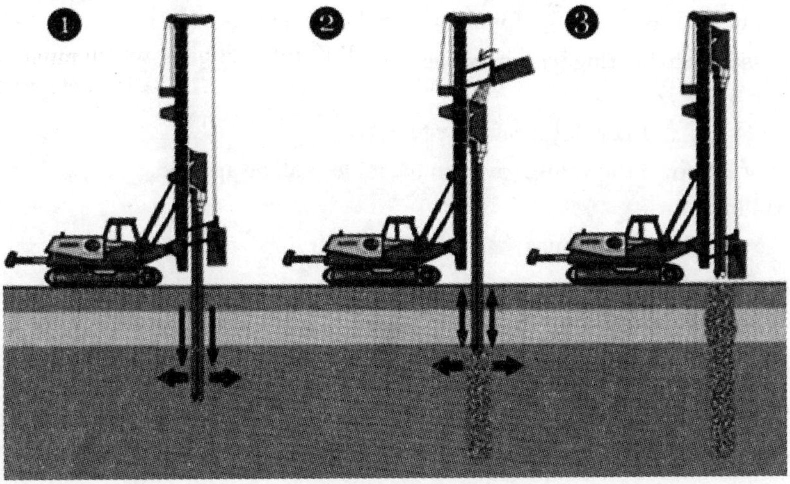

(b)

Figs 3.8(a and b): Illustration of construction of stone column (also called as **granular pile**) at site

Bulb diameter = 75 cm
Pile length = 2.5 m
The pile is driven in the medium compact sand or clay of medium consistency.

Solution:

a. Safe load in uplift:

From column 12, load for 3.5 m, 30 cm diameter pile = 8.00 tonnes
From column 15, decrease for 1.00 m length at the rate of 0.85 t/30 cm = 2.83 tonnes.
Thus, safe load in uplift = 8.0 – 2.83
= 5.16 tonnes **Ans.**

b. Safe load in compression (bearing):
From column 8, load for 3.5 m long pile of 30 cm diameter = 16.0 tonnes
From column 11, decrease for 1.0 m less length at the rate of 1.1 t/30 cm = 3.70 tonnes
Thus, safe load in compression = 16.0 – 3.70
= 12.30 tonnes **Ans.**

Example 3.5: Referring the Table 3.1 evaluate the safe load in compression and uplift for the single under-reamed pile of following details considering that the pile medium is:
(a) Medium compact sand ($10 < N < 30$)

Loose sand ($4 < N \leq 10$) or clay strata of soft consistency ($2 < N \leq 4$)
Stem diameter of pile = 30 cm
Bulb diameter = 75 cm
Pile length = 5.0 m
Solution:

a. Medium compact sand ($10 < N < 30$)
   i. *Safe load in uplift*:
   From column 12, load for 30 cm Uplift dia, 3.5 m long pile = 8.0 t
   From column 14, increased load for extra 1.50 m length @ 1.05t for each 0.30 m = 5.25 tonnes. Thus, safe load in uplift = 8 + 5.23 = 13.25 tonnes
   ii. *Safe load in bearing:*
   From column 8. load for 25 cm diameter and 3.5 m long pile = 12 t.
   From column 10, increase load for extra 1.5 m length @ 1.15t/0.30 m $\left(\frac{1.5}{0.3}\right) \times 1.15 = 5.75$ tonnes
   Thus, safe load in bearing (compression)
   = 12 + 5.75 = 17.75 tonnes **Ans.**

b. Loose sand ($4 < N \leq 10$) or clay strata ($2 < N \leq 4$)
   Reduction of 25% over the values given in the table shall be applied.
   i. Safe load in uplift
   = 0.75 × 13.25 = 9.94 tonnes
   ii. Safe load in bearing
   = 0.75 × 17.75 = 13.31 tonnes **Ans.**

Note: (a) For very loose sand strata ($N \leq 4$) or clay strata of very soft consistency ($N \leq 2$), a reduction of 50% over the table values shall be applied.

3.8 CAPACITY OF PILES IN INTERMEDIATE GEO-MATERIAL AND ROCK

(As per IRC: 78–2014, 'Standard specifications and code of practice for road bridges)

3.8.1 Axial Load Carrying Capacity

Piles in rocks and weathered rocks of varying degree of weathering derive their capacity by end bearing and socket side resistance. The ultimate load carrying capacity may be calculated from one of the two approaches given below:

Where cores of the rock can be taken and unconfined compressive strength is directly established using standard method of testing, the approach described in **method 1** shall be used. In situations where strata is highly fragmented, where RQD is nil or (CR + RQD)/2 is less than 30%, or where strata is not classified as a granular or clayey soil, or when the crushing strength is less than 10 MPa, the approach described in **method 2** shall be used. Also, for weak rock like chalk, mud stone, clay stone, shale and other intermediate rocks, method 2 is applicable.

METHOD 1

$$Q_u = R_e + R_{af} = K_{sp} \cdot q_c \, d_f \cdot A_b + A_s \, C_{us} \quad \ldots (3.28)$$

and,

$$Q_{allow} = (R_e/3) + (R_{af}/6) \quad \ldots (3.29)$$

where,

Q_u = Ultimate capacity of pile socketed into rock (Newtons)

Q_{allow} = Allowable capacity of Pile

R_e = Ultimate end bearing

R_{af} = Ultimate side socket shear

K_{sp} = An empirical co-efficient whose value ranges from 0.3 to 1.2 as per the table below for the rocks where core recovery is reported, and cores are tested for uniaxial compressive strength.

(CR + RQD)/2	K_{sp}
30%	0.3
100%	1.2

here,

CR = Core Recovery (%)

RQD = Rock Quality Designation (%)

For Intermediate values, K_{sp} shall be linearly interpolated

q_c = Average unconfined compressive strength of rock core below base of pile for the depth twice the diameter/least lateral dimension of pile (MPa)

A_b = Cross-sectional area of base of pile

$$d_f = \text{Depth factor} = 1 + 0.4 \times \frac{\text{Length of socket}}{\text{Diameter of socket}}$$

However, value of d_f should not be taken more than 1.2.

A_s = Surface area of socket

C_{us} = Ultimate shear strength of rock along socket length,

$= 0.225\sqrt{q_c}$, but restricted to shear capacity of concrete of the pile, to be taken as 3.0 MPa for M35 concrete in confined condition, which for other strength of concrete can be modified by a factor $\sqrt{(fck/35)}$

METHOD 2

This method is applicable when cores and/or core testing results are not available, or when geo-material is highly fragmented. The shear strength of geo-material is obtained from its correlation with extrapolated SPT values for 300 mm of penetration as given in table below:

Shear strength/consistency	*Moderately weak*	*Weak*	*Very weak*
Approx. N–Value	300–200	200–100	100–60
Shear strength/cohesion in MPa	3.3–1.9	1.9–0.7	0.7–0.4

(*Note*: 1 MPa ≃ 10 kg/cm^2 ≃ 100t/m^2)

$$Q_u = R_e + R_{af} = C_{u.b} \, N_c A_b + C_{us} A_s \quad \ldots (3.30)$$

and Q_{allow} is same as Eqn. 3.29. Thus,

$$Q_{allow} = (Re/3) + (R_{af}/6)$$

where,

$C_{u.b}$ = Average shear strength below base of pile, for the depth equal to twice the diameter/least lateral dimension of pile, based on average 'N' value of this region

C_{us} = Ultimate shear strength along socket length, to be obtained from table, based on average 'N' value of socket portion. This shall be restricted to shear capacity of concrete of the pile, to be taken as 3.0 MPa for M 35 concrete in confined condition, which for other strengths of concrete can be modified by a factor intermediate values C_{ub} and C_{us} can be interpolated linearly.

L = Length of socket.

$N_c = 9$.

Q_{allow} = Allowable capacity of pile.

The extrapolated values of 'N' greater than 300 shall be limited to 300 while using this method.

General notes common to Method 1 and Method 2

1. For the hinged piles resting on rock proper seating has to be ensured. The minimum socket length should be 300 mm in hard rock, and 0.5 times the diameter of the pile in weathered rock.
2. The allowable end bearing components after dividing by factor of safety shall be restricted to 5 MPa.
3. For calculation of socket friction capacity, the top rock 300 mm depth of rock shall be neglected. The friction capacity shall be further limited to depth of six times diameter of pile.
4. For the termination of working piles in the rocky strata methodology given in sub-clause 3.9 can be used as a quality control tool.

3.8.2 Moment Carrying Capacity of Socketed Piles

For the socketed pile, the socket length in the rock may be calculated from following equation:

$$L_s = \frac{2H}{\sigma_1 D} + \sqrt{\frac{4H^2}{\sigma_1^2 D^2} + \frac{6M}{\sigma_1 D}} \quad \ldots (3.31)$$

where,

L_s = Socket length

H = Horizontal force at top of the socket

M = Moment at the top of the socket

D = Diameter of the pile.

σ_1 = Permissible compressive strength in rock which is lesser of 30 kg/cm^2 or 0.33 q_c.

In case of socketed piles, for the satisfactory performance of the socket as fixed tip, the rotation at the top of the socket for the fixed condition (θ) should be less than or equal to 5% of the rotation for the pinned condition at the top of the socket (θ).

3.9 PILE TERMINATION CRITERIA AS A QUALITY CONTROL TOOL IN ROCKS

For establishing the similarity of soil strata actually met while advancing the pile-bore with the strata selected for terminating the pile on the basis of N values equivalent energy method can be used.

The concept of pile penetration ratio (PPR) is used in this method.

The PPR reflects the energy in tonne-metre required to advance the pile bore of 1 m^2 cross-sectional area by 1 cm.

1. In case of SPT test its PPR can be worked out as follows:
   Energy E spent for N blows = 63.5 kg × 75 cm × N blows (in kg - cm units) = E × 10^{-5} tonne metre. Area of samples is 0.786 × $(5.2)^2$ cm^2 = 21.24 cm^2, penetrating 30 cm.

   $$\text{Hence PPR} = 63.5 \times 75 \times N \times 10^{-5}/(21.24 \times 10^{-4} \times 30) = 0.747\, N \quad \ldots (3.32)$$

PPR for $N = 50 = 37.35$ tm/m^2/cm (Here, tm = Energy)
and for $N = 200 = 149.4$ tm/m^2/cm

where,
m^2 = Area
cm = Penetration

2. *PPR* (*P*), (for percussion piles) $= \dfrac{W.h.n}{A.P}$ ... (3.33)

where,
W = Weight of chisel in MT
h = Fall of chisel in 'm'
n = Number of blows of hammer
A = Area of pile in 'm^2'
P = Penetration in 'cm'

3. PPR (R), (for rotary piles) $= \dfrac{2\pi NTt}{AP}$ ... (3.34)

where,
N = Revolution per minute
T = Torque in '*tm*' for corresponding 'N'
t = Time in minutes
A = Area of piles in 'm^2'
P = Penetration in 'cm'

Chapter

4

Group of Piles

4.1 INTRODUCTION

Piles are seldom installed singularly but instead they are in groups. There are generally three piles under a foundation element because of alignment problems and inadvertent eccentricities. Therefore, building codes do not permit the use of less than three piles to support a major column and less than two piles to support a foundation wall. The group of piles may be any of the groups shown in Fig. 4.1.

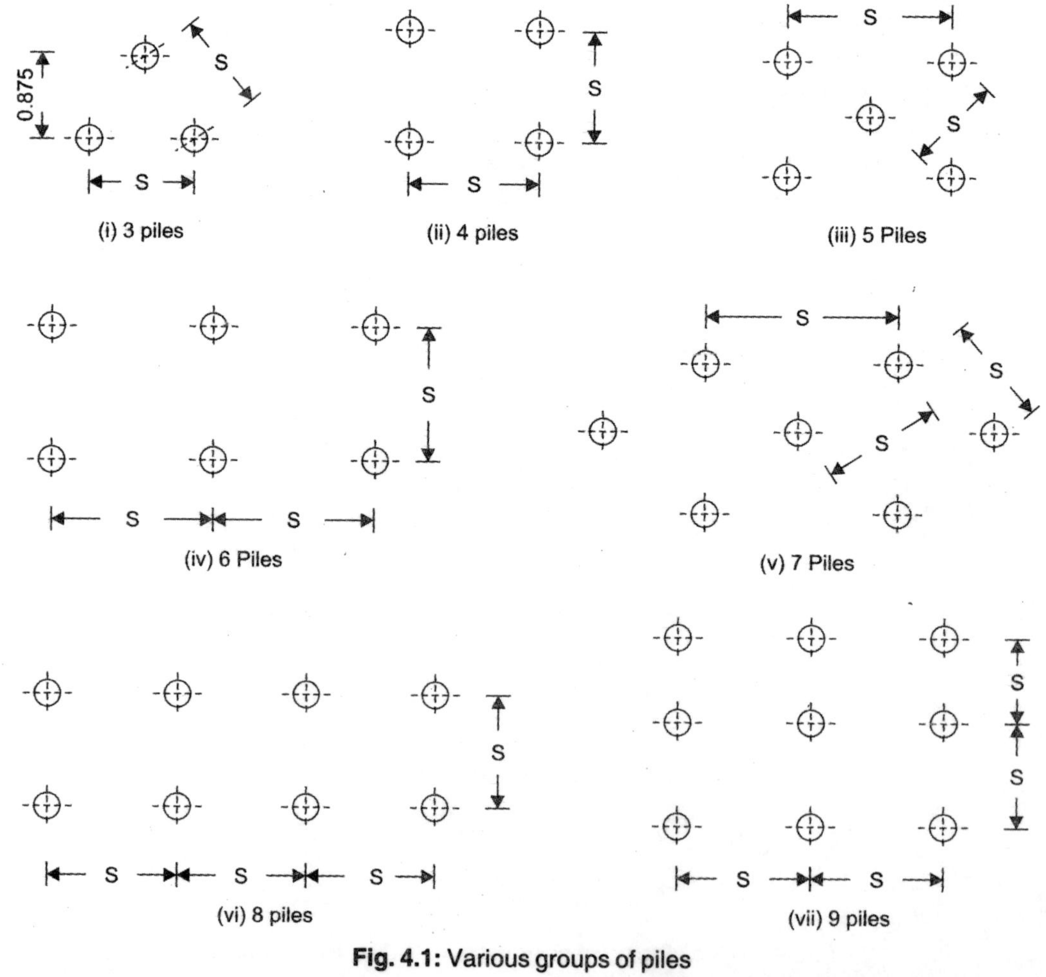

Fig. 4.1: Various groups of piles

The safe load capacity of individual pile is multiplied with the number of piles in a group to determine the group capacity of pile. This is perfectly true for friction pile, cast or driven into progressively stiffer materials or in end bearing piles. However, in case of friction piles in soft and clayey soils it is normally smaller. For driven piles in loose sandy soils, the group value may be higher due to the effect of compaction. In such a case, for determination of bearing capacity of the group, the load test should be made on a pile from the group after all the piles in the group are installed.

Figure 4.2 explains the overlapping of soil resistances that will develop in case of group of piles.

The bearing capacity of a pile group considering the piles spaced far apart and thus acting individually, may be taken as the sum of the individual pile capacities, i.e.

$$Q_{ug} = n.Q_u \qquad \text{... (4.1)}$$

where, n = No. of piles

Q_u = Ultimate bearing capacity of the individual pile

Q_{ug} = Ultimate bearing capacity of a pile group

From group action, the load bearing capacity of a group of piles is computed as:

$$Q_{ug} = \text{P.L.C.} + N_c.C_t.A \qquad \text{... (4.2)}$$

where, P = Perimeter of group of piles

A = Area of the pile group

C_t = Cohesion of soil at pile tip

C = Cohesion

N_c = Bearing capacity factor

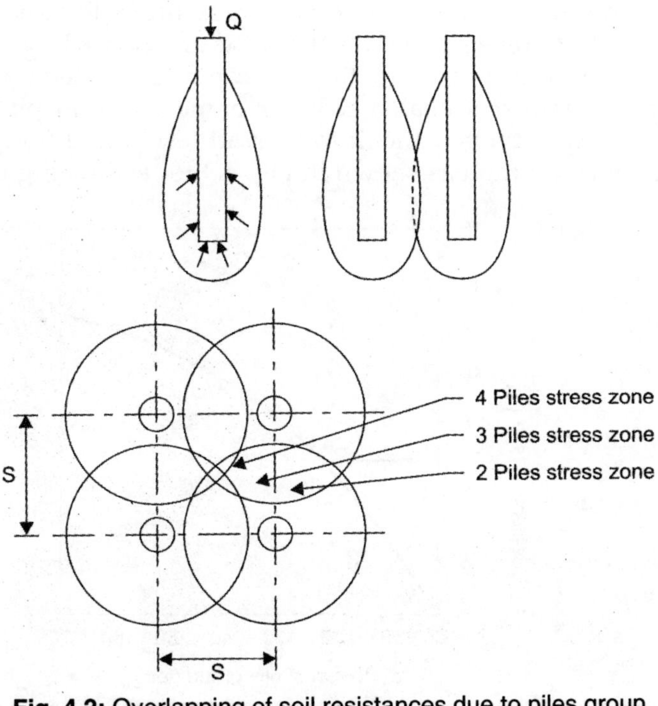

Fig. 4.2: Overlapping of soil resistances due to piles group

4.2 EFFICIENCY OF PILE GROUPS

The efficiency of a pile group or group efficiency of piles is the ratio of the actual group capacity to the sum of the individual pile capacities. The group efficiency is expressed as:

$$\eta_g = 1 - \theta \frac{(n-1)+(m-1)n}{90 \times m \times n} \quad \ldots (4.3)$$

where, m = Number of rows

n = Number of piles in a row

$\theta = \tan^{-1}(d/s)$

d = Diameter of pile

s = Spacing of pile centre to centre

4.3 SETTLEMENT OF PILE GROUPS

The total settlement of a single pile under axial load may consist of several components, *viz.*

1. Elastic compression of pile
2. Movement of pile relative to the surrounding soil
3. Settlement of surrounding soil due to the pile load
4. Settlement of soil under the pile tip (immediate and consolidation settlements)
5. Creep of pile material under constant axial load.

The closely spaced piles exhibit the overlapping stresses as is evident from Fig. 4.2. Therefore analysis of settlement of pile group becomes complex. Under equal axial load per pile, the pile group generally settles more than a single pile due to the stress overlapping. Therefore, the settlement of a pile group is larger than the settlement of a single pile and may be as large as 5 to 10 times, depending upon the size of the pile group. From elasticity considerations, the concept of pressure bulb indicates that, larger the group, the deeper the stresses penetrate the strata and hence larger the settlement.

The settlement of single pile under the load can be evaluated by load test. In case of group, the settlement must be evaluated by other means, as the pile group owing to be of its larger load capacity, is difficult to be load tested and much more expansive. Skempton presented a graph as shown in Fig. 4.3 for estimating the settlements of a

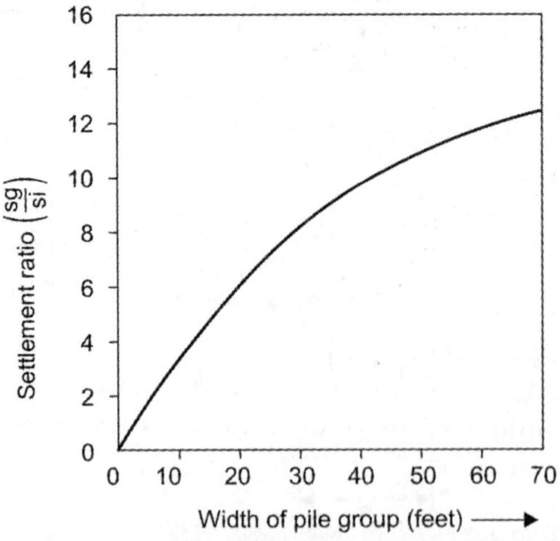

Fig. 4.3: Settlement of point bearing piles group (after Skempton)

point bearing pile group in sand having a usual pile spacing from the load settlement relationship of a single pile driven in granular soils. The settlement of a group of friction piles may be roughly estimated by assuming that the total load is distributed over the gross area of the group at a depth of two thirds of the pile length from the pile top. The soil above this two-thirds level is assumed to be incompressible and the presence of piles below is neglected. Settlement is calculated for the soil strata below this level.

The settlements of the pile groups may be computed by assuming the pile cap to be sufficiently rigid like rigid footing.

4.4 DESIGN OF PILE CAP

The pile cap is provided over the piles which supports the superstructure. The cap has to be designed with full skill so that there is no failure to structure. In general it is designed like a footing on soil but with the difference that instead of uniform reaction from the soil, the reactions in this case are concentrated on the piles which may be considered either point loads or distributed.

As per IS: 2911 (Part *I/Sec.* 3)-1979, the pile cap may be designed by assuming that the load from column is dispersed at 45° from the top of the cap up to the mid-depth of the pile cap from the base of the column or pedestal. The reaction from piles may also be taken to be distributed at 45° from the a edge of the pile, upto the mid-depth of the pile cap. On this bases the maximum bending moment and shear forces should be worked out at critical sections.

The pile cap accommodates deviations from the intended positions of pile sand by rigidly connecting all the piles in one group by a massive block of concrete, the ill effect of one or more defective piles are overcome by redistributing the loads. Caps for single piles must be interconnected by grade beams in two directions and for twin piles by grade beams in a line transverse to the common axis of the pair.

A deep cap is suitable for four piles. By adopting this arrangement the column load is transferred directly to the pile heads in compression.

A minimum recommended depth of cap is 30 cm but it should be thoroughly checked from shear criteria for diagonal tensions. The cap has to play a vital role for providing anchorage of column and piles through reinforcement.

When the pile reactions are considered as point load, the critical section for diagonal tension is located at a distance equal to one-half of the pile cap depth from the face of the column or pedestal.* In case of bending moment and shear for bond, the critical section is at the face of the column or pedestal.

According to IS: 456–1978, in computing the external shear on any section through a footing supported on piles, the entire reaction from any pile of diameter *D*, whose centre is located at *D*/2 or more outside the section shall be assumed as product shear on the section; the reaction from any pile whose centre is located *D*/2 or more inside the section shall be assumed as producing no shear on the section. For intermediate positions of the pile centre, the portion of the pile reaction to be assumed as producing shear on the section shall be based on straight line interpolation between full value at *D*/2 outside the section and zero value at *D*/2 inside the section:

In general, adopting one-half of the pile diameter instead of 15 cm as adopted by some designers, would be adequate. This is a very important point and should be kept in mind while designing a pile cap. As mentioned earlier, as per code, for the

**Sometimes pedestals are not provided. In such cases, the distance is taken from the edge of the column.*

case of distributed pile reactions, one method states to consider the load from column or pedestal to disperse at 45° to the mid-depth of the pile cap . The reaction offered by the piles is also distributed at 45° from the face of the pile to mid-depth of the cap. In this case also , the critical section for shear is located at a distance equal to half the pile cap depth from the face of column or pedestal and for the maximum bending moment, it is generally at centre. In this method, the resultant line of action of the distributed reaction from the pile does not coincide with the centre line of the pile.

The method considering the half of the pile diameter for shear is more in use. The reinforcement may be calculated using ultimate load theory. The clear overhang of the pile cap beyond the outermost pile in the group should normally be 10 to 15 cm depending upon the pile size.

The cap is generally cast over a 75 mm thick leveling course of concrete. The clear cover for main reinforcement in the cap slab shall not be less than 60 mm. The pile should project more than 50 mm into the cap concrete. Figure 4.4 shows the distribution pattern of maximum shear force and the reaction offered by the pile.

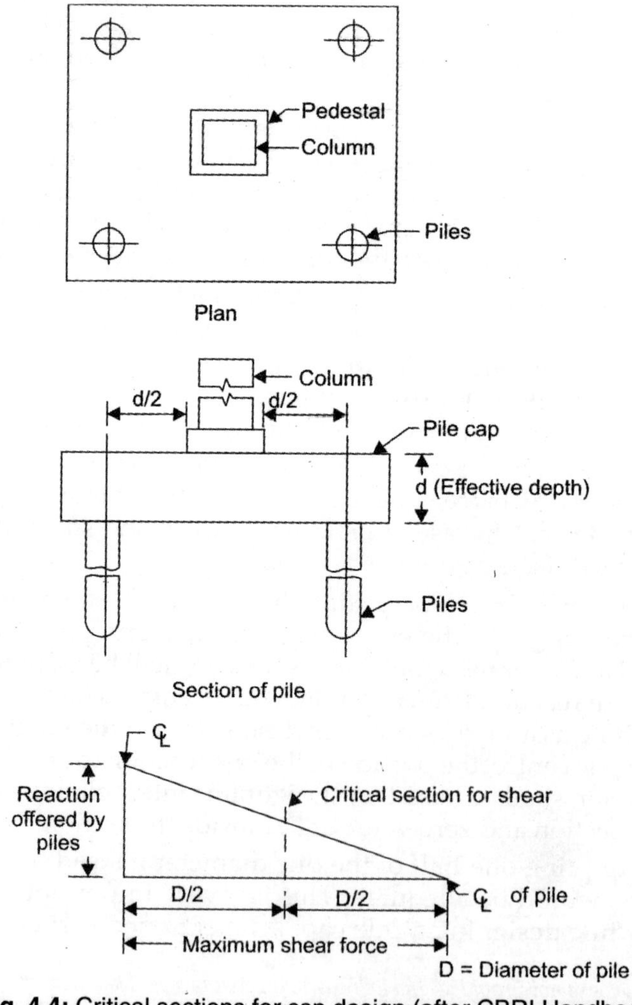

Fig. 4.4: Critical sections for cap design (after CBRI Handbook)

It also shows the plan and section of pile, pile caps and column (superstructure) assemblies. Whittle and Beattie have developed the relationship between dimension of the pile cap and the size of the pile. The dimension of some caps are shown in Fig. 4.5.

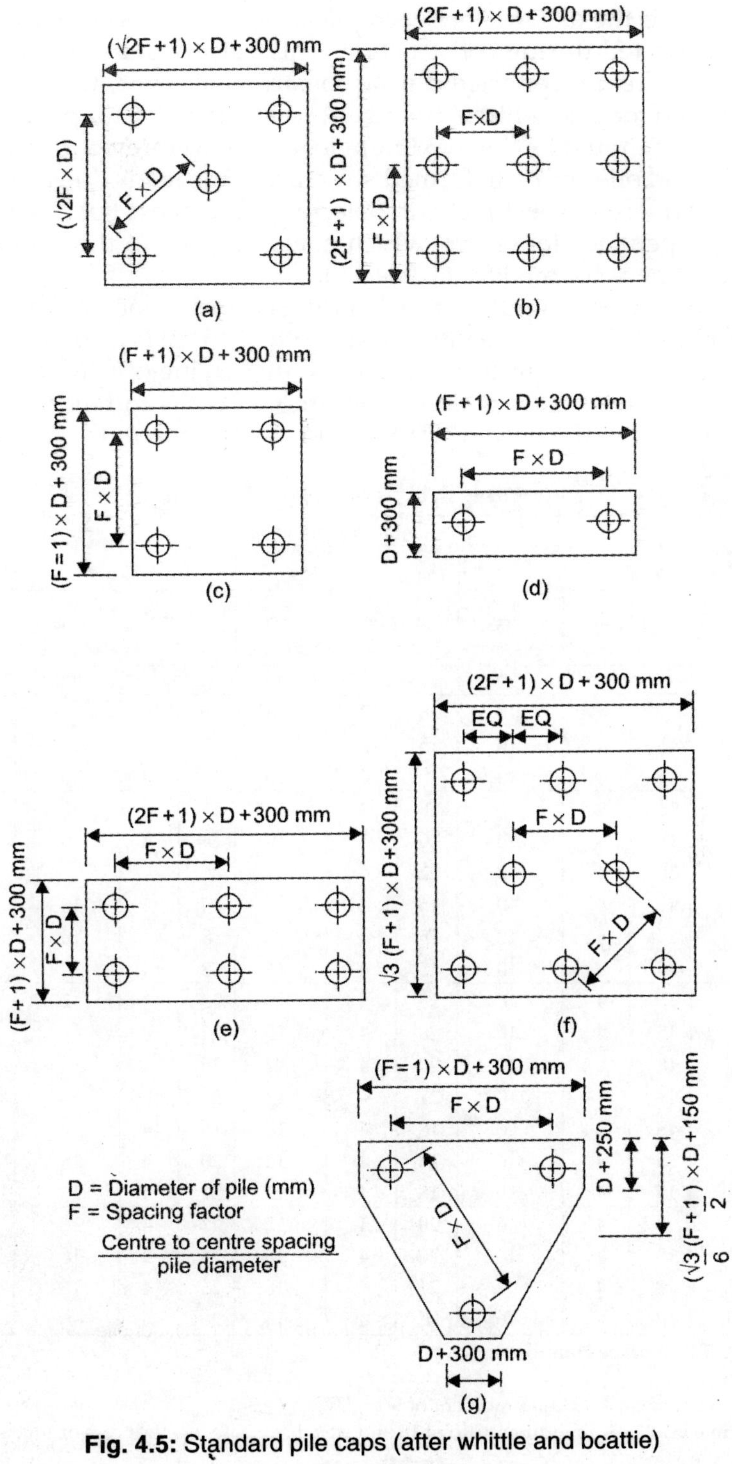

Fig. 4.5: Standard pile caps (after whittle and bcattie)

4.5 DESIGN OF GRADE BEAMS

Grade beams are provided over the piles to support the walls. The walls are usually made of bricks hence grade beams should be designed to take into account the panel action of masonry. The beam is designed for a bending moment of $wl^2/50$ where w is the intensity of uniformly distributed load per unit length (taken upto a maximum of 2 storeyes) and l is the distance between pile centres. The value of $wl^2/50$ is taken for the beams supported during construction. This value has to be enhanced to $wl^2/30$ when the beams are unsupported during construction. The ratio of wall height to the distance between the pile centres should not be less than 0.6 for panel action.

The value of w should be the maximum load of two storeyes in structure with load bearing walls and one storey in framed structures. The loads due to additional height or storey is considered to be directly transferred to the piles . For concentrated or other loads from suspended floors, etc. which come directly to the beams, full bending moment should be accounted for (Table 4.1).

The depth of beams computed from bending moments may be small in some cases. from practical considerations, a minimum depth of 15 cm is recommended. The top of the grade beam should be perfectly leveled so that laying of DPC, etc. may be proper. The beams are generally cast-*in situ* as continuous beams and the main reinforcement is kept continuous at the bottom throughout the beam length. An equal amount of

Table 4.1: Details of grade beams

		Spam of Beam, m											
Width of Beam	*Load on Beam Depth*	*1.5*			*2.0*			*2.5*			*3.0*		
		Depth		*Steel No. dia*	*Depth*		*Steel No. dia*	*Depth*		*Steel No. dia*	*Depth*		*Steel No. dia*
cm	*t/m*	*cm*		*mm*	*cm*		*mm*	*cm*		*mm*	*cm*		*mm*
1	2	3	4	5	6	7	8	9	10	11	12	13	14
	1.5	15	4	10	15	4	10	15	4	10	18	4	10
	3.0	15	4	10	18	4	10	18	4	10	21	4	10
23	4.5	15	4	10	18	4	10	21	4	10	24	4	12
	6.0	18	4	10	20	4	10	24	4	10	28	4	12
	7.5	18	4	10	22	4	10	26	4	12	30	4	12
	1.5	15	4	10	15	4	10	15	4	10	15	4	10
	3.0	15	4	10	15	4	10	18	4	10	22	4	10
	4.5	15	4	10	15	4	10	18	4	10	21	4	12
35	6.0	15	4	10	18	4	10	20	4	12	23	4	16
	7.5	15	4	10	20	4	10	22	4	12	25	4	16
	9.0	18	4	10	20	4	12	25	4	16	29	4	16
	10.5	18	4	10	21	4	12	25	4	16	29	4	16
	6.0	15	4	10	18	4	10	18	4	12	21	4	16
	7.5	15	4	10	18	4	12	20	4	16	23	4	16
	9.0	15	4	10	18	4	12	21	4	16	25	4	16
45	10.5	18	4	10	20	4	12	23	4	16	26	4	16
	12.0	18	4	10	20	4	16	24	4	16	28	4	16
	13.5	18	4	12	21	4	16	25	4	16	29	4	16
	15.0	18	4	16	22	4	16	26	4	16	31	4	16

Source: CSIR-CBRI Roorkee Handbook

Note:

1. The table is based on a bending moment of w/²/50
2. The concrete is M 150 of modular ratio of 18 and steel is for safe tensile stress of 1400 kg/cm².

negative reinforcement is provided to a distance of quarter span both ways from the pile centre. At least two bars of the top reinforcement should be kept continuous. Longitudinal reinforcement bars should not be less than 10 mm diameter and the minimum no. of bars both at the pile top and the bottom of the beam should be at least three on each side. The cranking up of the bars at the bottom is not considered very necessary. Stirrups of 6 mm diameter should be provided at a distance of 30 cm. and it will be sufficient for shear which is only nominal. No shear connectors are provide in the panel. In case of exterior beams over piles in expansive soils a ledge projection from the beam is provided on the outer side as shown in Fig. 2.1. It should be 75 mm thick and extend 80 mm into the ground. It is provided with nominal reinforcement.

The provision of reinforcement in beam for various loadings and spans is given in Table 4. 1.

4.6 STRUCTURAL DESIGNS OF PILES AND PILE GROUPS IN VARIOUS CONDITIONS

(with useful input from Prof. Bhupinder singh, IIT Roorkee)

Application of piles in bridge foundation (Fig. 4.6):

Diameter range	*Nomenclature*	*Applications*
(a) <175 mm	Micro-piles or grouted piles and seismic rehabilitation of Mini-piles	underpinning of building bridges, ground improvement, etc. Slopes protection, etc
(b) 175–300 mm >600 mm	Large diameter pile or drilled shafts. maximum practical diameter may be in the range 2~2.5 m	Bridges and tall buildings when rock deposits are available at resonance depths, if *not* adopt conventional piles!

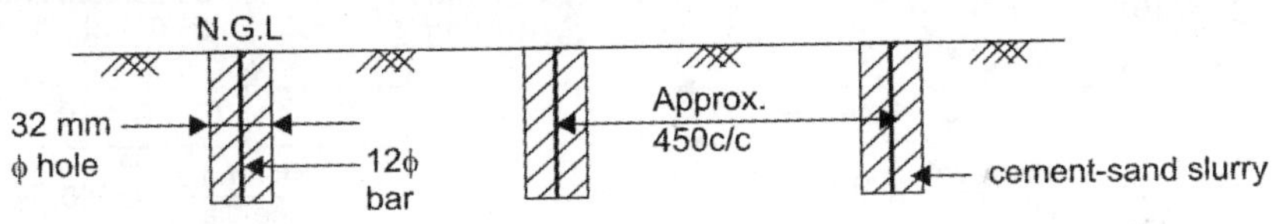

Fig. 4.6: A c/s of ples

Advantages of large diameter piles

1. Large diameter piles (typically 1500 mm φ, etc.) do not ned a pile cap for transferring loads to the piles.
2. Large diameter piles are easier to install in dense sand and gravel.
3. Large diameter piles resist lateral loads more efficiently.

Large diameter piles are most efficiently used when they can be terminated in rocks or socketed in weathered rocks.

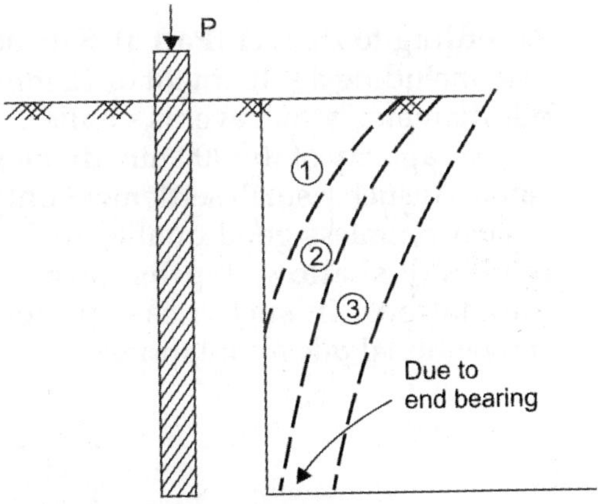

① : Load transfer at mobilisation of 10% frictional capacity
② : @ 30% frictional capacity
③ : @ 80% frictional capacity

Fig. 4.7: Load transfer mechanism

4.6.1 Load Transfer Mechanism in Large Diameter Piles (Fig. 4.7)

From static formulae, let allowable load = Q_{all}.

$$Q_{all} = \frac{Q_{ult.friction}}{(F.O.S)_f} + \frac{Q_{ult.bear}}{(F.O.S)_b}$$

$(F.O.S)_f = 1.5$, $(F.O.S)_b = 2.5$
overall (F.O.S) = 3.0

- Load is first resisted by friction and then by bearing
- Upon loading, the applied load is first resisted by upper part of pile by friction.
- Gradually the lower layers get stressed and finally end bearing resistance comes into action.

Consider a 900 mm ϕ bored cast—*in situ* circular pile
For 100% frictional, mobilization, required settlement is approx.

$$= \frac{0.6}{100} \times 900 = 5.4 \text{ mm}$$

For 100% bearing mobilization, required settlement is approx.

$$= \frac{6}{100} \times 900 = 54 \text{ mm}$$

$\therefore$ If approx. 54 mm settlement is allowed then the full frictional and bearing resistance is mobilized (Fig. 4.8).

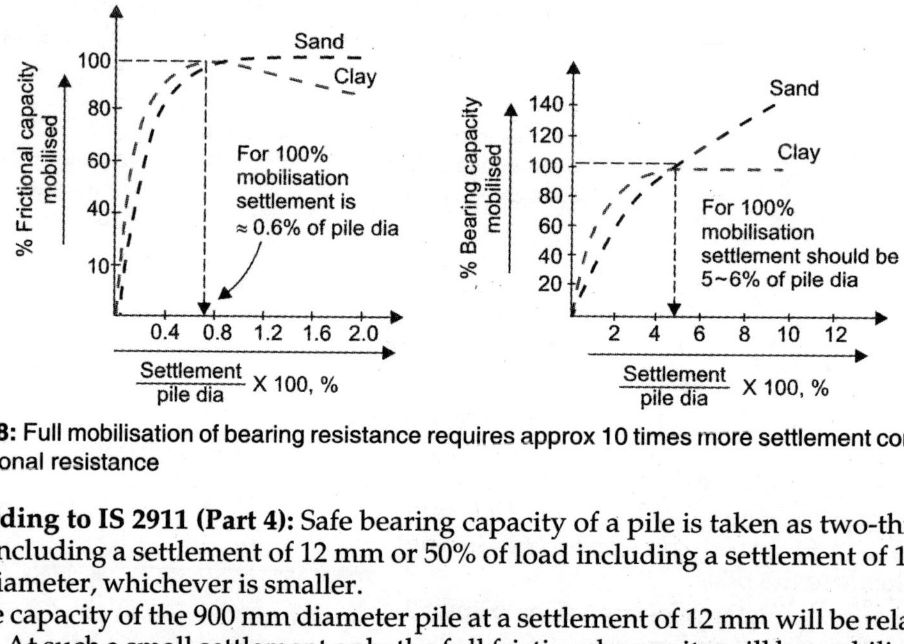

Fig. 4.8: Full mobilisation of bearing resistance requires approx 10 times more settlement compared to frictional resistance

According to IS 2911 (Part 4): Safe bearing capacity of a pile is taken as two-thirds of load including a settlement of 12 mm or 50% of load including a settlement of 10 % of pile diameter, whichever is smaller.

The capacity of the 900 mm diameter pile at a settlement of 12 mm will be relatively small. At such a small settlement only the full frictional capacity will be mobilized.

Hence, unless good quality rock or even weathered rock is available (Fig. 4.9) at relatively shallow depths, large diameter piles should *not* be used in bridge foundations. In such a case preference should be given to well foundations are conventional pile foundations!

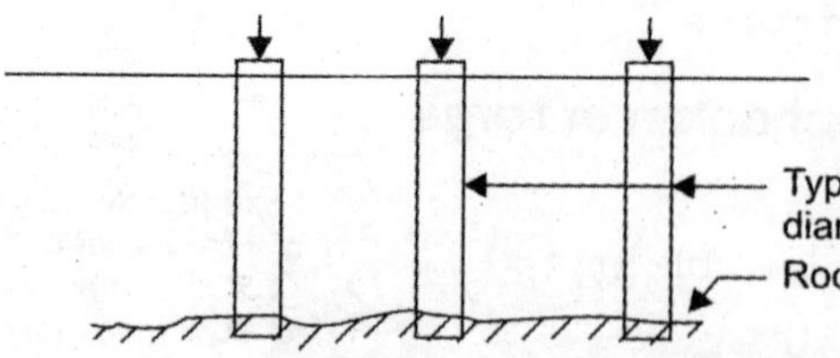

Fig. 4.9: Piles on rock

Rock quality for supporting large diameter piles

- Extract cores from the rock mass with core diameter being atleast 54 mm. Use diamond-tipped core-drilling maschine.
- Rock quality designation (RQD) is expressed as:

$$(RQD) = \frac{\text{Sum of length of cores of recovered length } \geq 100 \text{ mm}}{\text{Total length of drilling}}$$

RQD rock quality

<25	Very poor	
25 ~ 50	Poor	Preferable supporting large diameter piles
50 ~ 75	Fair	
75 ~ 90	Good	
90 ~ 100	Excellent	

- Find cylinder strength of rock cores with H/D = 2.0

Unconfined comp. strength v/s type of rock

>80 MPa	Good rock	Preferable supporting large diameter piles
~40 MPa	Moderately weak	
~20 MPa	Weak	
~8MPa	Very weak	

4.7 PRACTICAL SELECTION OF PILE PARAMETERS

i. Selection of type of pile:

- The soil profile at site representing the results of exploratory boring at site should be available so as to ascertain the nature of soil at site.
- Driven piles can be precast and installed according to specifications. The piling work can be planned in advance. They cause vibrations during driving operations and may affect adjoining structures. They are not suitable in clayey soils as build-up of pore-water pressure during operations is not desirable.
- Cast-*in situ* piles offer flexibility in use. Suitable where vibrations can be avoided and in clayey soils.
- Spacing between piles in a group: IS 2911 (Part1) – 1979

Length of pile	*Friction piles in sand*	*Friction piles in clay*	*End bearing piles*
Less than 12 m	38	48	38
12 to 24 m	48	58	48
> 24 m	58	68	58

where 'B' is pile diameter or larger edge of pile c/s.

- Minimum clear projection of pile cap beyond edge of outermost pile = 100 mm or diameter of pile whichever is larger (measured from centre of pile) (Fig. 4.10).

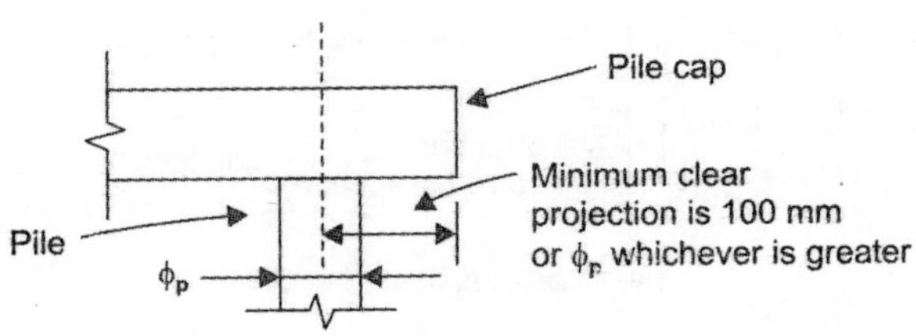

Fig. 4.10: Minimum clear projection of pile cap

Practical suggestions for construction of bored cast—*in situ* piles.

1. Piling work shall be carried out as per procedure laid down in IS-2911(Part-I, section 2) 1979 for large diameter bored cast—*in situ* piles.
2. The pile shall preferably have a clear cover of 100 mm. Cover shall be provided using 50 mm thick and 200 mm diameter rollers made of cement: Coarse sand mortar 1 : 1.5.
3. Desirable clear cover at top of pile cap: 50 mm, at bottom of pile cap: 50 mm, at bottom of pile: 75 mm (Fig. 4.11).
4. Concreting of pile shall be done in one continuous operation using a tremie pipe. Desirable slump is 100 to 150 mm.
5. For overlap/splicing of pile longitudinal reinforcement shall be staggered. Longitudinal distance between any two lap zones shall not be less than one lap length (Fig. 4.12).
6. Very long reinforcement (r/f) cage for pile cannot be fabricated as one unit. The cage may be fabricated in two to four units with appropriate splicing.

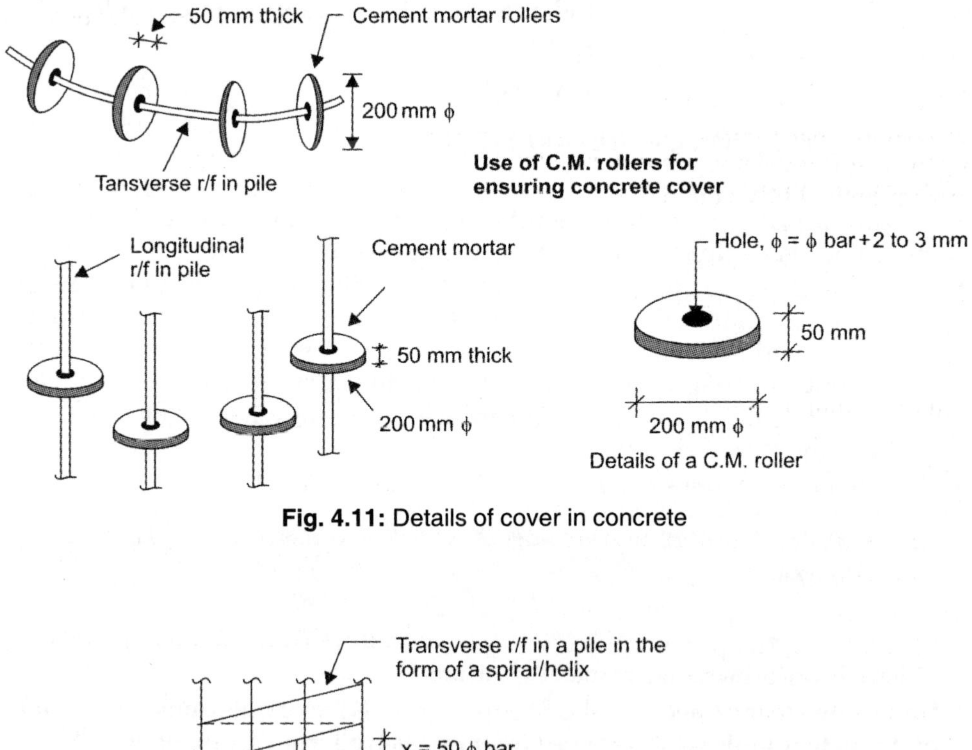

Fig. 4.11: Details of cover in concrete

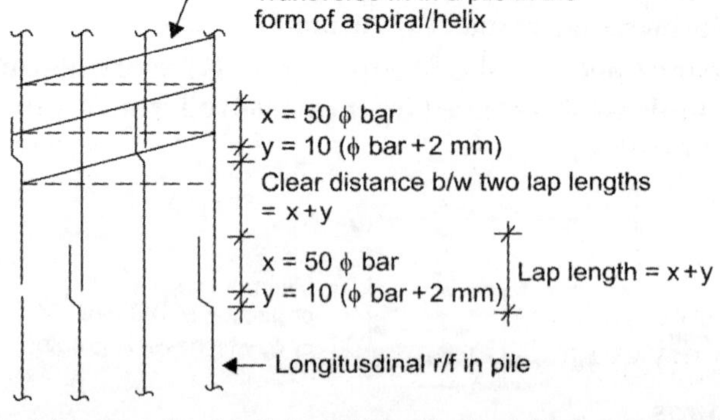

Fig. 4.12: Details of splicing of longitudinal reinforcement (r/f)

7. To impact rigidity to very long r/f cages, for handling and placing purposes, 16 T rings @ 5000 mm c/c may be provided along the length of the cage. These rings may be tack welded with alternate longitudinal bars and rigidly tied with binding wire with the remaining bars (Fig. 4.13).

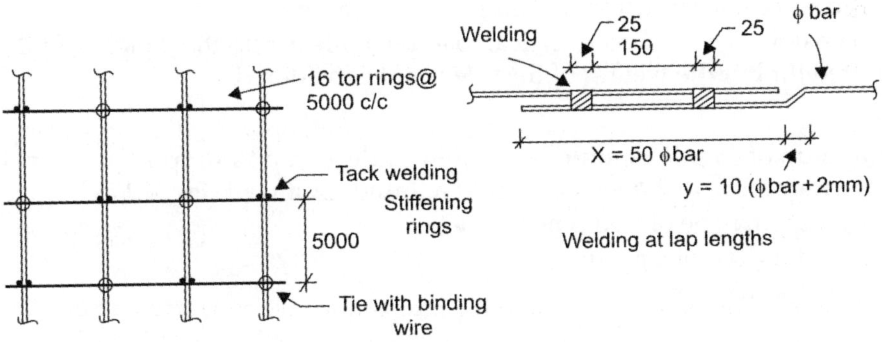

Fig. 4.13: Details of rings

8. Overlapping/splieed bars shall be welded in 25 mm lengths at two places 150 mm apart in the middle of the lap length.
9. When the cut-off level of the pile is below the ground/working level, the pile shall be cast at least 1.0 m above the cut-off level and bad concrete is to be chipped-off upto cut-off level such that hard, strong and pure concrete pile head is available at cut-off level. Only concrete above cut-off shall be removed leaving the *r/f* bars intact (Fig. 4.14).
10. The cut-off level shall be 50 mm above the bottom of the pile cap to ensure proper grip of pile inside the cap for transferring lateral loads.
11. All loose material shall be flushed out from the bore hole after lowering the r/f cage and before starting concreting.
12. For piles to be concreting in river beds having sandy soils and under waters, direct mud circulation method of boring shall be used.

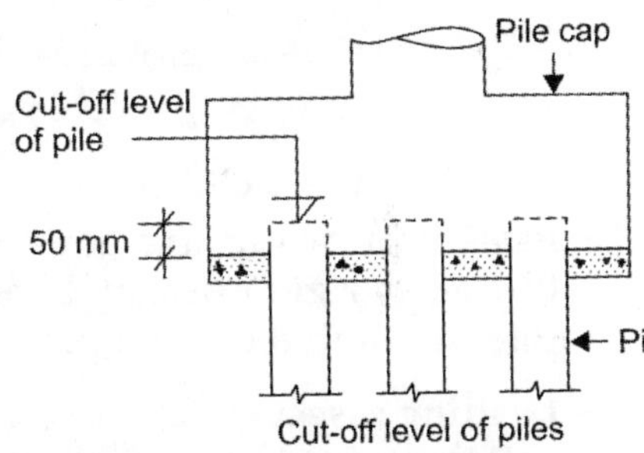

Fig. 4.14: Details of cut-off level

13. When concreting is done under water/drilling mud, the pile shall be cast 300 mm above G.L. After 3 days the concrete in the top portion which is usually contaminated by drilling mud etc. and is weak, shall be chipped off until hard and sound concrete surface is obtained. The pile is then extended upto cut-off level after making proper cold joint for which the pile head is cleaned properly and 1:1 cement coarse sand mortar layer 12 mm thick is applied and concrete is poured and vibrated in shuttering.

4.8 BATTER PILES

For inclined loads or heavy lateral loads under-reamed piles can be provided a maximum batter of 30° under dry ground conditions and 10–15° under submerged conditions.

Design of pile groups.

Pile groups may be subjected to (i) Vertical loads (ii) Moments (iii) Lateral loads.

Assumptions:

i. Connection between pile and pile cap is considered to be hinged. Hence no moment is transferred to the pile. Hence the effect of moment is considered in terms of an equivalent axial load (tension or compression on the pile).

ii. The lateral load on the pile group is assumed to be equally shared by all the piles. [Justified since lateral stiffness of all pile is the same.]

iii. When considering the direct vertical load on a pile group the weight of pile cap is accounted for but the weight of the pile is not considered.

Note:

i. Minimum of 03 piles should be used in a pile group to support a major column and a minimum of 2 piles to support a foundation wall (Fig. 4.15).

ii. Pile group may be of two types:

a. Free standing pile group

b. Piled foundation.

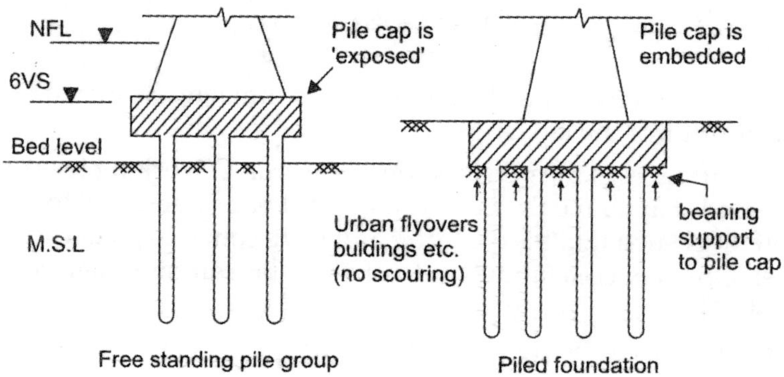

Fig. 4.15: Pile group

- Piles in a group are connected to each other through a rigid pile cap. Loads are transferred to piles through the pile cap.
- **Loading cases:**

  i. Pile cap subject to axially applied concrete load (Figs 4.16 to 4.18).

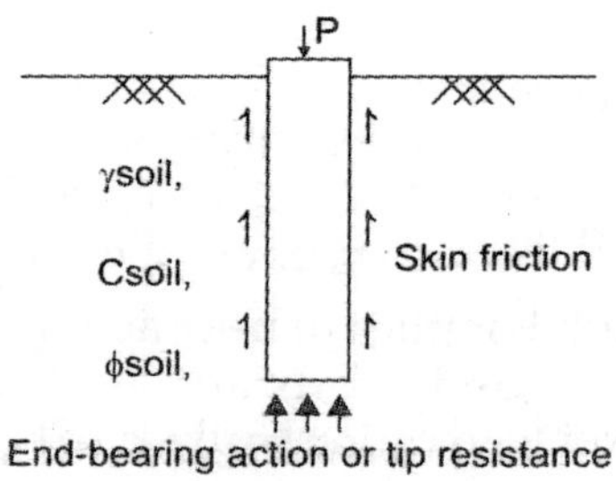

Fig. 4.16: Pile cap subjected to axially applied load

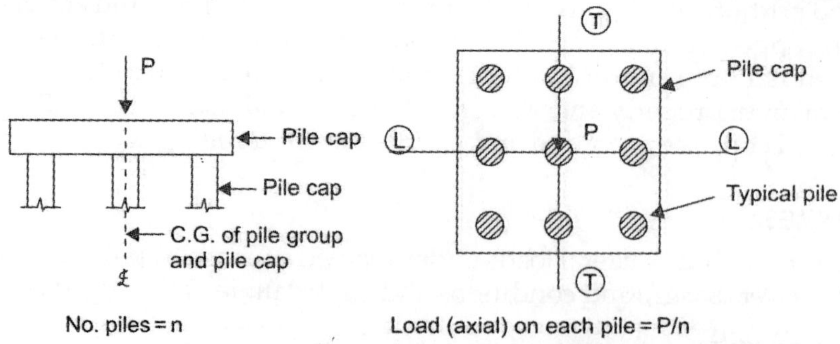

Fig. 4.17: Pile cap on group of piles with axial load

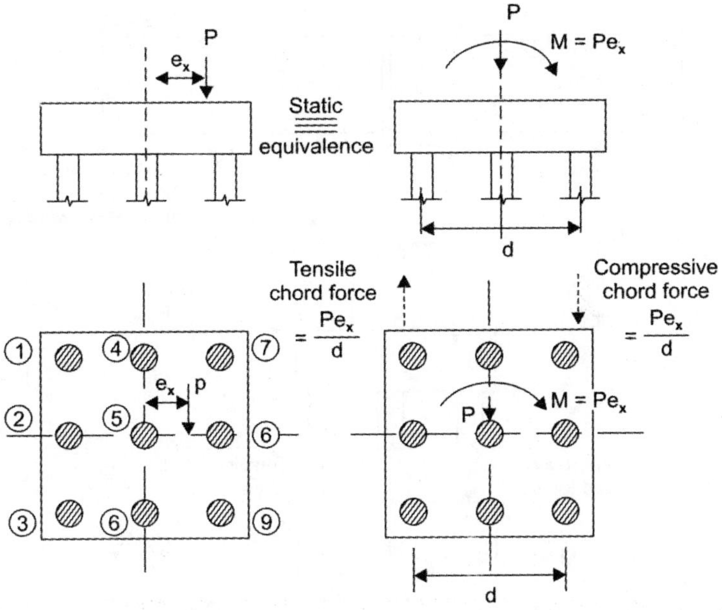

Fig. 4.18: Pile cap subjected to eccentric load

Following relationships will hold good:

Load with uniaxial eccentricity $e_x \neq 0,\ e_y = 0$

Load on each pile due to axial force $= \dfrac{P}{n}, n = 9$ (Comp).

Chord force due to moment $(M = Pe_x) = \dfrac{Pe_x}{d}$ (Comp/tension)

Tensile load on each of piles 1, 2, 3 $= \dfrac{Pe_x}{3d}$ due to moment 'M'

Compressive load on each of piles 7, 8, 9 $= \dfrac{Pe_x}{3d}$ due to moment 'M'

Load on each of piles 4, 5, 6, due to moment 'M' = zero

Net load on each pile 1, 2, 3 $= \dfrac{P}{n} - \dfrac{Pe_x}{3d}$ [compression positive]

Net load on each of piles 7, 8, 9 $= \dfrac{P}{n} + \dfrac{Pe_x}{3d}$

Net load on each of piles 4, 5, 6 $= \dfrac{P}{n}$

In a similar manner, axial loads on each pile in a group due to biaxial eccentricity $(M_x \neq 0, M_y \neq 0)$ can be computed by considering 'M_x' and 'M_y' separately.

Pile group subjected to a lateral load (Figs 4.19 and 4.20):

Following are the issues in the case:

- How is the lateral force F_x to be distributed amongst piles of the group?
- How is the proportioned lateral force resisted by each pile in a pile group?
- Lateral load analysis and design of piles.

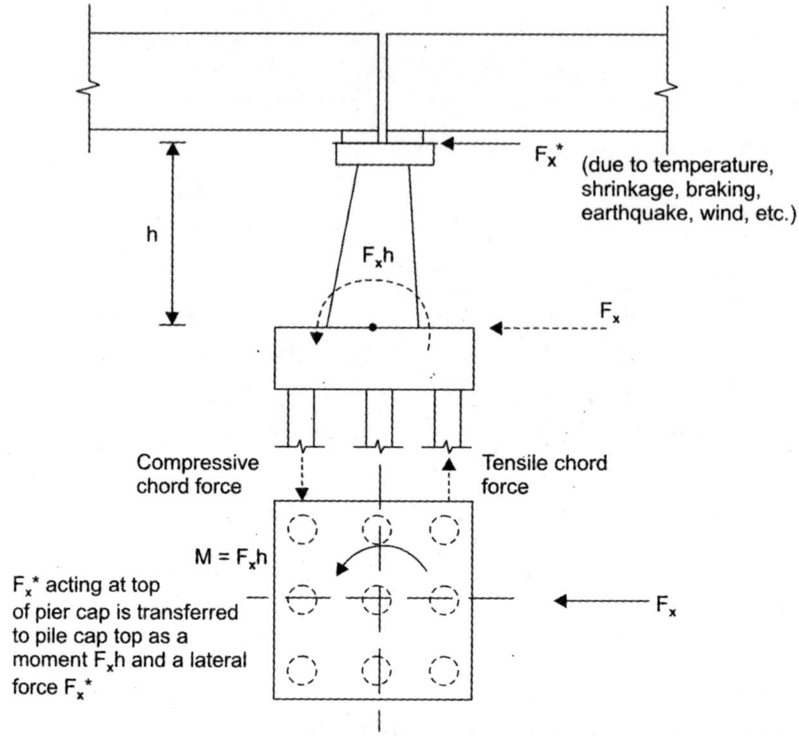

Fig. 4.19: Pile cap subjected to lateral load

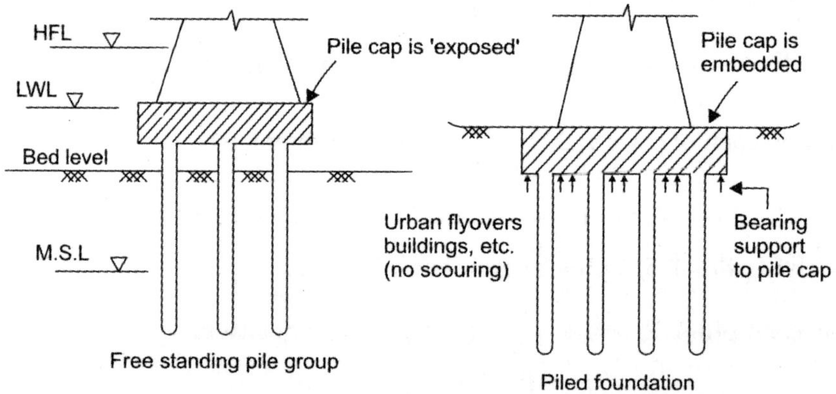

Fig. 4.20: Distribution of load between vertical piles of a pile group

Generalised expression

a. Eccentricity of vertical load about one axis (Fig. 4.21)

$$P_{ip} = \left(\frac{P}{n}\right) \pm \left(\frac{P.e.x_i}{I_g}.A\right)$$

where,

P_{ip} = Axial load on i[th] pile

P = Total vertical loading acting on pile group

n = No. of piles in pile group.

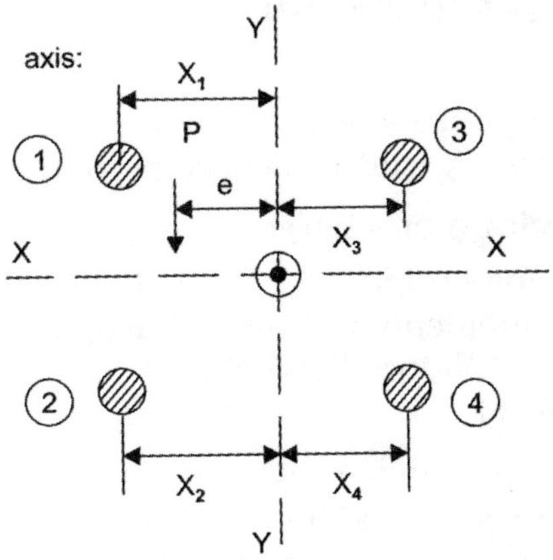

Fig. 4.21: Eccentricity of vertical load about and axis

e = Amount of eccentricity w.r.t centre of pile group.

x_i = Distance of centre of ith pile from centre of pile group measured parallel to 'e'

I_g = Moment of inertia to the piles about axis normal to the direction of eccentricity

$= Ax_1^2 + Ax_2^2 + ------ + Ax_n^2$

A = Area of pile c/s,

Since all piles in the group are assumed to be identical,

$$P_{ip} = \left(\frac{P}{n}\right) \pm \left(\frac{P.e.x_i}{\Sigma x_i^2}\right)$$

b. Eccentricity about two axes (Fig. 4.22)

$$P_{ip} = \frac{P}{n} \pm \frac{P.e_x.x_i}{\Sigma x^2} + \frac{P.e_y.y_i}{\Sigma y^2}$$

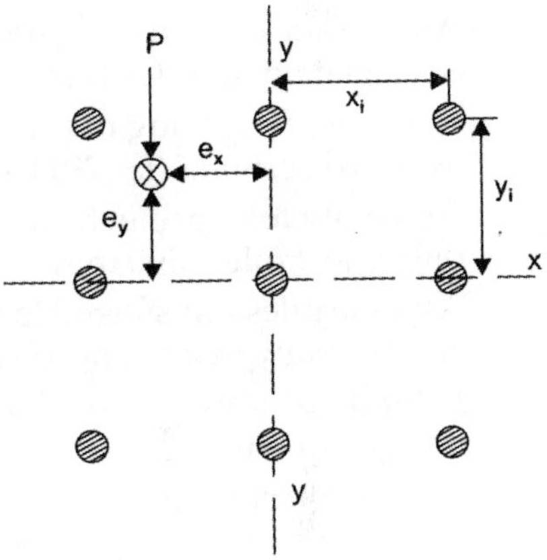

Fig. 4.22: Eccentricity about two axes

4.9 BOUNDARY CONDITIONS FOR PILE ANALYSIS

Three types of boundary conditions occur in practice:

i. **Fixed head pile:** A fixed head pile is free to move laterally but rotation is prevented completely. If longitudinal steel of piles is adequately embedded in the pile cap, the pile may be assumed to be fixed at the top and at the same time atleast three piles are connected to a rigid pile cap (Fig. 4.23).

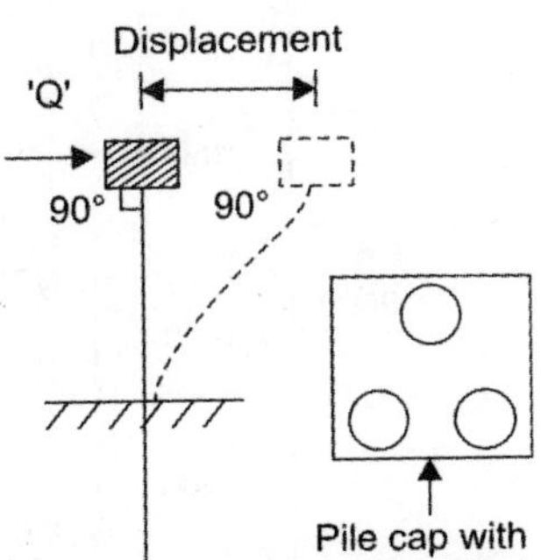

Fig. 4.23: Fixed head pile

ii. **Free head pile:** The pile head is free to rotate and displace without any lateral restraint. This condition is obtained in Jetties etc. At best, in such structures, the behaviour may be that of a partially restrained head pile (Fig. 4.24).

iii. **Partially restrained head pile: as above:** A pile cap containing either one or two piles only should be lateral restrained by orthogonal grained to be considered for case of fixed head pile.

4.10 SAMPLE DESIGN OF PILE FOUNDATION FOR BRIDGE PIER

Case 1 Piles in non-river bridge crossings.
Case 2 Piles in river bridge crossing (consider effect of scour)

4.10.1 Piles in non-river bridge crossings

Design the pile foundation for a bridge pier located in an urban area. The direct load, moments and lateral forces at the soffit of the pile cap are as follows (Fig. 4.25):

Total vertical load = 5929 kN.

Total moment about L-L axis = 1092 kNm

Total moment about T-T axis = 530 kNm

Total horizontal force along L-L axis = 83 kN

Total horizontal force along T-T axis = 172 kN

(practical pile sizes: 300,350, 400, 450, 500, 550, 600 - - - - 1500 mm ϕ)

- Assume all piles to be driven cast—*in situ* with ϕ = 500 mm

  Assume a total of 20 piles under the pier proposed arrangement of piles is as follows:

  Assume c/c spacing of piles to be 3 ϕ = 3 × 500 = 1500 mm in both directions. (IS : 2911(Part 1) -1979 recommendations)

  Assume clear projection of pile cap beyond centre of outermost pile = 500 mm.

  Assume piles transferred load partially due to skin friction and partially due to end-bearing action

  Lateral loads are assumed to be equally shared by all piles in a group (Fig. 4.26).

  ϕ for soil at site = 33.5°

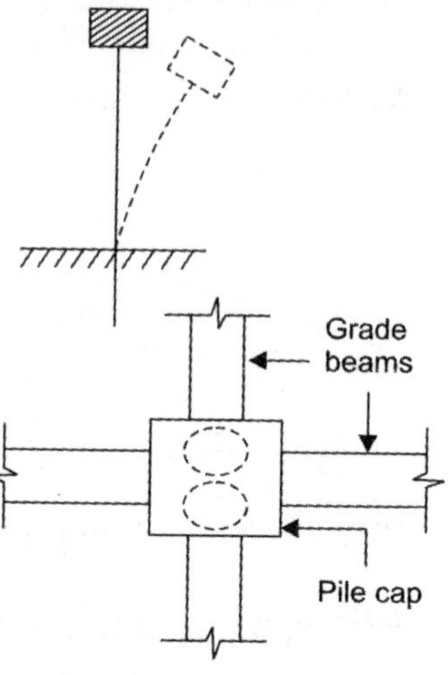

Fig. 4.24: Free head pile

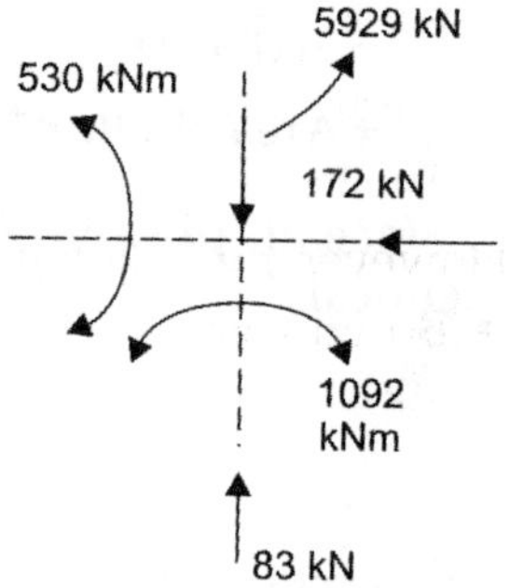

Fig. 4.25: Loads acting on pile cap

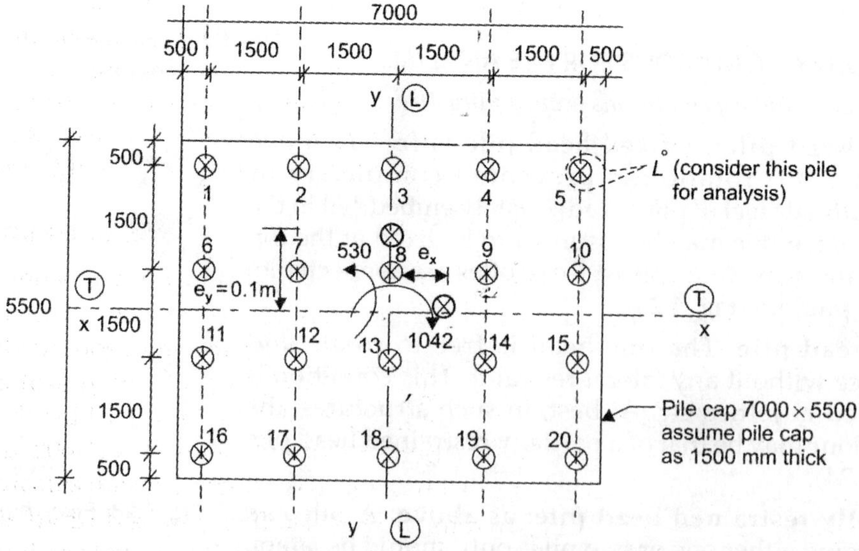

Fig. 4.26: Lateral loads acting on pile group

$$e_x = \frac{1092}{5929} = 0.184 \text{ say } 0.19\ m,\ e_y = \frac{530}{5929} = 0.089 \text{ say } 0.10\ m$$

Maximum vertical load on pile 'i' = P_i

$$P_i = \frac{P}{n} \pm \frac{P.e_x.x_i}{\Sigma x^2} + \frac{P.e_y.y_i}{\Sigma y^2}$$

$$\Sigma x^2 = 2[4 \times 1.5^2 + 4 \times 3^2] = 90 \text{ m}^2$$

$$\Sigma y^2 = 2[4 \times 0.75^2 + 4 \times 2.25^2] = 45 \text{ m}^2$$

$$\therefore \quad P_i = \frac{5929}{20} \pm \frac{5929 \times 0.19 \times 3}{90} + \frac{5929 \times 0.10 \times 2.25}{45}$$

or $\quad P_i = 296.45 + 37.55 + 29.64 = 363.64$ kN

Lateral load per pile in (L) –(L) direction = $\frac{83}{20} = 4.15$ kN

Lateral load per pile in (T) –(T) direction = $\frac{172}{20} = 8.6$ say 9 kN
(critical lateral load)

Under-reamed piles (Features, Analysis and Design):

The details of under-reamed pile foundations are given in other chapters also. However, for maintaining the continuity, some details are reproduced below (Fig. 4.27):

- Bored cast-*in situ* piles having one or more bulbs formed by enlarging the pile system

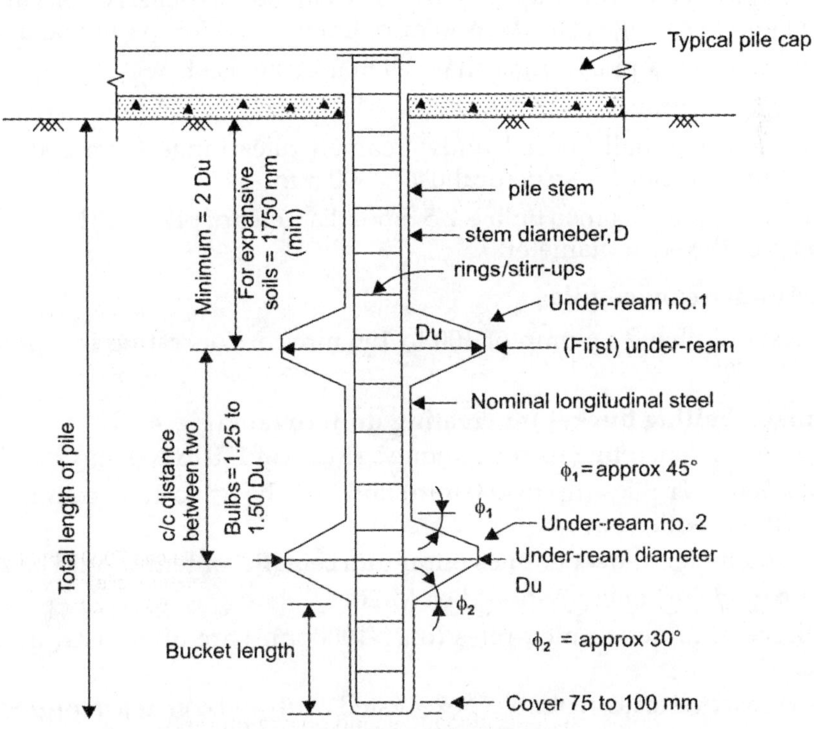

Fig. 4.27: Typical elevation of a double under-reamed pile

- Suited for expansive soils [These are subjected to significant ground movements due to seasonal moisture variations.]
- Also suited for filled up grounds and loose strata and foundations subject to uplift pressures.
- Used where strata of adequate bearing capacity is very deep.
- Load is carried by under-reamed pile through skin friction, end-bearing at pile tip and bearing action at the under-reams (Fig. 4.28).

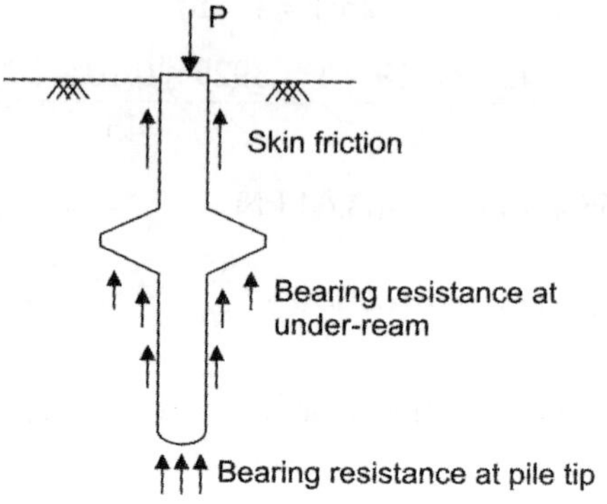

Fig. 4.28: Load carried out by under-reamed pile

- *Length of under-reamed piles*
  - In deep deposits of expansive clayey soils minimum length of pile: From ground level to the pile tip shall be 3500 mm so that piles are not affected by seasonal ground movement.
  - Piles should bypass filled up grounds or weak non-expansive soils and made to rest on good bearing strata. Hence, piles longer than 3.5 m may be used.
  - In sloping ground piles longer than 3.5 m may be used.
- *Stem and bulb diameter*
  - Diameter of manually bored under-reamed piles range from 200 to 500 mm. Minimum diameter for structural use = 300 mm.
  - Diameter of under-reamed bulb is 2.5 times diameter of pile stem but can vary from 2 to 3 times the stem diameter.

Features of under-reamed piles

- Concrete should have a slump of 100 to 150 mm for concreting in water free bore holes.

Bottom hinged belling bucket for creating under-ream (Fig. 4.29)

- For underwater concreting using a tremie a slump of 150 to 200 mm should be used.
- For small diameter piles (up to 400 mm) and lengths up to 10 m minimum cement content should be 350 kg/m^3.
- For larger diameter or deeper piles minimum cement content should be 400 kg/m^3.
- Minimum grade of concrete should be M-20.
- Large diameter under-reamed piles (dia >1000 mm) are often used in urban area flyovers.
- Savings in cost due to less volume of excavated earth and concrete required to replace the earth.

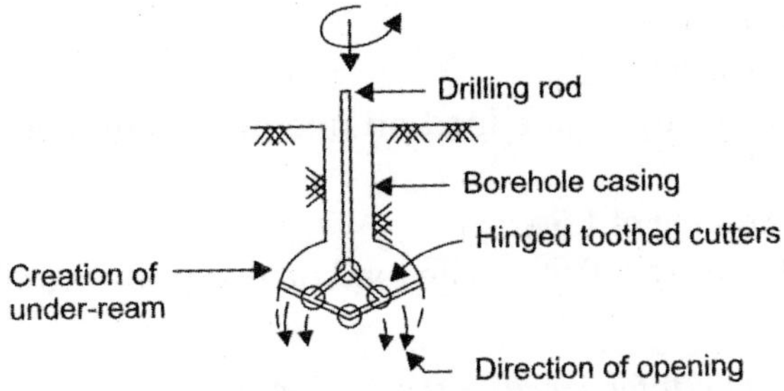

Fig. 4.29: Bottom hinged belly bucket

- Under-reamed piles are different to construct in bouldery soil or in cohesion-less soil where it may be difficult to form a large bulb.

  Spacing of bulbs along pile stem (Fig. 4.30)

  □ For pile stem diameter ≤ 300 mm, spacing of bulbs ≯ (not more than) 3.5 times diameter of bulb.

  For example consider pile stem diameter = 400 mm

  ∴ Diameter of under-ream bulb = 2.5 × 400 = 1000 mm

  ∴ Spacing of bulbs = 3 × 1000 = 3000 mm < 3.5 × 1000. OK.

- For pile stem diameter > 300 mm bulb spacing can be reduced to 1.25 bulb diameter.
- Top most bulb should be at a minimum depth of 2 × bulb diameter. In expansive soils top-most bulb should be at a minimum depth not less than 1.75 m below G.L.
- Minimum clearance below soffit of pile cap embedded in ground and bulb should be 1.50 × bulb diameter.
- Minimum stem diameter for bore holes needing stabilization by bentonite slurry shall be 250 mm.
- Minimum stem diameter for strata containing harmful chemicals like sulphates, etc. shall be 300 mm.

Spacing and positioning of piles:

- c/c spacing of under-reamed piles in a group should be $2\phi_b$ but not less than $1.5\,\phi_b$. For piles supporting grade beams maximum spacing of piles shall be 3.0 m.
- There is no upper limit on pile spacing but larger spacing means large size of pile caps.

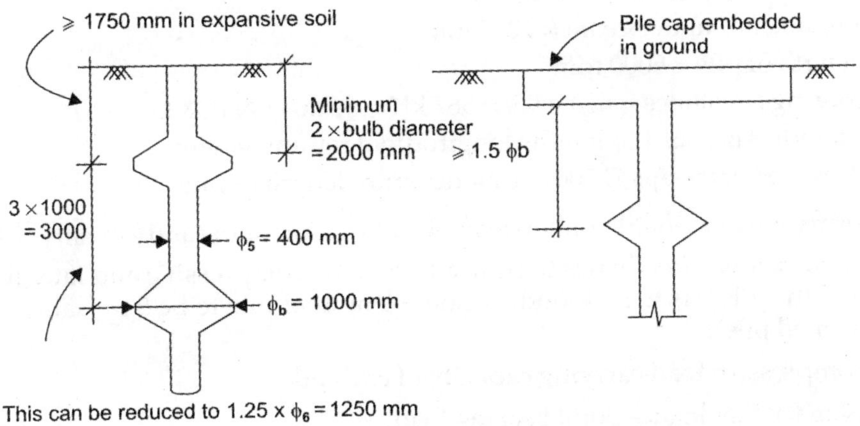

Fig. 4.30: Spacing between bulbs

Group efficiency of piles

i. For piles at a c/c spacing of 2ϕ_b:

Safe load on pile group = safe load on one pile × number of piles in the group. (100% efficiency)

ii. For piles at a c/c spacing of 1.5ϕ_b:

Safe load on pile group = 0.90 × safe load on one pile × number of piles in group.

Reinforcement

Minimum area of stem reinforcement = 0.4% of c/s area

- Reinforcement shall be provided for full length of pile.
- Minimum 3 nos. 8φ or 3 nos. 10ϕ M.S bars shall be provided as longitudinal steel.
- Minimum transverse steel shall be 6ϕ at a spacing not exceeding 300 mm or stem diameter whichever is less.
- Minimum cover to longitudinal steel shall be 40 mm.
  for soil containing sulphates, etc. minimum cover shall be 75 mm.
  (For taking preliminary guidance about safe loads, etc. on piles, readers may refer another chapter given in this book.)

If under-reamed pile has to be used then, as per IS: 2911 (Part III)–1980, for a single under-reamed pile length 3.5 m, the load carrying is (2.5 × 500 = D_u = 1250)

Bearing resistance = 420 kN > 363.64 kN IS: 2911 (PartIII)–1980

Uplift resistance = 210 kN > pile will be compression even under over turning moments

Lateral load resistance = 45 kN > 9 kN

Hence the pile considered is ok.

Note:

1. The provided spacing of piles = 1500 mm < 2 bulb dia = 2 × 1250 = 2500 mm. Hence, reduced allowable load to 90% = 0.90 × 420 = 378 kN to account for group behaviour. However, 378 > 363.64 kN. Hence OK.
2. Margin of safety is less, i.e. 378 ≃ 363.64 kN.
   Adopt double under-reamed pile of same specifications.

   Thus, revised specifications are:
   i. Diameter of pile stem = 500 mm
   ii. Diameter of under-ream = 1250 mm
   iii. Length of pile = 5000 mm
   iv. Bearing resistance = 630 × 0.9 = 567 kN > 363.64 kN (OK).
   v. Provide 9 nos. of 12φ bars as longitudinal reinforcement
   vi. Provide 6 stirr-ups @ 300 c/c along entire length of pile.

Note: Formation of a double under-ream may be difficult in sandy or silty soils.

If a bored cast-*in situ* pile has to be used, then the compressive and lateral carrying capacity of the pile has to be found. Assume length of pile to be 5 m (same as double under-reamed pile).

Safe compressive load carrying capacity of each pile:

Q_p = Skin friction load + point bearing load.
(Vertical stress changed to horizontal stress)

$= K\,\sigma_v\,Tan\,\delta\,A_s + \sigma_v\,@\,\text{tip} \times (N_q - 1)A_p$

Due to skin friction   Load resisted due to end bearing

where,

K = Co-efficient of horizontal pressure = 0.35 for bored piles

σ_v = Effective vertical stress at mid-grip length of pile

δ = Angle of wall friction between pile and surrounding soil $\ngtr$ (ϕ) 2/3 of diameter

A_s = Curved surface area of the pile over its grip length

A_p = Cross-sectional area of pile

N_q = Terzhagi's bearing capacity factor (a function of ϕ)

A factor of safety 03 shall be applied

Note: If critical, the uplift capacity of a bored cast-*in situ* pile is determined as follows:

$$Q_u = \left\{\frac{f_s A_s (RF) + W}{f.o.s}\right\}$$

where,

f_s = Average skin friction = K σ_v Tan δ

RF = Reduction factor = 0.60 for piles in sand.

W = (Submerged) weight of total length of pile

Apply a factor of safety of 3 to get safe uplift load. The Terzhagi's bearing capacity factors are given readily as follows:

$\phi°$	N_c	N_q	N_r	$\phi°$	N_c	N_q	N_r
0	5.7	1.0	0	34	52.6	36.5	30.0
5	7.3	1.6	0.5	35	57.8	41.4	42.4
10	9.6	2.7	1.2	40	95.7	81.3	100.4
15	12.9	4.4	2.5	45	172.3	173.3	297.5
20	17.7	7.4	5.0	48	258.3	287.9	780.1
25	25.1	12.7	9.7	50	347.5	415.1	1153.2
30	37.2	22.5	19.7				

$$Q_p = K\,\sigma_v\,Tan\,\delta\,A_s + \sigma_v\,@\,\text{tip} \times (N_q - 1)\,A_p$$

$$= 0.35 \times \frac{5}{2} \times 16 \times \tan\frac{2}{3} \times 33.5 \times \pi \times 0.5 \times 5.0 + 5 \times 16 \times (40-1) \times \frac{\pi \times 0.5^2}{4}$$

$$= 45 + 612.6 = 657.61 \text{ kN}$$

$$\text{Safe compressive load} = \frac{657.61}{3} = 219.20 < 363.64 \text{ kN}$$

Not OK.

Revise length of pile to 10 m and cross-check pile capacity.

$(Q_p)_{\text{safe}} = 438.4 \text{ kN} > 363.64 \text{ kN}$ **Hence OK.**

4.11 ANALYSIS FOR LATERAL LOAD CAPACITY OF PILES

Assumptions for analysis:

1. Laterally loaded pile is treated as a beam on elastic foundation. Lateral load is resisted by horizontal sub-grade reaction (Fig. 4.31).

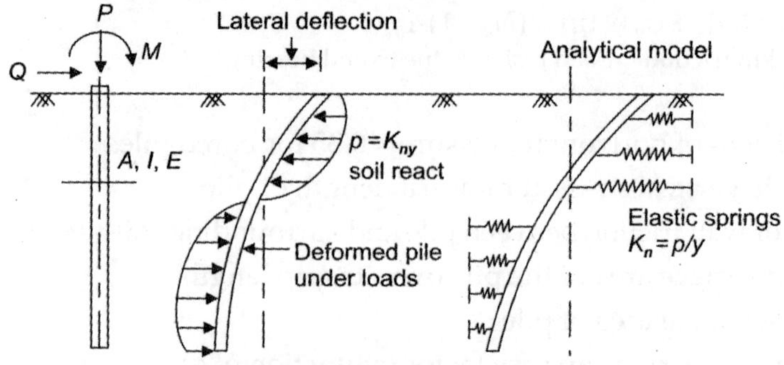

Fig. 4.31: Laterally loaded pile

2. The lateral load carrying capacity of a pile is that horizontal load applied at top edge of pile which causes a limiting lateral deflection.
3. Soil is replaced by a series of infinitely close independent elastic springs.
4. Stiffness of springs = K_h = modulus of horizontal sub-grade reaction = p/y.

   where,
   p = Soil reaction per unit length of pile
   y = Pile deformation

5. Soil pressure at any point along length of pile (deflection at that point.)

   $\therefore \qquad EI\, d^4y/dx^4 = -p = -K_h y.$

   Y is deflection of pile at depth 'x' from top of pile.

Note: How to find 'K_h'?

1. 'K_h' may be assumed to be constant with depth $K_h = K$ for piles in preloaded clays (clay has been loaded in past in its geological history)
2. $K_h = n_h \times$ for piles in sand and normally consolidated clays ("first time" loaded clay)

   where n_h = Co-efficient of modulus of sub-grade reaction expressed in kN/m^3
   x = Any point along pile length measured from ground

Values of 'K_h' for pre-loaded clays: (IS: 2911-Part 1)

Unconfined compression strength (kN/m^2)	*Range of values of K_h, kN/m^2*	*Probable value of K_h, kN/m^2*
20–40	700–4200	775
100–200	3200–6500	4880
200–400	6500–13000	9770
>400	–	19546

Source: Values of co-efficient of modulus of sub-grade reaction, n_h (IS: 2911-Part 1)

The lateral load capacity of a pile is that load which causes a limiting lateral deflection at the top of the pile. The limiting lateral deflection is on empirical serviceability criteria.

Soil type	n_h *in kN/m³*	
	Dry	*Sub-merged*
• Loose sand	2600	1460
• Medium sand	7750	5260
• Dense sand	20760	12450
• Very loose sand under repeated loading	–	–
• Very soft organic soil	–	110–270
• Very soft clay	–	–
– Static loads	–	450
– Repeated loads	–	270

6. The allowable lateral loads are calculated on the basis of limiting deflections by analysing an "equivalent cantilever" (Figs 4.32 to 4.35).

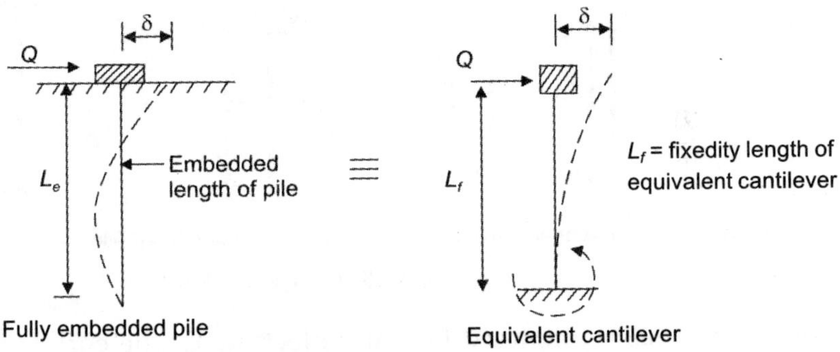

Fig. 4.32: Piled foundation

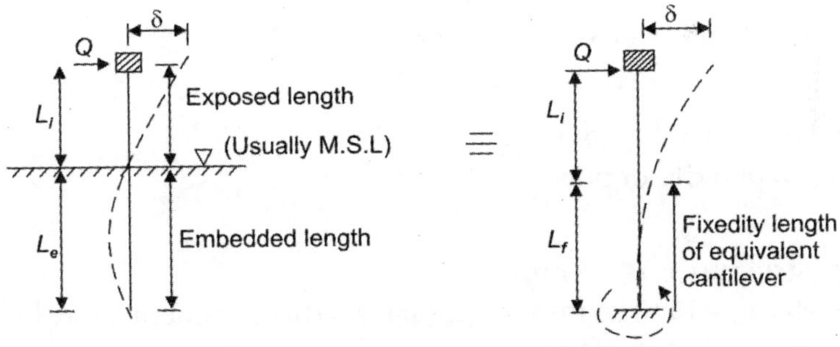

Fig. 4.33: Free-standing pile group

Case (1) fully embedded (non-river crossing, no possibility of exposure of pile due to scouring of soil).

Relative stiffness 'R' and 'T' are defined as:

$R = [EI/K]^{1/4}$ for piles in cohesive soils

$T = [EI/n_h]^{1/5}$ for piles in cohesionless granular soils

E, I are Youngs' modulus of pile material $= 5700\sqrt{f_{ck}}$ MPa

I = second moment of area of pile section.

Knowing the values of 'R' and 'T' the depth of fixity, L_f, for a fully embedded or partially embdedded pile can be found.

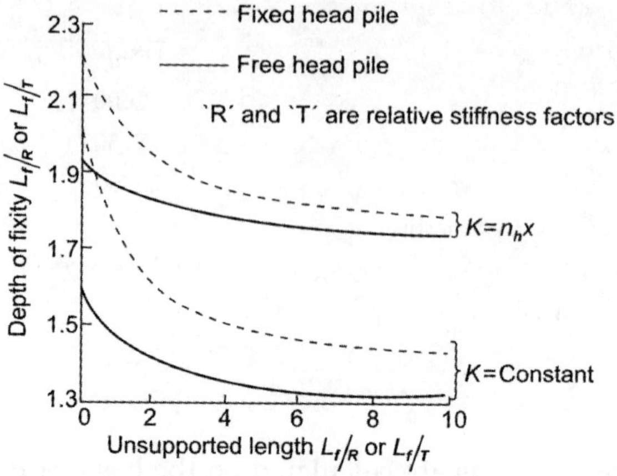

Fig. 4.34: IS: 2911 (Part 1/Sec 2)—1997, amendment No. 3, Appendix 'c'

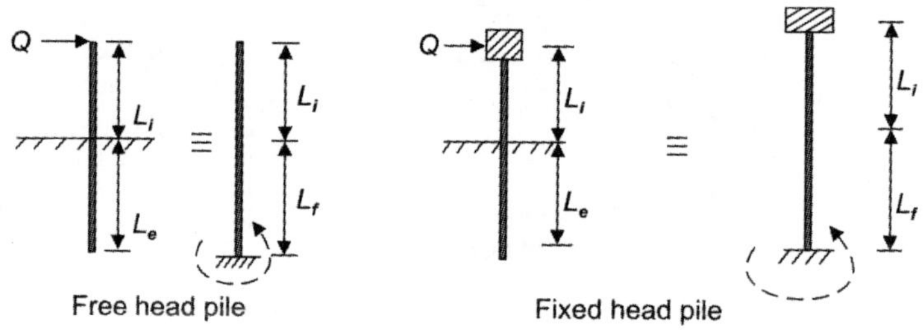

Fig. 4.35: Analysis of loads

For a permissible pile head lateral deflection Y_o, the corresponding permissible lateral load can be found as:

$Q = 3\, Y_o EI/(L_i + L_f)^3$ for a free head pile.

$Q = 12\, Y_o EI/(L_i + L_f)^3$ for a fixed head pile.

where

L_i = unsupported length of pile

L_f = depth of fixity of pile.

Note:

1. For bridge structures, Y_o = 5 mm
   For Jetties etc, Y_o = 15 to 25 mm [larger lateral displacements may be allowed]
2. Calculated value of Q should be compared with actual critical value of lateral load, (Q_{act}). If $Q < Q_{act}$ increase pile section.

4.12 STRUCTURAL DESIGN OF PILES FOR LATERAL LOADS (Fig. 4.36)

- Check cross-section of pile
- Provide longitudinal and lateral confining reinforcement
- Pile section is designed as a column subject to axial load and moment
- Design moment is the fixed end moment for the equivalent cantilever.

$M_f = Q(L_1 + L_f)$ for free head piles.

$M_f = Q(L_1 + L_f)/2$ for fixed head pile.

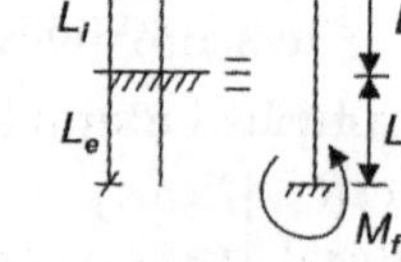

M_f = fixed end moment

Fig. 4.36: Free body diagram of lateral loads

A correction factor, m has to be applied to 'M_f' (Figs 4.37, 4.38)

∴ Corrected moment = $M_c = m\ M_f$ [Pile section has to be designed as a column subject to axial load and its moment].

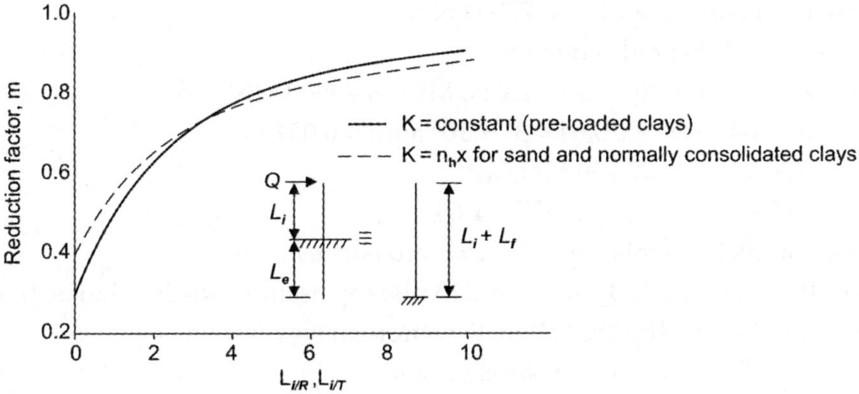

Fig. 4.37: Reduction factor for free head piles
Source: IS: 2911 (Part I/Sec. 2)—1979

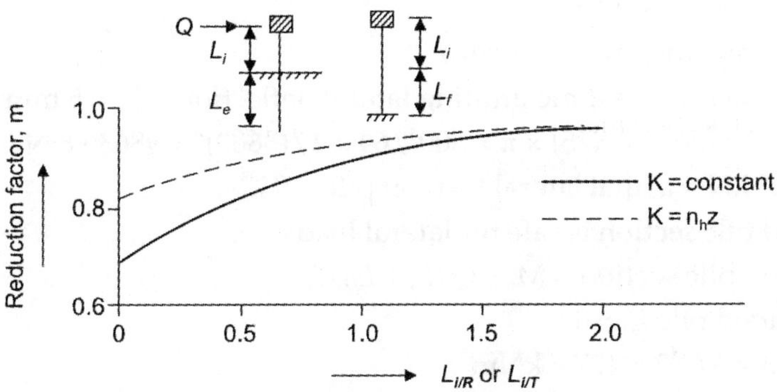

Fig. 4.38: Reduction factor for fixed head pile
Source: IS 2911 (part 1)

Note:

1. Knowing M_f, P design the pile section as a column section and provide the same reinforcement throughout
2. Pile has to be checked for axial load capacity of soil through skin friction and point bearing resistance
3. Procedure outlined above is approximate. For reliable prediction field lateral load tests shall be carried out on test or working piles.
4. From observed load—deflection behaviour of test pile the safe load on pile is estimated based on limiting—deflection criterion.

Example: Lateral load capacity of pile (for a group of 20 piles)

Given:

Lateral load on each pile = 9 kN

Diameter of each pile = 0.50 m

Length of each pile = 10 m [completely embedded]

Assume piles are founded in dry medium sand.

To determine the depth of fixity of pile.

Determine relative stiffness factor.

$R = [EI/K]^{1/4}$ for piles in cohesive soils.

$T = [EI/n_h]^{1/5}$ for piles in cohesionless soils.

For piles in dry medium sand, $n_h = 7750\ \text{kN/m}^3$

Assume pile is of M-25 grade concrete.

$$E = 5000\ \sqrt{f_{ck}} = 5000 \times \sqrt{25} = 28500 \text{ MPa or } 28500 \text{ MN/m}^2$$

$$I = \pi D^4/64 = \pi \times 500^4/64 = 31 \times 10^8 \text{ mm}^4 = 0.031 \text{ m}^4$$

$$n_h = 7750 \text{ kN/m}^3 = 7.750 \text{ MN/m}^3$$

$$\therefore \quad T = \{28500 \times 0.0031/7.750\}^{1/5} = 1.63$$

Since pile is completely embedded L_1/T = zero (since $L_1 = 0$)

Because of 20 nos of piles [>3] connected to pile cap assume pile head to be fixed.

For $L_1/T = 0$, $L_f/T = 2.2$ [IS: 2911 (Part 1/section 2)]

$$\therefore \quad L_f = 2.2 \times T = 2.2 \times 1.63 = 3.586 \text{ say } 3.6 \text{ m.}$$

For a fixed head pile,

Lateral load corresponding to a lateral displacement Y_0 is

$$= Q = 12\ Y_o EI/(L_1 + L_f)^3$$

Note: For fully embedded pile, L_1= zero

For bridge substructures assume limiting lateral deflection = Y_o = 5 mm.

$$\therefore \quad Q = 12 \times 5 \times (5000 \times \sqrt{25}) \times \pi \times 500^4/64 \times 1/(3600)^3 = 98635.6 \text{ N}$$

Or $\quad Q = 98.63$ kN > actual lateral load on pile = 9 kN

Hence, assumed pile section is safe for lateral loads.

Fixity moment on pile section = $M_f = Q\ (L_1 + L_f)/2$

For fully embedded pile $L_1 = 0$

$$\therefore \quad M_f = 98.63 \times 3.6/2 = 177.6 \text{ kNm}$$

Axial load on pile = 363.64 kN (calculated)

Corrected moment on pile section = $M_c = m\ M_f$

where 'm' is the reduction factor.

For fixed head piles in cohesionless soils,

$m \approx 0.82$ [IS: 2911 (Part 1/section 2)]

$\therefore$ Corrected moment = 0.82 ×177.6 = 145.63 kNm

Design the circular pile section of 500 mm diameter for an axial load of 363.64 (say 364 kN) and moment of (146 kNm).

Assume $d'/D = 0.10$

$P_u/f_{ck}D^2 = 1.5 \times 364 \times 10^3/25 \times 500^2 = 0.09$

$M_u/f_{ck}D^3 = 1.5 \times 146 \times 10^6/25 \times 500^3 = 0.07$

Since the pile is a fully embedded pile in a reasonably good soil, it is designed as a short column under axial load and moments.

$$p/f_{ck} = 0.06 \text{ or}$$

$$p = 1.5\%$$

$$A_{sc} = 2945 \text{ mm}^2$$

Provide 6 bars of 25φ . Thus, A_{sc} provided = 2940 ≈ 2945 mm²

Hence OK.

As per IS: 2911 (Part 1)

- Minimum diameter of link or spirals = 6 mm
- Minimum spacing of links or spirals = 150 mm
- Practical spacing of links or spirals = 300 mm

Provide 6 mm spirals at 300 mm c/c (centre to centre) throughout the length of the pile.

Group action of piles (Fig. 4.39):

$$\eta = \text{group efficiency} = Q_{ug}/nQ_u$$

group; Q_u: Ultimate load capacity of one pile.

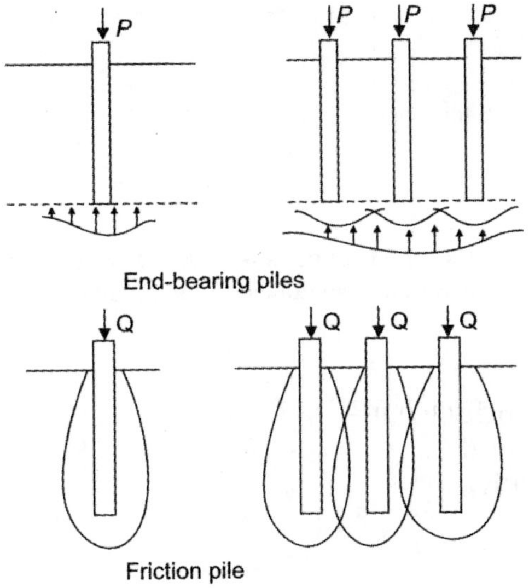

- Disturbance of soil. During installation of piles and overlap of stresses may result in $\eta < 1$.

- Gencrally for smaller pile spacing $\eta < 1$
- For larger pile spacing effect of pile interaction diminishes and $\eta \rightarrow 1$

Fig. 4.39: Group action of pile

Pile group in cohesionless soil

- For driven free standing pile-groups $\eta > 1$ for closer spacings and $\eta \rightarrow 1$ when pile spacing is increased to 5 to 6 times the pile diameter. An efficiency factor of 1.0 is commonly used for such piles in design.
- In piled-foundation additional support is available because of bearing action of pile-cap on soil. However, to mobilize this support excessive settlement of pile cap may be required.
- Unlike in driven piles, method of installation of bored piles results in general loosening of soil around the piles and no compaction of soil takes place. An efficiency factor of 1.0 is commonly used for such piles in design.

Behaviour of piles and pile groups in clayey soils.

Case 1: Hard clays, very stiff clays, stiff clays

How to classify the clay?

a. *Hard clay*: Unconfined compressive strength (UCS) >400 kN/m^2
{The material effectively behaves as a rock.}
b. *Very stiff clay*: UCS in the range of 200–400 kN/m^2
c. *Stiff clay*: UCS in the range of 100–200 kN/m^2

General shear failure occurs (load is carried through end-bearing action and skin friction. Major portion of the load is resisted by end-bearing action) *see* Fig. 4.40.

- Generally, shear failure also occurs in pile foundations located in very dense sands (relative density 85 to 100%) and dense sands (relative density 65 to 85%)

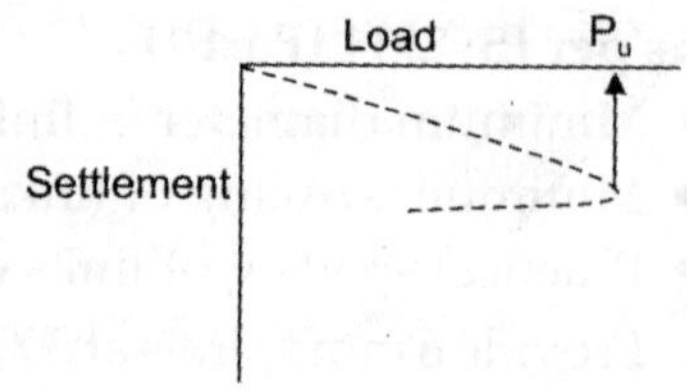

Fig. 4.40: Brittle mode of failure; sudden and drastic loss of load-carrying capacity after reaching peak load

Case 2: Medium clays {UCS : 50 to 100 kN/m^2} some settlement of the soil takes place under applied loads and both end bearing action and skin friction are significant (Fig. 4.41).

Local shear failure occurs: There is no sudden and drastic loss of load-carrying capacity. The resisted load gradually varies with increasing settlements.

- Local shear failure also occurs in sands with relative density in the range of 30% to 50%.

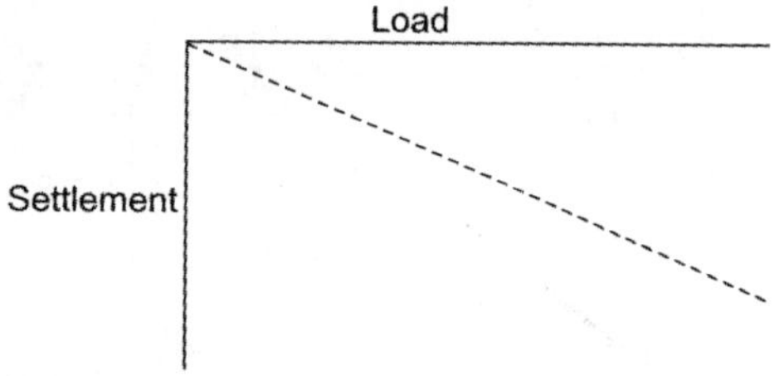

Fig. 4.41: Local shear failure occurs. This is a relatively ductile mode of failure

Case 3: Soft clays (UCS: 25–50 kN/m^2) and very soft clays (UCS < 25 kN/m^2). Because of the highly compressible nature of the soil significant settlements take place under the applied loads (Fig. 4.42).

- Block failure can also take place in a pile foundation supported on very loose sand (relative density <15%) or loose sand (relative density: 15 to 35%).

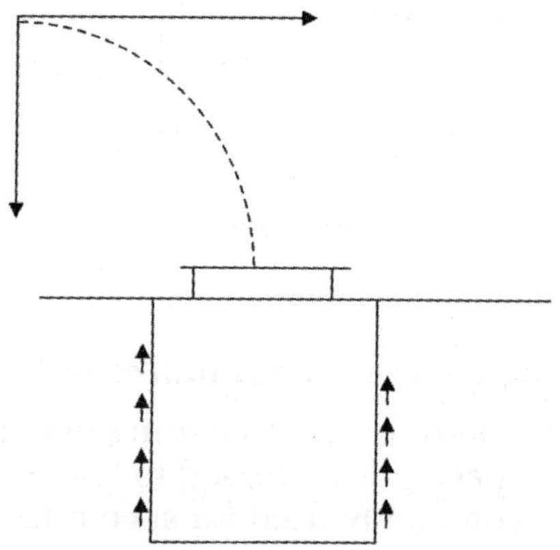

Fig. 4.42: Pile in soft clays

Design of pile foundation in alluvial beds

Design the pile foundation for a bridge pier in scourable cohesive soil. Maximum design discharge of scour depth calculation is 7500 m^3/sec. Following data is provided:

i. Bearing capacity factor $N_c = 9.0$
ii. Unconfined compression strength of clay at 0.5 m depth may be taken as 50 kN/m^2
iii. Adhesion factor for medium clays for bored cast-*in situ* piles = α = 0.45
iv. Unconfined compression strength of clay may be assumed to linearly increase at rate of 15 kN/m^2
v. Average size of particles in river bed = 0.23 mm.
vi. Total vertical loads = 5929 kN
vii. Total moment about L–L axis = 1260 kNm
viii. Total moment about T–T axis = 628 kNm
ix. Total horizontal force along L–L axis = 128 kN
x. Total horizontal force along T–T axis = 165 kN
xi. Elevation of HFL = +475.50 m
xii. Elevation of LWL = +472.00 m

xiii. Elevation of river bed = +470.6 m

- Assume the foundation is made-up of bored cast-*in situ* piles of diameter 50 mm.
- Assume a total of 20 piles under the pier
- Proposed arrangement of piles is as follows:
  - Assume c/c spacing of piles to be 3 times of dia = 3 × 500 = 1500 mm
  - Assume 04 rows of 05 piles each (Fig. 4.43)

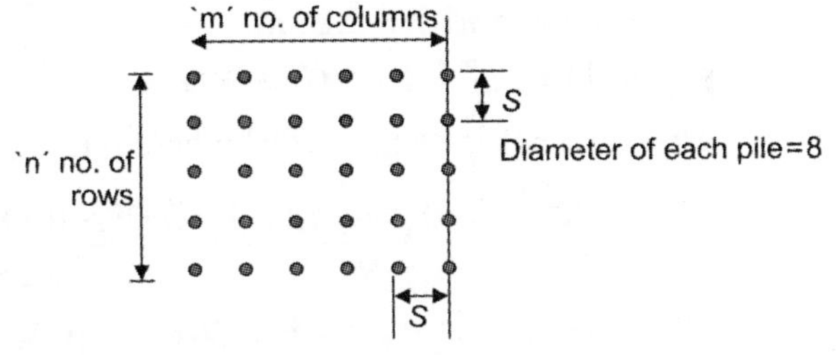

Fig. 4.43: Schematic of piles in rows

Note: The spacing of piles may also be decided on the basis of efficiency of a pile group especially for friction piles.

Consider the arrangement of piles shown below.

Pile group capacity = sum of individual pile capacities

Spacing of piles = $S = (1.57\text{ B m-2B})/(m + n - 2)$

For the present case; $B = 0.5$ m, m = 5, $n = 4$

$$S = 2.10 \text{ m}$$

[Not strictly applicable in this case since the piles are not friction piles.]

The above equation is more readily applicable for piles in cohesion less soils.

Minimum clear overhang of pile cap beyond outermost pile in group = 150 mm [IS: 2911 Part 1/Section 2].

Assume clear projection of pile cap beyond centre of outermost pile = 500 mm.

Hence clear overhang of pile cap = 500–500/2 = 250 > 150 mm.

Hence OK.

Assume piles transfer load to soil partly by skin friction and partly by end bearing action (Fig. 4.44).

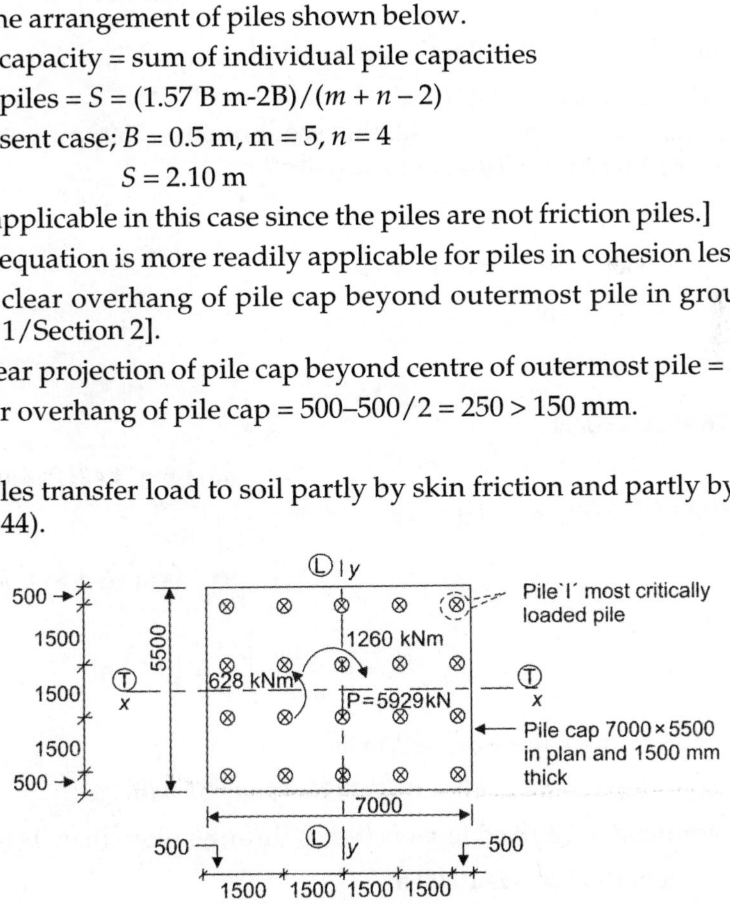

Fig. 4.44: Load transfers through piles

Geometry of pile group is:

$$e_x = \frac{1260}{5929} = 0.212 \text{ m}, \qquad e_y = \frac{628}{5929} = 0.11 \text{ m}$$

Maximum vertical load on pile 'i' $= P_i = \dfrac{P}{n} + \dfrac{Pe_x x_i}{\varepsilon x^2} + \dfrac{Pe_y y_i}{\varepsilon y^2}$

$$P = 5929 \text{ kN}, n = 20$$

For the critical pile $i, x_i = 3.0$ m and $y_i = 2.25$ m

$\varepsilon x^2 = 2[4 \times 1.5^2 + 4 \times 3.0^2] = 90 \text{ m}^2$

$\varepsilon y^2 = 2[5 \times 0.75^2 + 5 \times 2.25^2] = 56.25 \text{ m}^2$

$$P_i = \frac{5929}{20} + \frac{5929 \times 0.212 \times 3}{90} + \frac{5929 \times 0.11 \times 2.25}{56.25}$$

or $P_i = 296.45 + 41.9 + 26.08 = 364.43$ say 365 kN

Lateral load on each pile along $L-L$ direction $= \dfrac{128}{20} = 6.4$ say 7 kN

Lateral load on each pile along $T-T$ direction $= \dfrac{165}{20} = 8.25$ kN

{Assuming all piles share the lateral load equally}

Hydraulic calculations (Fig. 4.45):
Maximum design discharge for calculating scour depth = 7500 m^3/s
Effective linear water-way $= W = C\sqrt{Q}$

Take $C = 4.5$

$\therefore \quad W = 4.5 \times \sqrt{7500} = 389.71$ m say 390 m

R_{sf} = Silt factor $= 1.76\sqrt{M_r}$

$M_r = 0.23$ mm (given)

$\therefore \quad k_{sf} = 1.76\sqrt{0.23} = 0.84$

$\therefore$ Mean depth of scour below $HFL = d_{sm} = 1.34\left[\dfrac{D^2}{k_{sf}}\right]^{1/3}$

$$d_{sm} = 1.34\left[\frac{\frac{7500^2}{390}}{0.84}\right]^{1/3} = 10.19 \text{ m}$$

{Verify this value using Lacey's equation also}.

Maximum scour depth $= 2\,d_{sm} = 2 \times 10.19 = 20.38$ say 21.0 m

Pile head is assumed to be fixed {connection is through more than 03 piles}

The embedded length of pile is 10.00 m.

Hence, total length of bored cast-*in situ* pile = 10 + 16.10 = 26.10 m.

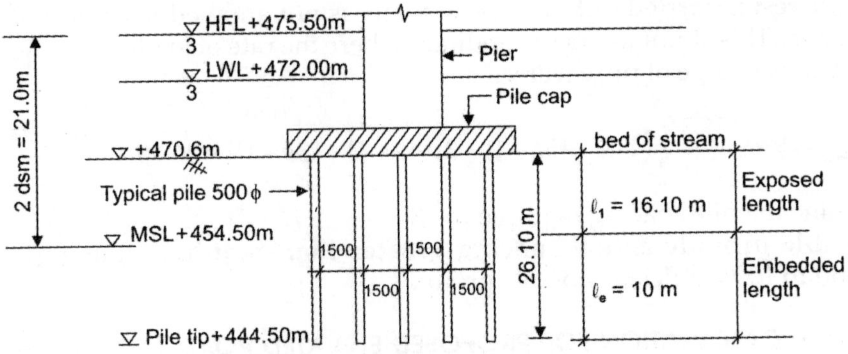

Fig. 4.45: Hydraulic parameters

Note: Group efficiency of friction piles in clay

- How to obtain optimum group efficiency with suitable spacing of piles.
- Consider load test on a single friction pile in clay.

  There is a sudden break in the load-settlement diagram when friction is overcome (point 'A') and the pile continues to settle without further gain in load (Fig. 4.46)
- Failure of a group friction piles in clay is not in the form of shear failure of individual piles at their interface with soil. Since the piles are connected by a rigid pile cap at the top the piles must fail as a group (Fig. 4.47).

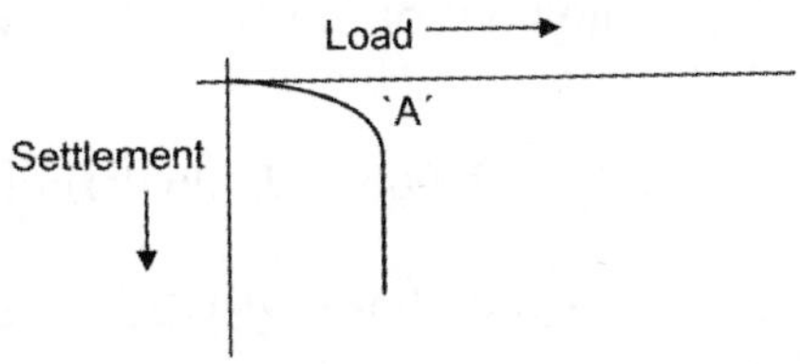

Fig. 4.46: Load-settlement graph

Failure of rectangular prism of soil takes place.

$\therefore$ Load carrying capacity of group = f [shear strength of soil over surface area of the group, bearing capacity of rectangular prism of soil at the base]

Ultimate load capacity of a pile group considering block failure

$$= Q_{ug} = N_c C_D A_{gb} + P_g \bar{C}_u$$

where N_c = Bearing capacity factor may be taken as 9.0 for clayey soils.

C_b = Shear strength of clay at bottom tip of block.

where σ_1 = Unconfined compressive strength of clay

C_u = Average shear strength of clay adjacent to the block shaft

A_{gb} = Area of cross-section of block

P_g = Lateral surface area of the block

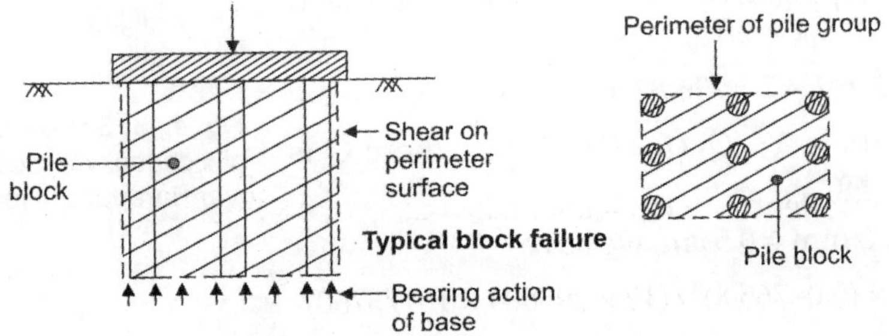

Fig. 4.47: Piles connection through cap

Since the test is carried out quickly so water is not allowed to drain out from the soil specimen. This simulates field condition where the rate of load application is faster than the rate of escape of pure water.

$$Q_{ug} = 9 \times \frac{441.5}{2} \times (4s+1)(3s+1) + 2[(4s+1)+(3s+1)] \times 26.10 \times \left(\frac{50+441.5}{2 \times 2}\right)$$

Determine the pile spacing required $S = 8$

If possible provide actual spacing greater than required spacing. This will correspond to 100% efficiency of pile group.

4.13 SAMPLE CALCULATION FOR PROPOSED ELEVATED ROAD ALONG A NALLAH IN NEW DELHI

Lateral Load Capacity Bored Cast-*in situ* RCC Pile (Fig. 4.48)

The lateral load carrying capacity of bored piled has been calculated in the following way, as per IS: 2911, Part 1/section-2-1979

$$R = \text{'}v\text{'}\ E.I/K_1$$

where K_1 is constant given in table, appendix c_1 of above code

'E' is Young's modulus of the pile material in kg/cm^2

'I' is the moment of inertia of the pile cross-section in cm^4.

As per IS: 456, 2000, $\quad E = 5000\sqrt{f_{ck}}$

where f_{ck} is the characteristics strength of concrete of pile.

Considering

$$f_{ck} = 35\ \text{N/mm}^2$$

$$E = 5000\sqrt{35} = 29580.4\ \text{N/mm}^2 = 295804\ \text{kg/cm}^2$$

For dia of pile, 'D' = 120 cm, $I = [1.64 \times (120)^4$

$K_1 = 0.525$ (for medium dense sub-merged sandy strata)

Putting these values into the above equation, we get

$$R = 356.2\ \text{cm}$$

From Fig. 2 of Appendix of above code for fixed head pile and $L_1 = 0.0\ L_1T = 2.15$

or $\quad L_f = 356.2 \times 2 = 756.8$ cm

And for equation, $Y = Q_1 (L_1+L_f)^3/12EI$, where Q is lateral load in kg.

Adopting $Y = 5$ mm $= 0.5$ cm, we have

or $\quad 0.5 = Q_1 \times (0.0–765.8)^3/(12 \times 295804 \times 10178760)$

or $\quad Q_1 = 40226$ kg $= 40.2$ T.

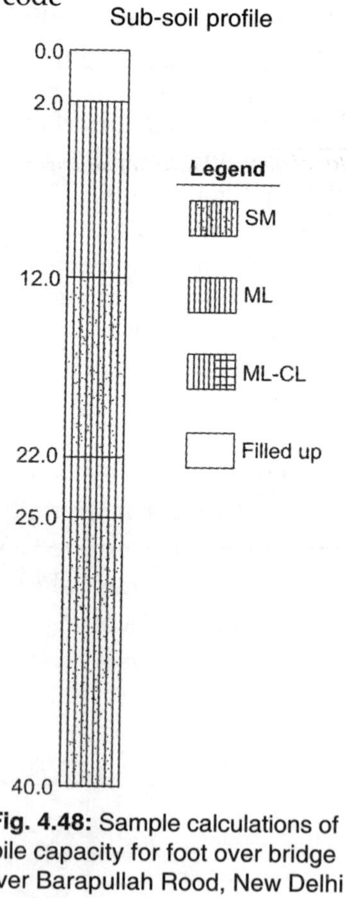

Fig. 4.48: Sample calculations of pile capacity for foot over bridge over Barapullah Rood, New Delhi

Design of pile caps:

- Pile caps are used to transfer column loads to the pile foundation
- Fundamental size requirement: Pile cap should be deep enough to provide adequate anchorage for pile reinforcement and column starter bars (Fig. 4.49).

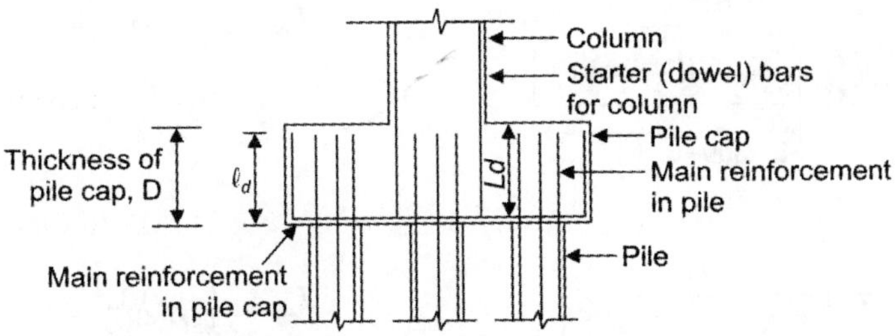

Fig. 4.49: Front elevation of a typical pile foundation

Design parameters for pile cap

- Shape of pile cap in plan
- Depth (thickness) of pile cap
- Amount and detailing of reinforcement

Recommended shapes of pile caps (Figs 4.50 to 4.53)

No. of piles	*Shape of pile cap*	*Principal tensile force by truss analysis*
2	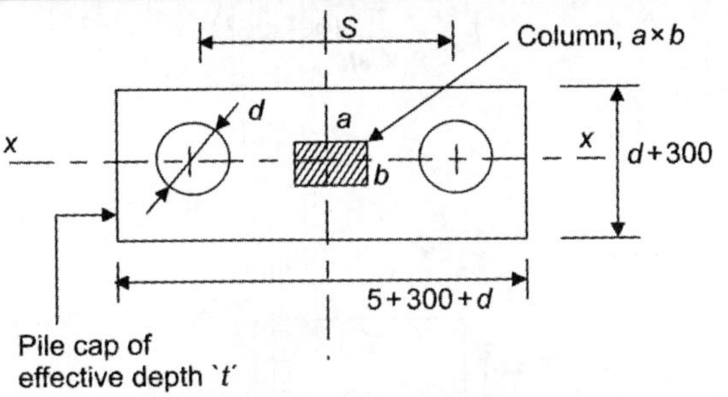	$T_{xx} = \frac{P}{12st.}(3s^2 - a^2)$

Fig. 4.50: Shapes of pile cap

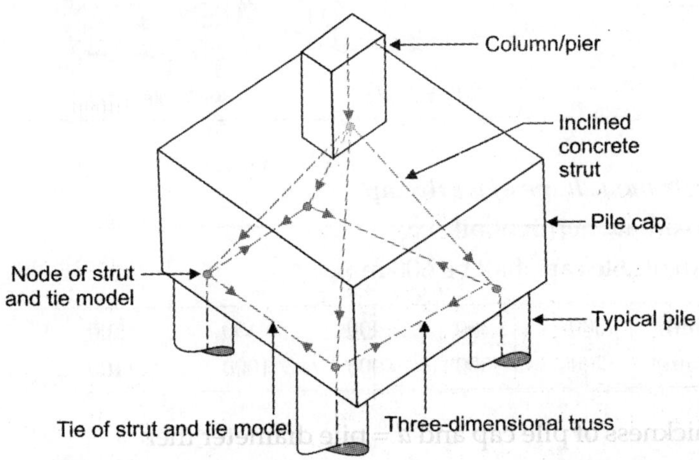

Fig. 4.51: Truss model for a pile cap supported by four piles

No. of piles	*Shape of pile cap*	*Principal tensile force by truss analysis*
3.	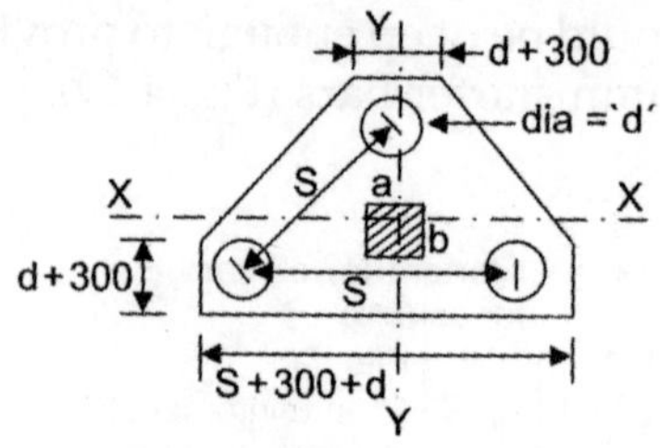	$T_{xx} = \frac{p}{36st} \times (4s^2 + b^2 - 3a^2)$ $T_{yy} = \frac{p}{18st}(2s^2 - b^2)$
4.	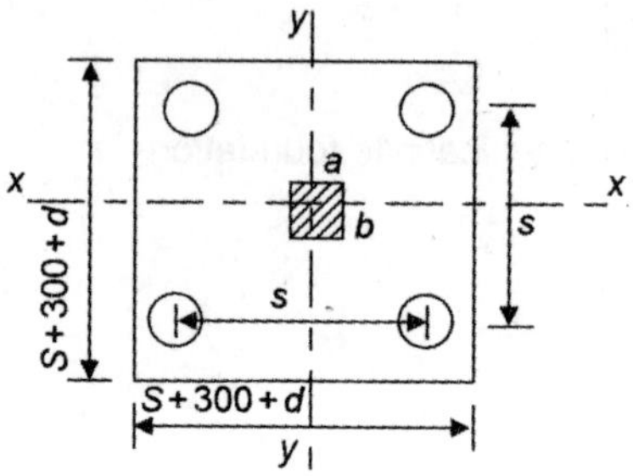	$T_{xz} = \frac{p}{24st} \times (3s^2 - a^2)$ $T_{yy} = \frac{p}{24st}(35^2 - b^2)$
5.	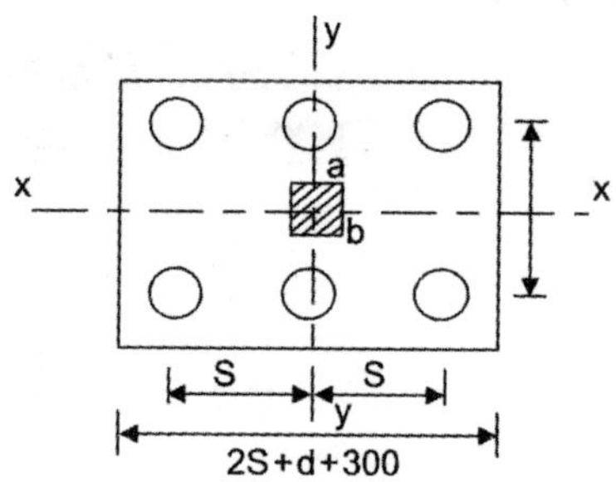	$T_{zz} = \frac{p}{18st} \times (3s^2 - a^2)$ $T_{yy} = \frac{p}{36st}(3s^2 - b^2)$

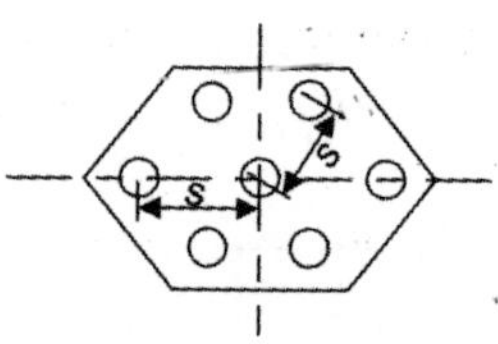

7-Pile group

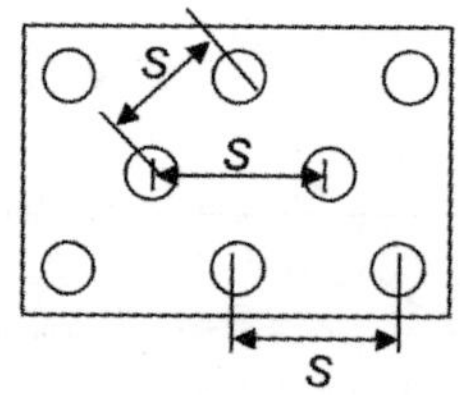

8-Pile group

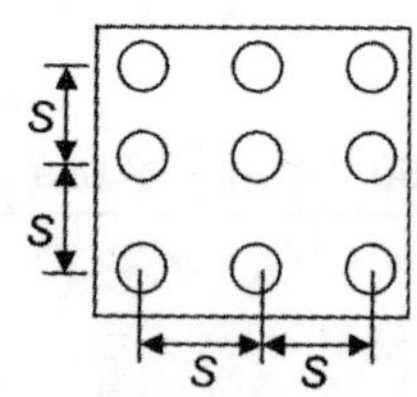

9-Pile group

Selecting the optimum shape of a pile cap

Selecting approximate depth of pile cap

Minimum depth of pile cap shall be 600 mm.

Pile diameter (mm)	300	350	400	450	500	550	600
Pile cap depth (mm)	700	800	900	1000	1100	1200	1400

If D = total thickness of pile cap and d = pile diameter then

$D = (2d + 100)$ for $d \ngtr 500$ mm

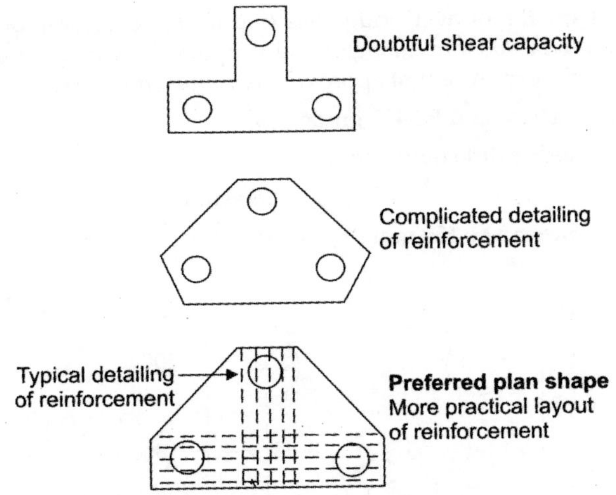

Fig. 4.52: Various shapes of pile caps

$$D = \frac{1}{3}(8d + 600) \text{ for } d \geq 550 \text{ mm}$$

'D' and 'd' are in "mm".

Design basis for pile caps (Fig. 4.53):

i. Truss theory
ii. Beam or bending theory

Illustrative design of a pile cap using truss analogy

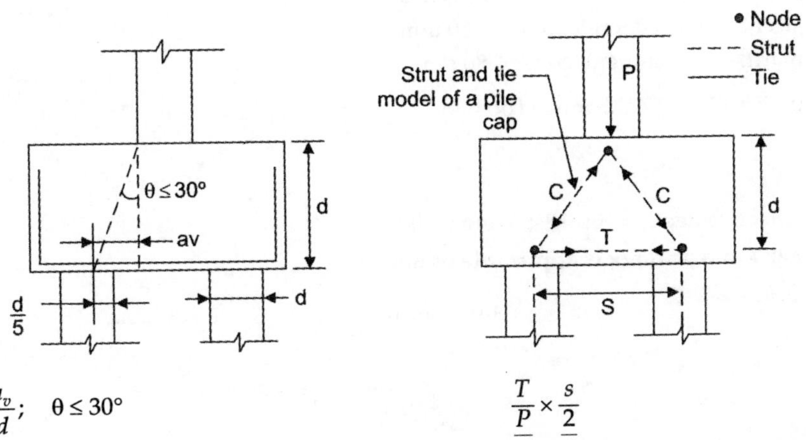

$$\tan\theta = \frac{a_v}{d}; \quad \theta \leq 30°$$

i.e. $\frac{a_v}{d} \leq 0.58$

Applied load is transferred to piles through truss action by formation of internal struts and ties.

$$\frac{T}{\frac{P}{2}} \times \frac{\frac{s}{2}}{d}$$

$$T = \frac{Ps}{4d}$$

Required area of steel = A_{st}

$$A_{st} = \frac{T}{0.87\,fy} = \frac{P_S}{0.87\,fy\,4d}$$

Fig. 4.53: Forces acting on pile cap

Example: The pier of a small foot over bridge has to transmit a factored design load of 3850 kN. Assuming the pier shaft to be 900 mm diameter and supported on four symmetrically placed piles of 600 mm ϕ, design the pile cap. As a first approximation, the effect of moment may be neglected.

- Assume M 25 grade concrete and Fe-415 grade steel.

  Let spacing of piles in either direction = 3ϕ

  $$= 3 \times 600 = 1800 \text{ mm c/c}$$

  Let clear overhang of pile cap be 150 mm (Fig. 4.54)

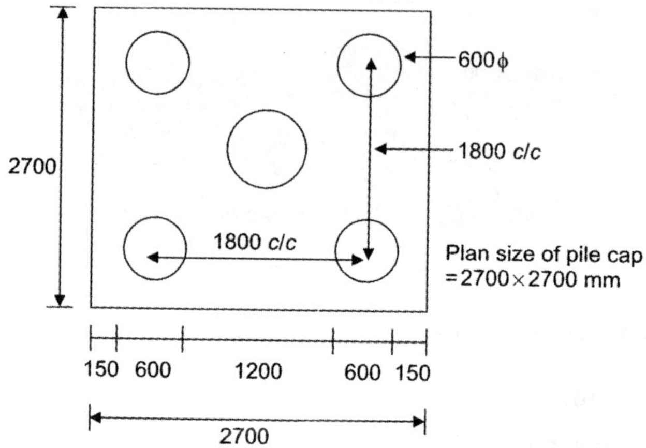

Fig. 4.54: Overhang in pile cap

Overall depth of pile cap = $D = 1/3\ (8 \times 600 + 600) = 1800$ mm

$\therefore$ Overall dimensions of pile cap are 2700 × 2700 × 1800 mm.

Calculation of effective depth: d:

For rebar adjacent to soil face, cover = 75 mm (clear)

For rebar adjacent to lean concrete, cover = 60 mm

For rebar in marine environment, cover = 80 mm

$\therefore d = 1800 - 75 - 12.5 = 1712.5$ mm. [Too high]

$$\frac{\phi}{2} = \frac{25}{2}$$

Adopt an effective depth = ½ pile spacing = 900 mm

Verify whether Truss analogy is applicable or not.

From Fig. 4.55, $a_v = 270$

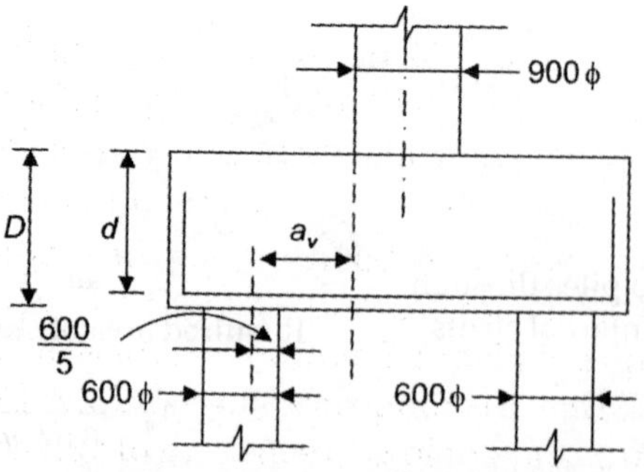

Fig. 4.55: Truss analogy in pile cap

$$\frac{a_v}{d} = \frac{270}{900} = 0.30 < 0.60$$

Hence, truss analogy is applicable for analyzing and design.

Tensile force in either direction $= T = \frac{P}{24st}(3s^2 - a^2)$

$P = 3850$ kN, $S = 1.80$ m c/c, $t = d = 900$ mm, $a = 797$ mm (size of equal area square column)

$$\therefore\ T = \frac{3850}{24 \times 1.80 \times 0.90}(3 \times 1.80^2 - 0.797^2) = 899.8 \text{ kN say } 900 \text{ kN.}$$

Area of steel required $A_{st} = \frac{900 \times 10^3}{0.87 \times 415} = 2492.72 \text{ mm}^2$

Provide 10 bars of 20 φ equally spaced in both directions near the soffit of the pile cap

A_{st} provided = 3140 > 22492.7 mm²

Hence OK.

Provide full anchorage to tension reinforcement

Beyond critical section

$$L_d \text{ reqd} = \frac{\phi\sigma_s}{4T_{bd}} = \frac{20 \times 0.87 \times 415}{4 \times 1.4 \times 1.60} = 806 \text{ mm}$$

Percentage of steel provided

$$= p_t = \frac{18 \times 314}{2700 \times 900} \times 100 = 0.13\% > 0.12\%$$

Check for shear

Critical section for shear is located at a distance 'd' from face of column.

All piles located at $\frac{\phi \text{ pile}}{2}$ or more outside this section will produce shear on the critical section.

No. of such piles = Nil

Hence, check for shear is *not* required.

Provide 12 φ bursting r_{einf} in the form of closed rectangular loops over full depth of pile cap. Provide full anchorage to pier and pile r_{einf} inside the pile cap (Fig. 4.56).

Design of a pile cap.

Q. A reinforced concrete column supporting a light pedestrian foot over bridge carries an axial load of 1000 kN. And a moment of 140 kNm in the X-direction (*see* Fig. 4.57). There is a seismic moment of 500 kNm in any direction at one time. The four bored cast-*in situ* piles are 300 mm φ and are arranged as shown. Design the pile cap using M25 grade concrete and Fe 415 rebars.

Assume pile cap to be 800 mm thick. $D = 800$ mm. Assume 50 mm bottom clear cover and 20 mm φ main bars.

$\therefore\ d = 800–50–10 = 740$ mm [Minimum thickness of pile cap shall be 600 mm.]

Critical section for one-way shear is located at a distance 'd' from face of column.

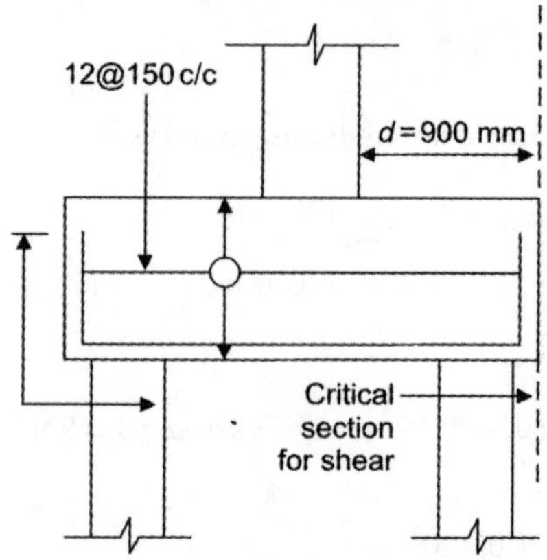

Fig. 4.56: Reinforcement details

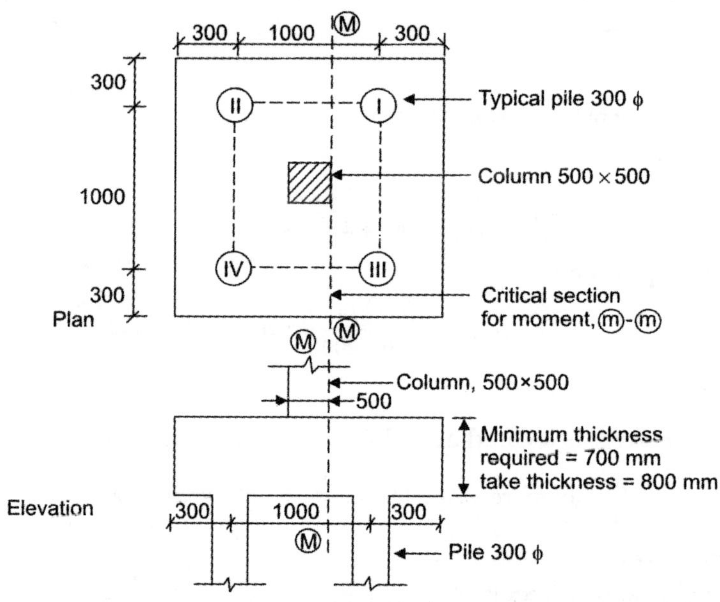

Fig. 4.57: Force and moment on pile

Critical section for two-way shear is located at a distance '$d/2$' from face of column (Fig. 4.58).

To find the loads on the piles: (equivalent axial load).

1. Due to axial load $= \dfrac{1000}{4} = 250$ kN

2. Due to regular moment $= \dfrac{140}{1 \times 2} = 70$ kN

3. Due to seismic moment $= \dfrac{500}{1 \times 2} = 250$ kN

$\therefore$ Equivalent axial load on critical pile = 250 + 250 + 70

= 570 kN

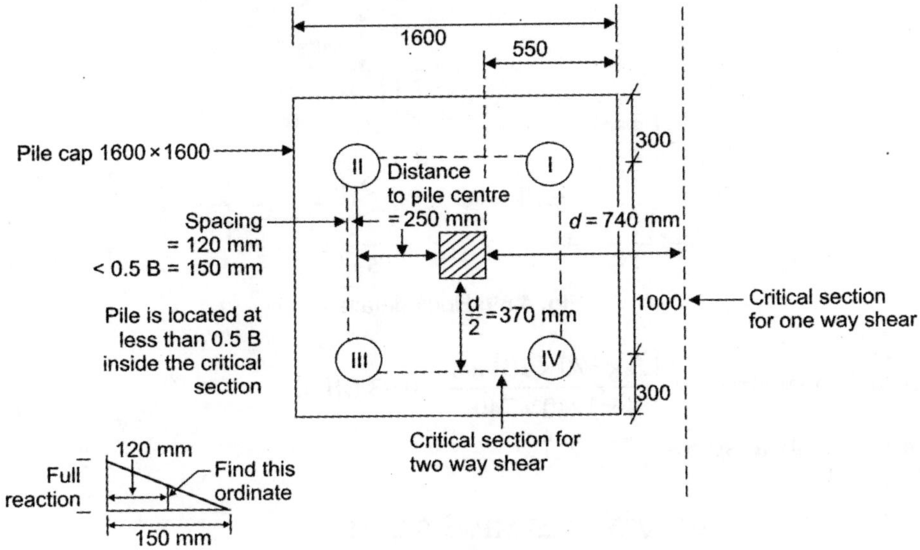

Fig. 4.58: Critical section for two-way shear

B.M. at face of column = 2 × 570 × 0.25 = 285 kNm

Design for moment:

$$\frac{M_u}{bd^2} = \frac{1.5 \times 285 \times 10^6}{1600 \times 740^2} = 0.49$$

Section can be designed as a singly reinforced one. Using SP-16, for $\frac{M_u}{bd^2} = 0.49, p_t \cong 0.142\%$

$$\therefore \qquad A_{st} = \frac{0.142}{100} \times 1600 \times 740 = 1681.28 \text{ mm}^2$$

$$\frac{A_{st}\text{ min}}{bd} = \frac{0.85}{f_y} \text{ or } A_{st}\text{ min} = \frac{0.85 \times 1600 \times 740}{415} = 2425 \text{ mm}^2$$

Provide 10 bars of 20 φ @ the bottom @ both ways.

A_{st} provided = 314 provided = 3140 > 2425 mm². **Ok.**

Check for development length.

Development length required for a 20 φ bars is:

$$L_d = \frac{0.87 f_y \phi}{4T_{bd}} = \frac{0.87 \times 415 \times 20}{4 \times 1.4 \times 1.60} = 806 \text{ mm}$$

Straight length available beyond critical section for moment = 550–50 (cover) = 500 mm < 806 mm.

Not OK, provide a standard 90° hook with an extension of 16 φ = 16 × 20 = 320 mm at both the ends of the main tension reinforcement (Fig. 4.59).

Standard 90° bend. Anchorage value for 20 φ bar = 160 mm

∴ Total anchorage length available = 500 + 160 + 12 × 20 = 900 > 806 mm. **Hence OK.**

Check for two-way shear:

Full pile reaction is assumed for two way shear check.

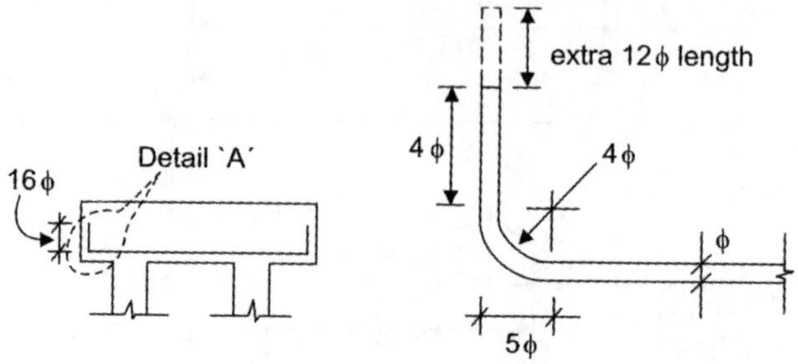

Fig. 4.59: Hook details

$$\text{Nominal shear stress} = \frac{1.5 \times 570 \times 10^3 \times 4}{4 \times 1240 \times 740} = 0.93 \text{ MPa.}$$

Permissible shear stress = $T_c' = K_s T_c$.
Take $K_s = 1$

$$T_c = 0.25\sqrt{f_{ck}} = 0.25\sqrt{25} = 1.25 \text{ MPa} > 0.93 \text{ MPa}$$

The detailing of pile cap shall be done as per recommendations of SP: 34 (Figs 4.60 and 461).

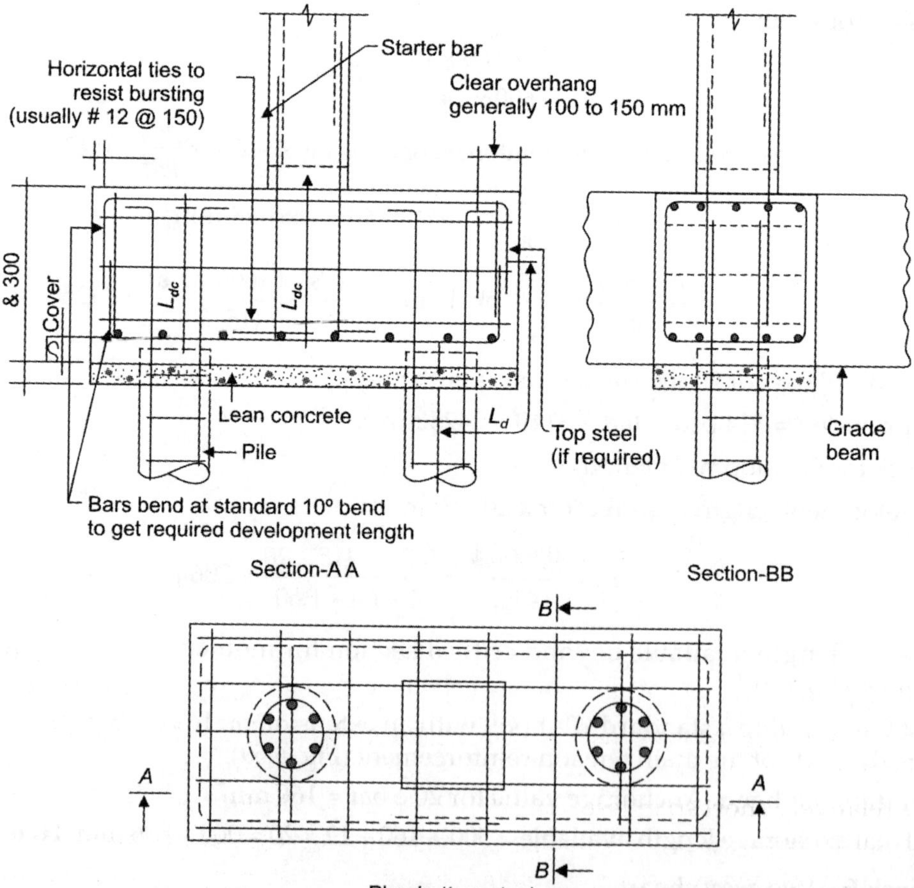

Fig. 4.60: Typical details of a 2-pile cap

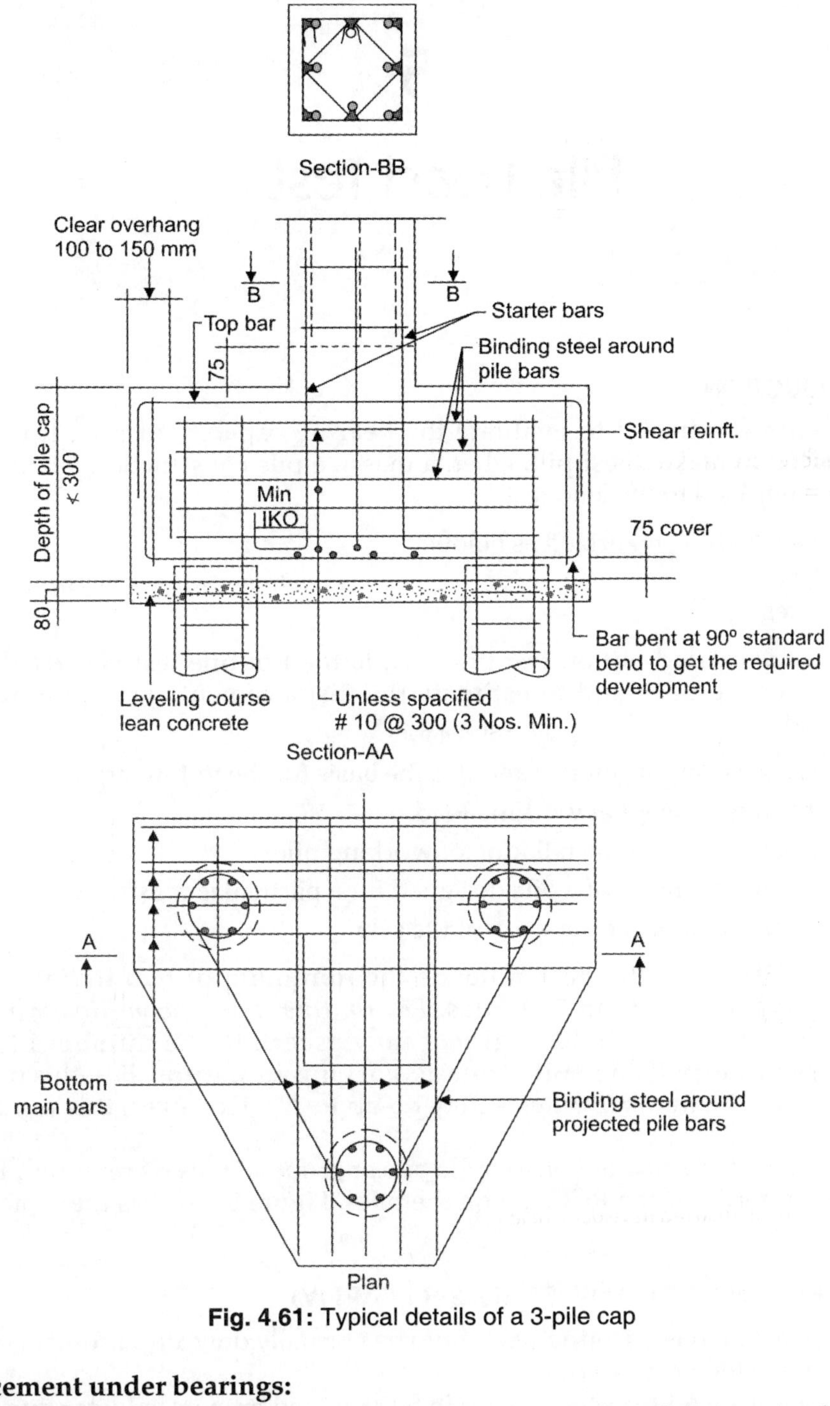

Fig. 4.61: Typical details of a 3-pile cap

Reinforcement under bearings:

i. Bridge bearing may be kept on a R.C bearing pedestal (for major long span bridges).

ii. For relatively short span less important bridges, the bridge are directly placed on the bridge seat.

iii. Very high bearing stresses in the concrete under the bearings. Provide local hairpin mesh reinforcement under all bearing locations.

Chapter

5

Pile Load Tests

5.1 INTRODUCTION

Load tests are conducted to evaluate the bearing capacity of pile. It is strongly recommended to make a test pile when a massive pile construction job takes place. This pile is subjected to the load test.

The load test may be categorized as below.

1. Initial test
2. Routine test

Initial test is carried out on test piles while the Routine test is carried out as a check on working piles and to estimate the displacement corresponding to the working load.

Initial test is a very important test and it is the basis for the following:

1. To establish settlement at working load.
2. To establish criteria for installation of working piles.
3. To check the suitability of a pile designed for a particular cause.
4. To estimate the safe load capacity of the pile.

IS: 2911 (Part IV)—1979 recommends the minimum of two initial tests for a sizeable work of more than 200 piles. *For routine tests, the minimum number of tests should be 0.5%.* The field engineer may ascertain the number of tests up to 2% or even more if the soil strata is varying in nature. For the important structures also, the number of tests may be up to 2%. However, this number may vary.

The tests may be carried out on a single pile or group of piles as required. The loads are applied in series over a RCC cap on the piles. Figure 5.1 shows the typical set-up for pile load test.

5.2 PREPARATION OF PILE FOR TEST (IS: 2911 PART IV)

1. For compaction piles, group tests should be normally done on a group of piles with their cap resting on ground.
2. In case a portion of the pile is likely to be removed or exposed later due to scour, dredging or otherwise, then the capacity contributed by the portion of the pile during load tests shall be duly accounted for. The pile groups in the above conditions shall be tested without their cap resting on the ground.
3. The tests should be carried out at cut-off level, wherever applicable. Otherwise the suitable allowance shall be made in the interpretation of test results/test load if the test is not carried out at cut-off level.

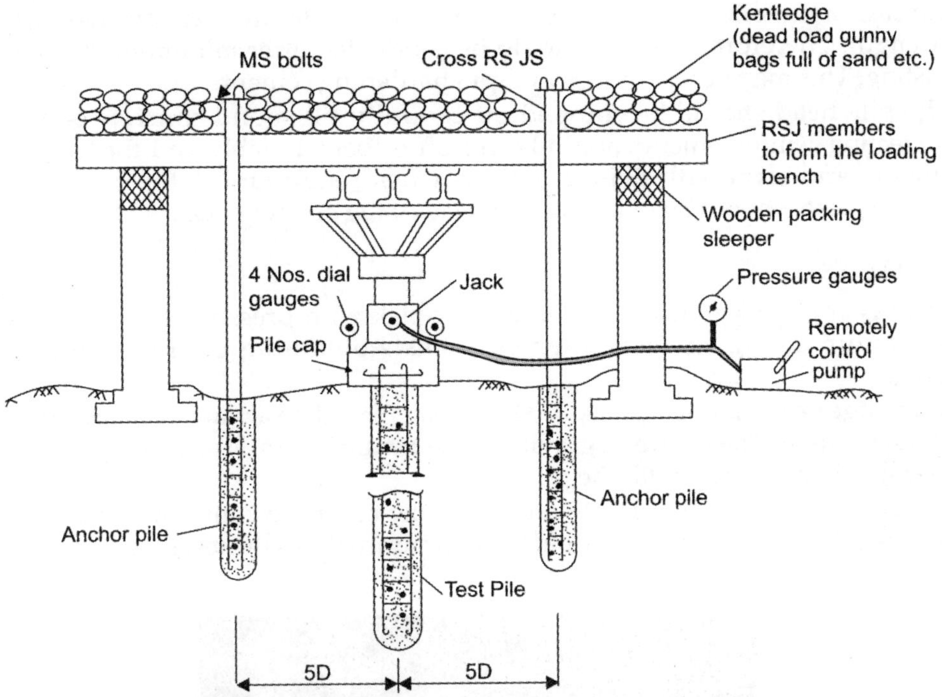

Fig. 5.1(a): Sectional layout of typical pile-load test set-up

Fig. 5.1(b): A view of typical pile-load test set-up
(Courtesy: SMEC, Roorkee, Ph : 9412071753)

4. For cast-*in situ* piles, the test should be conducted after the concrete had gained its strength. Therefore, The test should be conducted after minimum 28 days after casting. This may be less in case of rapid hardening cement.
5. The pile head should be chipped off carefully till sound concrete is met. The projecting reinforcement should be cut-off or bent suitably and the top finished smooth and level with plaster of Paris when required. For the jacks to rest, a bearing plate should be placed on the head of the pile (Fig. 5.2).

5.3 INITIAL TEST

The load is applied by means of a hydraulic jack with pressure gauge and remote control pump acting against rolled steel joists obtaining the reaction from the following:

a. Kentledge heavier than the required test load placed on a platform supported clear of the test pile. The centre of gravity of the kentledge and the load applied by the jack should be coaxial with the pile.
b. Anchor piles or other suitable anchors—the centre to centre distance between the test pile and anchor pile should be minimum 5 times the test pile shaft diameter.

Fig. 5.2: A view of the test pile (top made smooth), ready for test
(Courtesy: SMEC, Roorkee, Ph : 9412071753)

The test load should be applied in increments of about 20% of the assumed (or predicated from design equations) safe load. The reaction to be made available for the test should be 25% more than the final test load proposed to be applied. The settlements are recorded by four dial gauges of 0.02 mm sensitivity each positioned at equal distances around the piles and normally held by datum bars resting on immovable supports atleast 5 *D* (subject to a minimum of 1.5 m) away from the test piles periphery (*D* being the pile stem diameter in case of circular piles or diameter of circumscribing circle in case of square piles).

Each stage of loading is maintained till the rate of movement of the pile top is not more than 0.1 mm/h in sandy soils and 0.02 mm/h in case of clayey soils or a maximum of 2 hours whichever is greater. The estimated safe load may be maintained for 24 hours and settlements should be observed every hour during this period. For each increments, application of load shall be smooth as far as possible.

The loading shall be continued up to twice the safe load (safe load calculated from the static formula) or the load at which the total displacement of pile top equals the appropriate value specified in para 5.3. 1 and 5.3.2, whichever is earlier.

The load on the pile may be removed in one stage by releasing the jack steadily after completion of the test and the rebound observations are made for 2 hours.

5.3.1 The Safe Vertical Load on Single Pile for the Initial Test Should be Least of the Following

1. For piles up to and including 600 mm diameter
   a. Two-thirds of the final load at which the total displacement attams a value of 12 mm unless otherwise required in a given case on the basis of nature and type of structure in which case, the safe load should be corresponding to the stated total displacement permissible; and
   b. Fifty percent of the final load at which the total displacement equal to 10 percent of the pile diameter in case of uniform diameter piles and 7.5 percent of bulb diameter in case of under-reamed piles.
   c. Two-thirds of the final load at which the net settlement attains value of 6 mm.
2. For piles more than 600 mm diameter
   a. Two-thirds of the final load at which the total displacement attains a value of 18 mm or maximum of 2 percent pile diameter whichever is less unless otherwise required in a given case on the basis of nature and type of structure in which case, the safe load should be corresponding to the stated total displacement permissible and
   b. 50 percent of the final load at which the total displacement equal to 10 percent of the pile diameter in case of uniform diameter piles and 7.5 percent of bulb diameter in case of under-reamed piles.

5.3.1.1 However, routine test shall be carried for a test load of at least 1.5 times the working load; the maximum settlement at the test load being not greater than 12 mm for piles diameter up to and including 600 mm and 18 mm or 2 percent of pile diameter whichever is less for piles of diameter more than 600 mm.

5.3.2 The Safe Vertical Load on Groups of Piles for Initial Test shall be Least of the Following

1. Final load at which the total displacement attains a value of 25 mm unless otherwise required in a given case on the basis of nature and type of structure; and

2. Two-thirds of the final load at which the total displacement attains a value of 40 mm.

5.3.2.1 However, routine test on group of piles shall be carried for a test load of at least equal to the working load; the maximum settlement of the test loading in position being not exceeding 25 mm.

Note: For differential settlements, IS 1904 shall be referred.

5.3.3 Maintained Load Method

This is applicable for both initial and routine test. In this method, each stage of loading shall be maintained till the rate of movement of the pile top is not more than 0.2 mm/h or until 2 h have elapsed, whichever is earlier subject to a minimum of I h. The maximum test load shall be maintained for 24 h.

5.3.3.1 Duration of Vertical Loading

5.3.3.2 Vertical loading on single pile shall be continued till one of the following takes place:

1. **In case of initial load test:**
   a. Applied load reaches 2.5 times the safe estimated load; or
   b. Maximum settlement of pile exceeds a value of 10 percent of pile diameter in case of uniform diameter piles and 7.5 percent of bulb diameter in case of under-reamed piles.
2. **In case of routine load test:**
   a. Applied load reaches 1.5 times the working load; or
   b. Maximum settlement of pile exceeds a value of 12 mm for piles diameter up to and including 600 mm and 18 mm or maximum of 2 percent of pile diameter whichever is less for piles of diameter more than 600 mm.

5.3.3.3 Vertical loading on group of piles shall be continued till one of the following takes place:

1. **In case of initial load test:**
   a. Applied load reaches 2.5 times the safe estimated load; or
   b. Maximum settlement of pile exceeds a value of 40 mm.
2. **In case of routine load test:**
   a. Applied load reaches the working load; or
   b. Maximum settlement of pile exceeds a value of 25 mm.

5.3.4 Safe Load through Initial Test (Pile Group)

The safe load on groups shall be the smallest of the following:

a. The final load at which the total settlement attains a value of 25 mm unless a total settlement different from 25 mm is specified in a given case on the nature and type of structure.

b. Two-thirds of the final load at which the total settlement attains a value of 40 mm.

Readers are advised to refer IS:2911 (Part 4) for more details (latest version)

5.4 ROUTINE TEST

In this test, the loading shall be carried out up to one and a half times the safe load or upto the load at which the total settlement attains a value of 12 mm for single piles or 40 mm for group of piles whichever is earlier. The reaction provided may be 25% more than the test load. The procedure for the test is same as that for Initial test.

5.5 CYCLIC LOADING TEST

This load test is carried out if specially needed to divide the load on a pile top into skin friction and point bearing resistance. This test is particularly useful for piles passing through soft compressible layers likely to develop negative skin friction. The test load is applied in increments of about one-fifths of the assumed (or predicted by equations) allowable load in a test pile and the pile is completely unloaded at every stage of the load increase and the total elastic rebound of the pile top is also measured. The separation of the total pile load at any stage of loading into skin friction and point resistance is based on that the load on the pile too increases linearly with the elastic compression of the soil. And the straight line showing the relationship between point resistance and elastic compression of soil is parallel to the straight line portion of a curve drawn between the load on the pile top and the elastic compression of the soil. Fig. 5.3 shows the distribution pattern. As the value of skin friction which is required for the determination of the elastic compression of the pile is not known in advance, it is first assumed that there is no compression in the pile and the curve is plotted between the total elastic recovery and the load on the pile top (curve I). A straight line A_1, is drawn from the origin parallel to the straight portion of the curve I which is supposed to divide the total load on the pile into skin friction and point resistance.

By the approximate known value of skin friction, the elastic compression of the pile is calculated from the following equation:

$$\Delta = \frac{(T - F/2)}{AE} \times L$$

where,

Δ = Elastic compression of pile
L = Length of pile
T = Total load on pile
F = Frictional resistance
E = Modulus of elasticity of pile material
A = Cross-sectional area of pile

By subtracting the elastic compression of the pile from the total elastic recovery of the pile, the elastic compression of the soil is obtained corresponding to the load on the pile. Another curve (curve II) is plotted and a line A_2 is drawn from the origin parallel to the straight portion of curve II which gives the more accurate values of skin

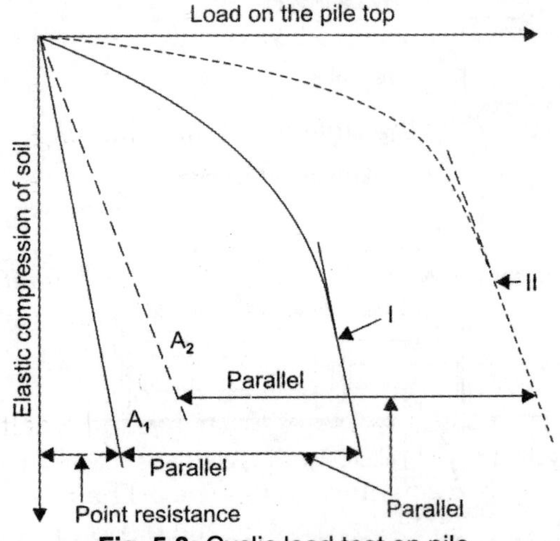

Fig. 5.3: Cyclic load test on pile

friction and point resistance than earlier values. From these new values of skin friction, the compression of the pile is recalculated and the process of plotting is repeated which gives reasonably accurate values of skin friction. The effective friction is taken as the total frictional resistance minus the friction over the length of the pile likely to be subjected to negative skin friction in compressible strata. The allowable load for the pile is obtained by dividing the ultimate values of the effective skin friction and point resistance corresponding to a total settlement equal to one-tenth of the diameter of the pile with factors of safety of 2 and 2.5 respectively.

5.6 LATERAL LOAD TEST ON PILE

The lateral load test shall be conducted as far as possible at the cut-off level of piles. In case of group of piles, the test shall be conducted after providing caps such that required conditions of actual use are fulfilled. The test may be conducted by introducing a hydraulic jack with gauge between two piles or group under test or the reaction may be similarly obtained otherwise. This test may also be conducted by applying the lateral pull by a suitable set-up. If it is conducted by jack located between two piles or groups, he full load imposed by the jack shall be taken as the lateral resistance of each pile or group. The loading should be applied in increments of about 20% of the estimated safe load. The next increments shall be applied after the rate of displacement is about 0.1 mm per 30 min in sandy soils and 0.02 mm/h in clayey soils or 2 h whichever is earlier. Displacements shall be recorded by using minimum two dial gauges spaced at 30 cm and kept horizontally one above the other on the test pile. Wherever it is not possible to locate one of the dial gauges in the line of jack axes, then the two dial gauges may be kept at a distance of 30 cm at a suitable height and the displacement is interpolated at load point from similar triangles. The schematic of lateral load test set-up is shown in Figs 5.4 and 5.5.

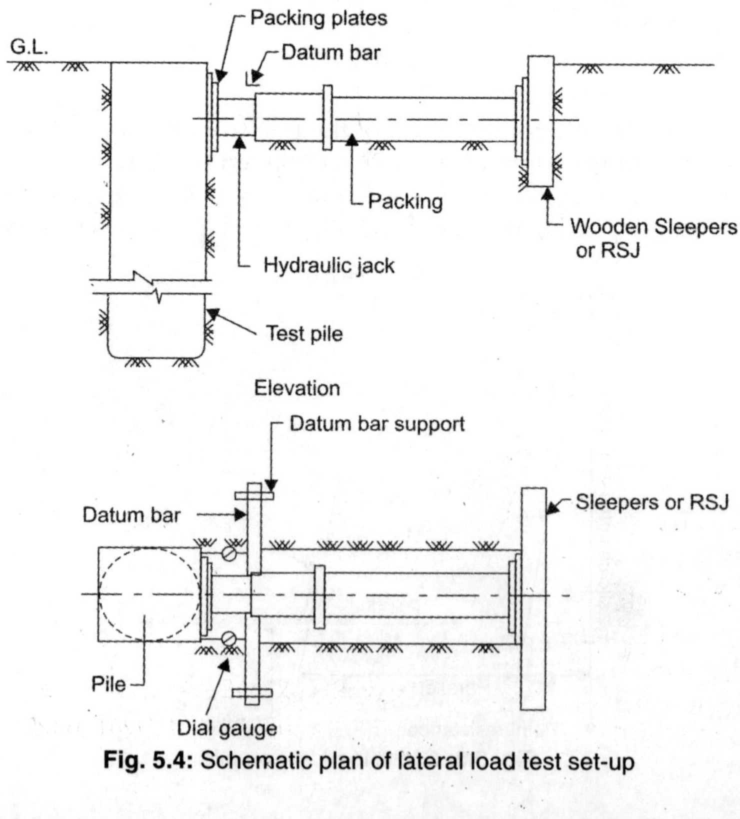

Fig. 5.4: Schematic plan of lateral load test set-up

Fig. 5.5(a): A view of lateral load test set-up in field
(Coutesy: SMEC, Roorkee, Ph: 9412071753)

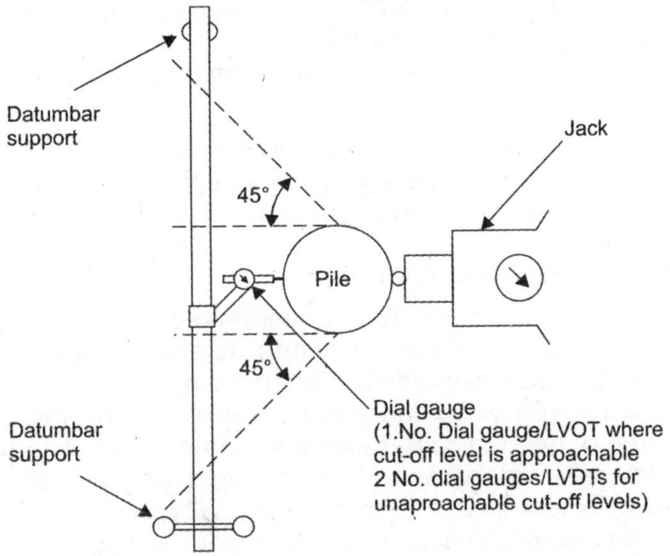

Fig. 5.5(b): A view of lateral load test set-up in field (in plan), IS: 2911(Pt.IV)

IS: 2911 (Part IV)—1985 (Re-affirmed 2015) recommends for keeping dial gauge on pile surface to chip-off uneven concrete on the side of the pile and to fix a piece of glass 20 to 30 mm^2. The dial gauges tips shall rest on the central portion of the glass plate.

The safe lateral load on the pile shall be taken as the least of the following:

a. Fifty per cent of the final load at which the total displacement increases to 12 mm.
b. Final load at which the total displacement corresponds to 5 mm.
c. Load corresponding to any other specified displacement due to performance requirements.

5.7 PULL OUT TEST ON PILE

Uplift force may be found out by means of hydraulic jack with gauge using a suitable pullout set-up.

According to the IS: 2911 (Part IV)—1985, one of the methods for pullout tests may be where hydraulic jack is made to rest on rolled steel joists (RSJ) resting on two supports on the ground. The jack reacts against a frame attached to the top of the test pile such that when the jack is operated, the pile gets pulled up and the react ion is transferred to the ground through the supports which are at least 2.5 D away (where 'D' is pile diameter) from the test periphery. The framework can be attached to the pile top with the reinforcement bars which may be threaded or to which threaded bolts may be welded. As an alternative, it is sometimes preferable to use a central rod designed to take pile load and embedded centrally in the pile equal to the bond length load required. It has the threads at top for fixing it to the framework. For larger loads, the number of rods may have to be more and depending on the set-up these may be put in a line or in any other symmetrical pattern. For routine test, the framework is normally attached to the reinforcing bars but a central rod may also be used in case the upper portion of the pile is required to be built up.

The test pile should have adequate quantity of steel to withstand the pulling during tests. Sometimes, it may be required necessary to provide the additional reinforcement to allow for neck tension in a pull out test.

The safe load from the pull out test shall be smallest of the following:

a. Two thirds of the load at which the total displacement is 12 mm or the load corresponding to a specified permissible uplift.
b. Half of the load at which the load-displacement curve shows a clear break (downward trend).

The Initial test shall be carried out up to twice the estimated safe load or until the load displacement curve shows a clear break. While the routine test shall be carried out to one and a halftimes the allowable load or 12 mm pull whichever is earlier.

5.8 CONSTANT RATE OF PENETRATION TEST

This test is conducted to estimate the ultimate load capacity of a pile. The test is conducted by penetrating the pile at constant rate and continuous measuring the corresponding load. The load versus deflection (penetration) curve does not represent an equilibrium relationship between load and deflection and therefore it does not necessarily give the deflection under any maintained load. The test is generally suitable for friction piles in clay and alluvium deposits For predominantly point bearing piles, the load versus penetration curve does not give a well-defined peak and as such some arbitrary chosen deflection criterion would be needed.

IS: 2911 (Part IV)—1985 describes the procedure of this test as below.

The load should be applied by a remote controlled hydraulic jack. The penetrations are measured by means of a dial gauge. The jack is operated in such a way so to cause the pile penetration at uniform rate which may be controlled by checking the time taken for small increments of penetration and adjusting the pumping rate accordingly. Readings of time penetration and load should be taken at sufficient close intervals to give adequate control of the rate of penetration. A rate of penetration of about 0. 75 mm per minute is suitable for predominantly friction piles. For predominantly end bearing piles in sand or gravel, the rate of penetration of 1.5 mm per minute may be used. The rate of penetration if steady, may be half or twice these values without significantly affecting the results. The test should be carried out for the penetration more than 10% of the diameter of the pile base.

5.8.1 Ultimate Load Capacity from Graphs

The curve of load versus penetration in the case of a predominantly friction pile will represent either a peak and the subsequent downward trend or a peak and then almost a straight line as shown in Fig. 5.6 (a). The peak load marked 'A' in Fig. 5.6 (a) will represent the ultimate load capacity of pile.

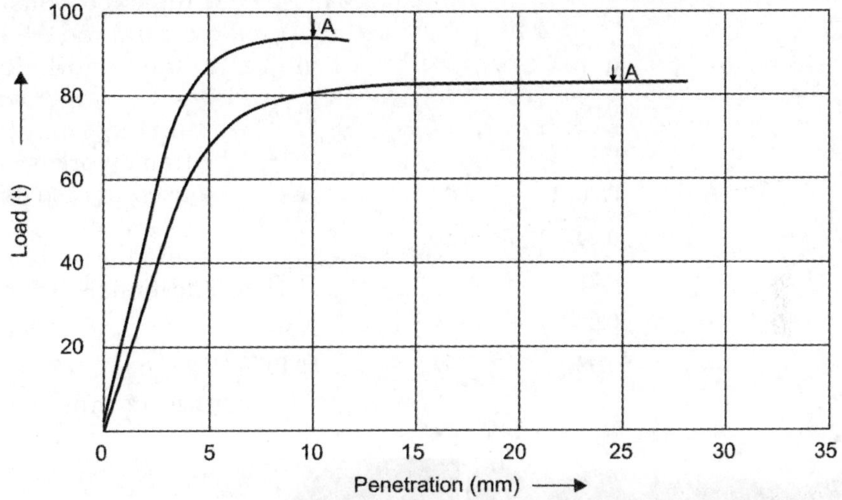

Fig. 5.6 (a): Illustrates load v/s penetration curve for friction pile and

In case of predominantly end bearing pile, the curve will be similar to that shown in Fig. 5.6 (b) and the ultimate load capacity may be taken as the load corresponding to the penetration equal to 10% of the diametric of the pile base.

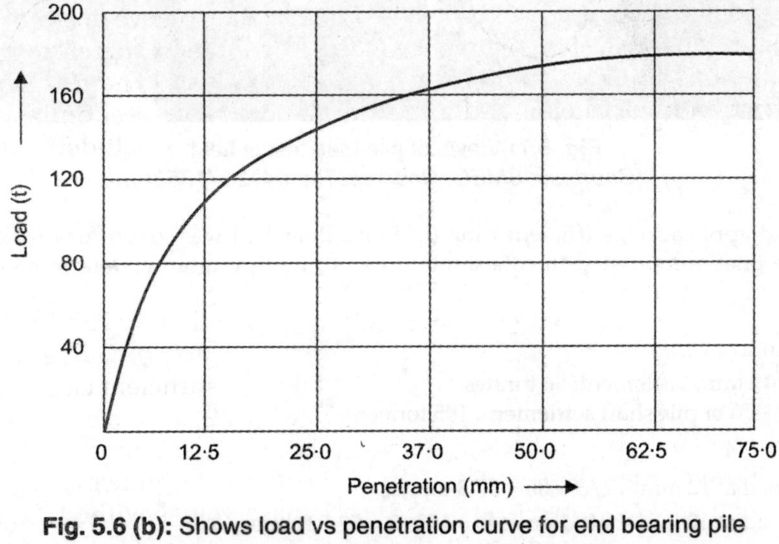

Fig. 5.6 (b): Shows load vs penetration curve for end bearing pile

Example 5.1: A RCC cast-*in situ* bored pile of 450 mm dia and 16.5 m length below its cut-off level was tested at Delhi border for a three storey building. The pile load test data is as below (Table 5.1). Two dial gauges were fixed during the testing. The settlement readings of both dial gauges are given as D1 and D2 respectively. The settlement of pile is computed by taking average of these 2 dial gauge readings. The load was applied through a calibrated remotely controlled hydraulic jack. The photographs taken during the tests are also given below for better understanding of test (Fig. 5.7). The load v/s settlement curve is shown in Fig. 5.8.

Table 5.1: Load v/s settlement data for 450 mm diameter 16.5 m long bored cast-*in situ* pile, pile no. 1, (GP11-2-2)

Load (t)	*D1*	*D2*	*Average (mm)*
0	0	0	0
10	0.06	0.68	0.37
20	0.12	1.12	0.62
30	0.20	1.81	1.005
40	0.24	3.0	1.62
50	0.27	4.32	2.295
60	0.53	6.62	3.575
70	3.98	11.10	7.54

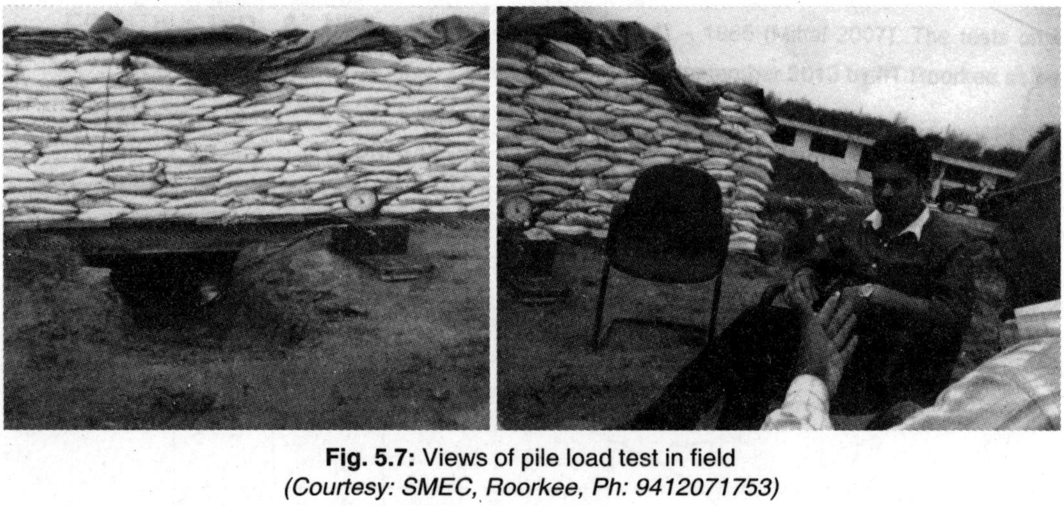

Fig. 5.7: Views of pile load test in field
(Courtesy: SMEC, Roorkee, Ph: 9412071753)

For every load application, sufficient time (of more than 1 h) was given for dial gauges to be constant. The piles were casted in cohesionless soil strata throughout. Determine the pile load capacity.

Observation:

a. Load at 12 mm, settlement: 86 tonnes
b. Load at 10% of pile shaft settlement: 105 tonnes

Results:

a. 2/3 of load at 12 mm = 2/3 × 86 = 57.3 tonnes
b. 1/2 of load at 10% of shaft diameter = 105/2 = 52.5 tonnes

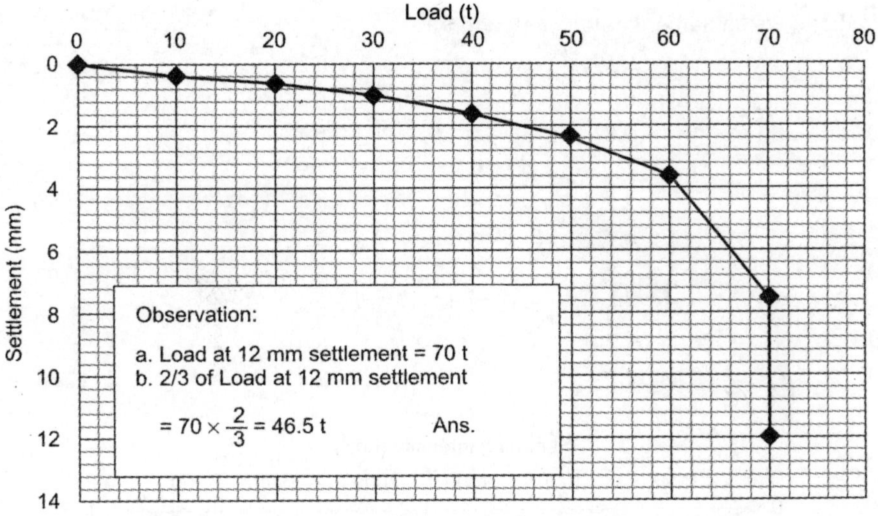

Fig. 5.8: Load v/s settlement curve for pile (Dia = 450 mm, Lenght =16.5 predicted capacity = 51t)

Example 5.2: At a thermal power station in India, it was decided to make granular piles (stone columns) to improve the flyash bed. Two test granular piles were made each of 4 m length and 300 mm diameter. First pile was made like conventional stone column, whereas other pile was made duly encased in geogrid. The site photographs also given here (Figs 5.9 to 5.13). The test results are given in tabular form (Tables 5.2 and 5.3). Compute the load bearing capacity for each case.

Table 5.2: Load v/s settlement data (without geogrid)

Load (kg)	*D1*	*D2*	*Average (mm)*
0	0	0	0
250	0.7	0.7	0.7
500	2.3	2.4	2.35
750	2.9	4	3.45
1000	4.5	4.9	4.7
1500	8.5	8.9	8.7
2000	13.5	13.9	13.7
2500	20	21	20.5

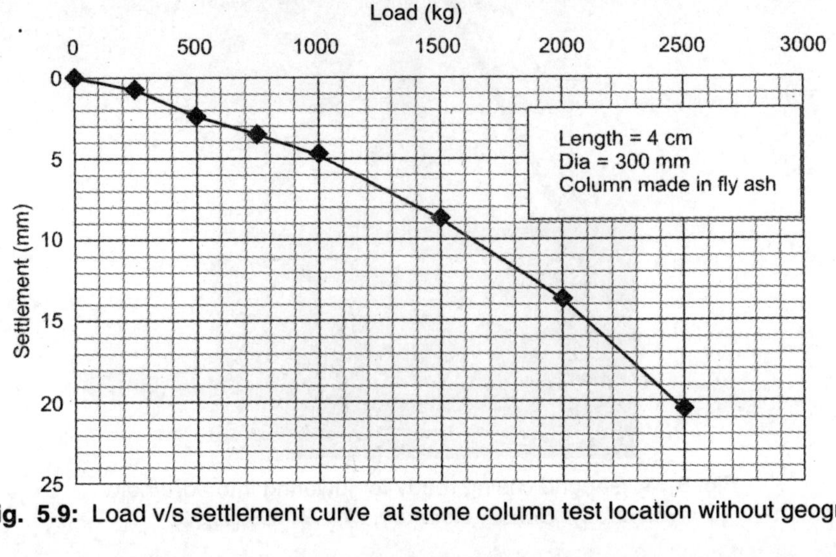

Fig. 5.9: Load v/s settlement curve at stone column test location without geogrid

Table 5.3: Load v/s settlement data (with geogrid)

Load (kg)	*D1*	*D2*	*Average (mm)*
0	0	0	0
250	0.8	0.7	0.75
500	1.1	1.2	1.15
750	2.2	2.4	2.3
1000	3.1	3.3	3.2
1500	5.1	5	5.05
2000	6.6	6.6	6.6
2500	9.2	8.9	9.05
3000	12.7	12.6	12.65

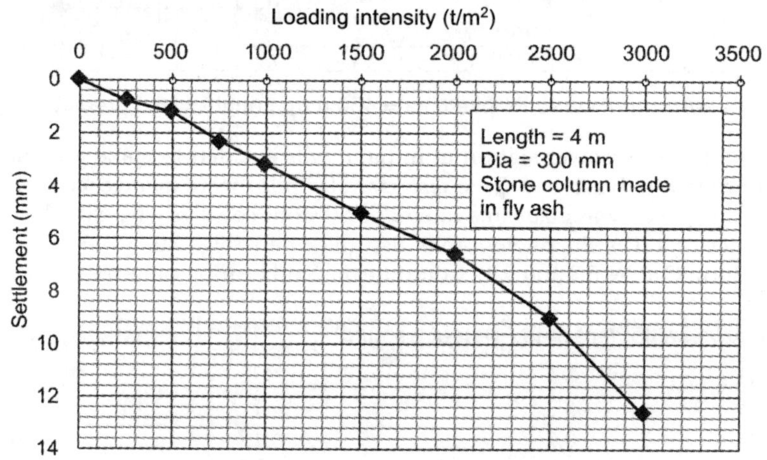

Fig. 5.10: Load v/s settlement curve at stone column test location with geogrid encasing

Fig. 5.11: Geogrid casing ready for lowering into bore hole
(Courtesy: SMEC, Roorkee, Ph : 9412071753)

Fig. 5.12: Top view of geogrid casing
(Courtesy: SMEC, Roorkee, Ph : 9412071753)

Fig. 5.13: Compaction of aggregates in granular pile (stone column)
(Courtesy: SMEC, Roorkee, Ph : 9412071753)

Solution:

a. From Fig. 5.9 (without geogrid), the load corresponding to 12 mm settlement = 1800 kg (1.8 tonnes). The two-thirds of this load = 1200 kg. Hence, load carrying capacity of this granular pile (4 m length, 300 mm diameter made in fly ash) = 1.2 tonnes.

b. From Fig. 5.10 (with geogrid), the two-thirds of load corresponding to 12 mm settlement of this granular pile = 1934 kg (1.9 tonnes). Hence, load carrying capacity of this granular pile = 1.9 tonnes (with geogrid encasing).

This example clearly shows that geogrid encasing improves load carrying capacity of granular pile *(for more details, readers can refer, "An Introduction to Ground Improvement Engg." by Dr. Satyendra Mittal, published from SIPL publications, New Delhi).*

Example 5.3: A stone column was tested at a site in Chennai (India). This test column was situated in a group of column. The c/c spacing of columns was 2.2 m (columns were installed in isosceles triangular pattern). This column had been designed to carry a load of 36 tonnes (Mittal, 2016). The diameter of stone column and length were 800 mm and 13.5 m respectively. The testing was done by providing a very strong Kentledge (loading platform) by a reputed multinational company. The stone column was made at site on 18.11.2016 and was tested on 29.11.2016. The stone column was made by vibroflot. The results and other details are given below for the benefit of readers.

a. Stone column ID and location : Col ID. TP 4—initial test location
b. Date of installation : 18/11/2016
c. Date of commencement of test : 29/11/2016
d. Finished diameter of stone column : 800 mm loading
e. Intensity : 90 kPa (9 t/m2)
f. Spacing and pattern of stone columns : 2.2 m c/c and triangle
g. Influence area of single column : 4.19 m2
h. Depth of the stone column : 13.5 m from cut-off level

Loading system

a. Diameter of steel plate used for testing : 2.31 m (circular)
b. Design load on the plate : 38.0 T
c. Test load (1.5 times design load) : 57.0 T
d. Kentledge load (1.3 times test load) : 75.0 T
e. Load per each increment : ~9.1 T
f. Effective area of ram : 452.50 cm^2
g. Least count of pressure gauge : 10 kg/cm^2
h. Least count of dial gauge : 0.01 mm

Preparation and layout

The layout of group of stone column is illustrated below in Fig. 5.14

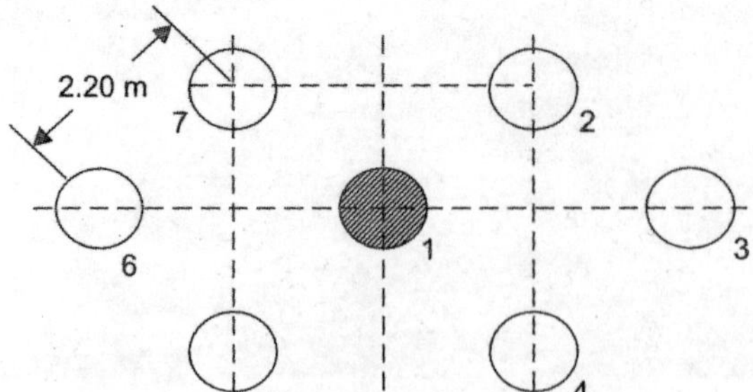

Fig. 5.14: Plan of group of stone columns

SUGGESTED ARRANGEMENT OF STONE COLUMN TEST

1. Pile to be tested : ● No. 1 (Refer plan/Fig. 5.15)
2. Capacity of each pile/column : 38 t
3. Loading platform/Kentledge : 1.5 × 38 = 57 t
4. Testing to be done after constructing all 7 columns (Fig. 5.15)

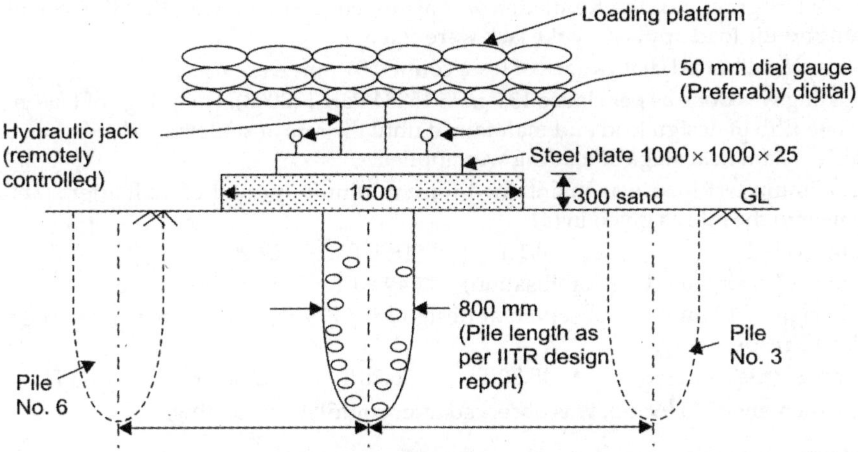

Fig. 5.15: Section of test pile No. 1

Procedure for stone column load test

The following systematic procedure was adopted during execution of initial single column load test at site:

1. Test column TP-4 was identified from the group of columns installed at test location.
2. The following procedure was adopted to prepare test column:
   i. The top surface of stone column was exposed and levelled using a thin layer of granular material of about 300 mm thick and lightly compacted to provide a firm horizontal-loading surface.
   ii. A 60 mm thick, 2.31 m diameter circular steel plate was placed over the effective tributary soil area of stone column.
3. A calibrated 200 T capacity hydraulic jack was placed over the centre of the steel plate for loading purpose.
4. Appropriate steel packing plates were used as per site requirements to provide necessary support between loading jack and stone column.
5. Two datum bars were installed on either side of the stone column (as per the convenience at the test stone column location) to support 4 nos. of calibrated dial gauges. The datum bar supports were driven well below EGL and bracing arrangements were provided in order to avoid deflections or local disturbances.
6. Stone column settlements corresponding to each stage of loading/unloading were monitored by recording the dial gauges.
7. All testing equipment and monitoring units were protected from disturbances prior to and throughout the testing.

Loading procedure

Loading of the column was carried out by means of jacking the primary beam against the reaction from Kentledge. The test was carried out by applying a series of load increments as per the steps given below.

1. The initial single column load test was carried out using kentledge system.
2. Brief details of kentledge are as below:
   a. Primary beam: 3 nos. ISMB 300, 8.3 m long
   b. Secondary beam: 14 nos. ISMB 300, 8.3 m long
   c. Weight of Kentledge: 85T (>1.3 times test load using 1.2 m × 1.2 m × 0.8 m concrete blocks)
   The centre of gravity of the Kentledge was arranged in such a way that the axis of the stone column and the load applied by the jack were coaxial.
3. Four nos. of calibrated dial gauges of least count 0.01 mm were used.
4. The loading was done as per clause 13.7 in IS 15284 Part 1 (2003). Each stage of loading was near about one-fifth of design load and maintained until the rate of settlement is less than 0.05 mm/h at which instant next stage of loading was applied.
5. The maximum test load was maintained for a minimum period of 12 h after stabilization of settlement to the rate as given in (4).
6. Loading cycle 1 9.1 T 18.1 T 27.20 T 31.70 T 40.70 T
   (Design load maintained for stabilisation) 49.80 T 54.30 T 58.80 T
7. A settlement of 10 mm was observed at design load (38 T) and 24.5 mm was observed at test load 57.0 T respectively.
8. Unloading cycle 45.30 T 36.20 T 22.60 T 13.60 T 0.00 T
9. A net settlement of 20.99 mm was observed after complete unloading.

Settlement measurements

Four calibrated dial gauges of least count 0.01 mm were mounted and recorded independently. For plotting purposes, the average of the recorded readings of the dial gauges was considered. Readings were recorded at 1 min, 2 min, 4 min, 8 min and 16 min up to stabilisation, i.e. the rate of settlement is equal to either 0.05 mm/hr.

Results and discussion

Field records

The brief summary of the load vs. settlement data is shown in Table 5.4 and load vs. settlement graph is shown in Fig. 5.16.

Table 5.4: Load vs. settlement values

S. No	*Load in %*	*Load, T*	*Settlement, mm*
1	0	0.00	0.00
2	20	9.10	2.47
3	40	18.10	3.83
4	60	27.20	6.32
5	80	31.70	8.28
6	100	40.70	11.05
7	120	49.80	19.32
8	140	54.30	20.43
9	150	58.80	29.63
10	120	45.30	29.63
11	90	36.20	29.49
12	60	22.60	28.05
13	30	13.60	27.18
14	0	0.00	20.99

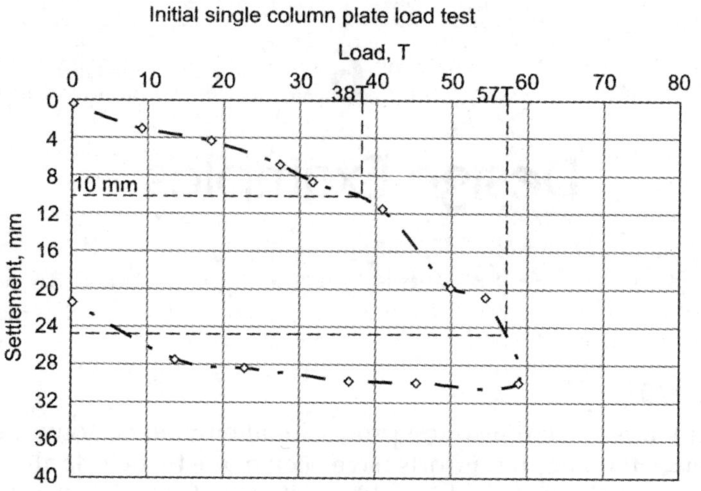

Fig. 5.16: Load vs settlement curve

Summary of results

Summary of the results of single column initial load test are presented in Table 5.5.

Table 5.5: Summary of results

Ref. No.	Initial single column load test
Date of testing	29.11.2016
Settlement @ design load, 38.0 T	10.0 mm
Settlement @ test load, 57.0 T	24.5 mm
Net settlement	20.99 mm
Rebound	8.64 mm
Allowable settlement as per IS15284 (Part 1): 2003	12 mm @ design load

Observations

Based on the Load vs. settlement plot, following points are observed and concluded.

1. The observed settlements are within the acceptable limits for the applied design load of 38.0 T (90 kPa). The proposed ground improvement design sufficing the bearing capacity requirements.
2. According to the criteria stipulated in IS: 15284 Part 1, the required safe bearing capacity of improved soil (9.0 T/m^2) is achieved as per the project requirements.
3. Based on the test results, it may be concluded that compaction efforts were adequate and quality of installed stone column was satisfactory.

Chapter

6

Design Examples

6.1 BACKGROUND

For the sake of students and also the professional engineers, some design examples have been given in this chapter. Efforts have been made to include all possible types of pile foundations in different possible field conditions. Thus, this chapter shall help for design of pile foundations.

Example 6.1: Determine the length and safe loads in uplift and compression for a pile of 3 bulb multi under-reamed stem diameter 40 cm and bulb diameter as 100 cm.

Solution: The length of the pile is decided by the following considerations:

i. The top bulb should not be placed closer than twice the under-reamed bulb diameter.

ii. The stem diameter of pile being greater than 30 cm, the preferred spacing between the two adjacent bulbs is 1.25 times the under-reamed bulb diameter.

iii. The distance between the centre of the lower most bulb up to the stem will be 0.55 times the stem diameter.

iv. The pile stem portion towards the toe which will accommodate the under-reamer bucket during construction will be 55 cm for a 40 cm diameter pile.

Therefore, the length of the pile

$$= (2 \times D_u + 2 \times 1.25\, D_u + 0.55\, D + 0.55) \text{ m}$$

where D_u = Diameter of under-ream bulb (m)

D = Diameter of pile (m)

$$= 2 \times 1.0 + 2 \times 1.25 \times 1.0 + 0.55 \times 0.4 + 0.55$$

$$= 5.27 \text{ m, (Say 5.30 m)}$$

Referring the Table 3.1, for medium compact sand ($10 < N < 30$) or clay of medium consistency ($4 < N < 8$).

a. *Safe load in uplift*:

From column 12, the safe uplift load of a single under-reamed pile is 14 tonnes. Increase due to each additional 30 cm length is 1.45 tonnes.

$$\text{Safe uplift load} = 14 + 2 \times 7.0 + \frac{(5.3 - 3.5) \times 1.45}{0.3}$$

$$= 36.7 \text{ tonnes}$$ **Ans.**

b. *Safe load in compression*:

From column 8, safe load of a 3.5 m long 40 cm diameter, single under-reamed pile is 28 tonnes.

Increase due to each additional bulb is 14 tonnes.

Increase for each additional 30 cm length from column 10 is 1.9 tonnes.

Thus, safe load in compression

$$= 28.0 + 2 \times 14.0 + \frac{(5.3 - 3.5) \times 1.9}{0.30}$$

$$= 67.40 \text{ tonnes}$$ **Ans.**

For strata of better bearing properties, an increase of 25%, for strata of poor bearing properties, a decrease of 25% and for under-water concreting, a decrease of 25% in safe loads as given in Table 3.1 is to be done.

Example 6.2: Nine piles of 25 cm diameter are arranged in a square form. The centre to centre spacing is 1.0 m. If the length of piles is 12 m and cohesion of the soil is 0.8 kg/cm^2, determine if the failure will occur with the piles acting individually or as a group. Neglect bearing at the top of piles (*Refer* Fig. 6.1).

Solution: The load carrying capacity of the pile for individual action is given by the equation

$$Q_u = cN_cA_p + m.c.A_s \quad \ldots (6.1)$$

Neglecting the first term, we may write

$$Q_u = m.c.A_s$$

where A_s = Area of surface of each pile.
for circular pile, $A_s = \pi d \times L$
Therefore, $Q_u = c.m.\ \pi d \times L$
For 9 piles $Q_u = 9 \times c.m.\ \pi dl$

$$= 9 \times \left(\frac{0.8 \times 100 \times 100}{1000}\right) \times 1.0 \times \pi \times 0.25 \times 12$$

(taking the value of $m = 1.0$)
Thus, $Q_u = 678.24$ tonnes
For group action, considering equation given in chap. 4

$$Q_{ug} = P.L.c. + N_c \cdot c_t.A \quad \ldots (6.2)$$

In this case, second term may be neglected,
Therefore, $Q_{ug} = P.L.c$
Here, P = Perimeter of group of piles
For 9 piles, the perimeter of the group will be:

$$4 \times \left(1 + 1 + \frac{0.25}{2} + \frac{0.25}{2}\right) = 9 \text{ m}$$

Thus, $Q_{ug} = 9 \times 12 \times 8 = 864$ tonnes
In group action piles will take more load than acting individually.
Hence the failure will occur with piles acting individually. **Ans.**

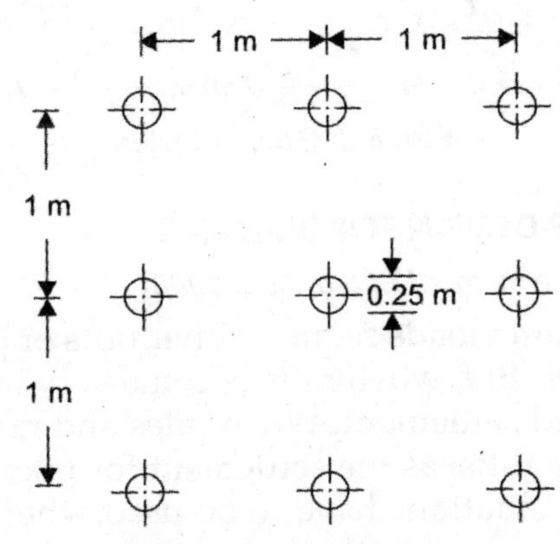

Fig. 6.1: Group of piles

Example 6.3: Determine the group strength of the piles arranged in a square form at 90 cm centre to centre. Total number of piles is 9 and diameter and depth of piles is 30 cm and 12 m respectively. The unconfined compressive strength of the surrounding soil is 150 g/cm² Refer Fig. 6.2.

Solution:

$$C = \frac{q_u}{2} = \frac{150}{2} = 75 \text{ gm/cm}^2 = 0.75 \text{ t/m}^2 \quad \ldots (6.3)$$

Surface area of the entire pile group acting as a unit

$$= 12 \times 4 \left(\frac{90 + 90 + 15 + 15}{100} \right) = 100.8 \text{ m}^2$$

Thus, total shear resisted = 100.8 × 0.75 = 75.5 tonnes

Safe soil pressure at base

$$q_u = 0.95q \left(1 + 0.3 \times \frac{b}{l} \right) \quad \ldots (6.4)$$

where l = length of footing

b = width of footing

q = unconfined compressive strength

$$= 0.95 \times 1.5 \left(1 + 0.3 \times \frac{2.1}{2.1} \right) = 1.852 \text{ t/m}^2$$

Load carrying capacity at base

= 1.852 × 2.1 × 2.1 = 8.16 tonnes

Group strength of piles = 75.5 + 8.16

= 83.66 tonnes **Ans.**

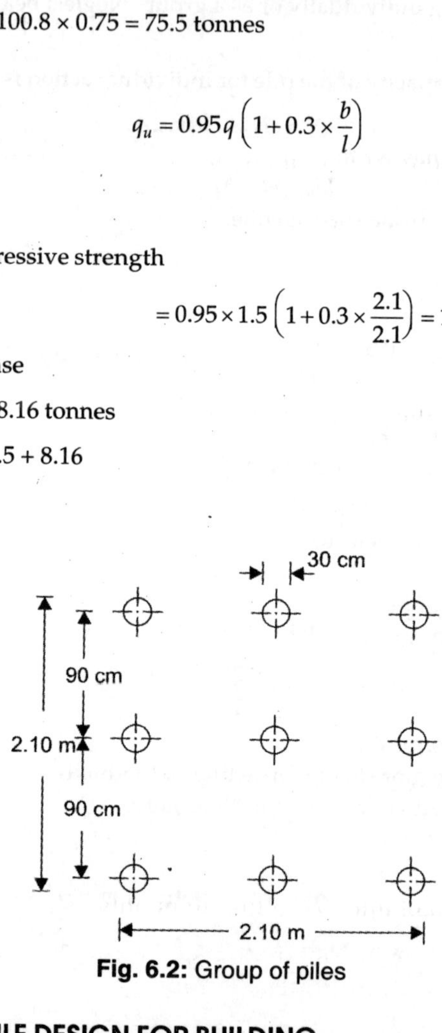

Fig. 6.2: Group of piles

6.2 DESIGN STEPS FOR PILE DESIGN FOR BUILDING

(Ref: UH Varyani, Journal of The Instn. of Engrs, Aug. 1993)

Weak soils with heavy column loads require either rafts or piles. Rafts, in general, are more economical than piles. But, when rafts become very large, piles have to be used for restricting both cost and settlement. When piles and rafts, are both equal in cost, then piles are preferable to rafts, as the settlement for piles is considerably less than that of rafts. Thus, pile foundations have to be used when raft foundations are not suitable on grounds of economy of settlement considerations.

Cast-*in situ* driven piles 300 to 600 mm in diameter are quite common for medium capacity piles of 40 to 120 t. Cast-*in situ* bored piles are available for high capacities (150 to 300 t) and cause no vibrations to the existing adjoining buildings unlike driven piles which are noisy and disturb the structures in the immediate vicinity. Cast-*in situ* under-reamed piles 200 to 450 mm in diameter, 3 to 6 m deep are low capacity piles of 20 to 40 t. These piles are ideal for foundations of low-rise buildings in black cotton soils, as these provide good resistance to upward lifting forces caused by the soil's expansion.

Economy of pile foundations is achieved when the cost of pile caps is minimized. This is possible if a few high capacity piles are used under each column. Capacity of a pile is increased by increasing either the diameter of the depth of a pile. By increasing the diameter of a pile, pile spacing (equal to three times the diameter) will increase and this will lead to a large pile cap, which will result in increase cost. So, pile depth should be increased to get high capacity piles. Thus, economy requires the use of a few high capacity deep piles under a column rather than a large number of low capacity shallow piles.

A minimum of three piles are required under a column in order to resist all the column loads and moments acting on it. If one or two piles are provided under a column, then grade beams have to be provided in one or more directions, which have to be designed to resist fully column base moments since piles will resist the vertical load only. Further, column shear will also be resisted by piles, which are assumed to have a horizontal load capacity equal to 5% of its vertical load capacity. It may be noted that horizontal load on a pile is the result of earthquake or wind loads only, for which 25% excess piles capacity is allowed by codes.

6.2.1 Design Steps

Design of a pile foundation consists of:

i. Design of pile group or pile layout
ii. Design of pile cap

6.2.2 Design of Pile Group or Pile Layout

A pile group should be so designed that pile load works out to be less than or equal to the pile capacity in both the vertical and the horizontal directions. Also, the pile layout should be close knit so that pile cap area is as small as possible.

The general loading cases met with in practice are:

i. $P + MX - MY$: Vertical load with moment in each principal

ii. $V + T$: Shear (V) with torque (T) in horizontal plane, resulting from earthquake and wind loadings only

There are two general layouts, one rectangular and the other circular, and also layout for a single column or for multiple columns.

i. Rectangular layout : For large lift-well rectangular shafts or shear walls, tubes, etc.

ii. Circular layout : Useful for circular shaft-like tall structures, e.g. chimneys and towers.

iii. Pile layouts for isolated columns : Analogy of combined footing.

iv. Pile layouts for multiple columns : Analogy of combined footing, strip footing or raft.

Rectangular Layout

Piles are arranged on a rectangular grid with equal spacing S in each direction so that total number of piles equals mn (Fig. 6.3), where m = number of rows of equal no. of piles in a rectangular grid, n = no. of columns of equal no of piles in a rectangular grid or total no. of piles in a circular layer.

S, the spacing of piles = 3 D for 100% efficiency of pile group.

Load Cases

i. Vertical load P alone:

$$\text{Pile load} = \frac{p}{mn} \quad \text{(on each pile)}$$

ii. $P + M_y$:

$$\text{Pile load} = \frac{p}{mn} + \frac{M_y}{mn\frac{(n+1)\cdot S}{6}} \quad \text{(valid for extreme piles).}$$

iii. $P + M_y + M_x$: Pile load

$$= \frac{p}{mn} + \frac{M_y}{\frac{mn(n+1)\cdot S}{6}} + \frac{M_x}{\frac{mn(m+1)\cdot S}{6}}$$

Here, M_y, M_x are moments about $y - y$ and $x - x$ axes respectively.

iv. $V + T$: Shear per pile $= \dfrac{V}{mn} + \dfrac{T}{I_z}$

(valid for corner piles)

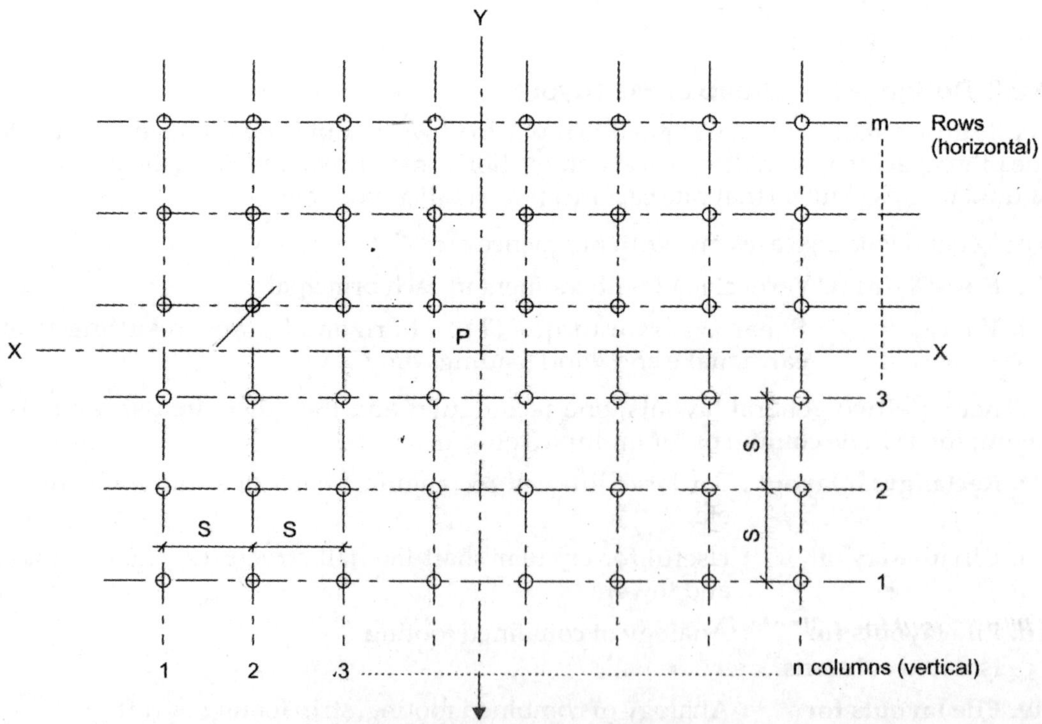

Fig. 6.3: Rectangular pile layout

where $$\frac{I_z}{Y} = \frac{mnS}{6} \cdot \frac{(m^2 + n^2 + 2)}{\sqrt{(m^2 + n^2 + 2m + 2n + 2)}}$$

where I_z = polar moment of inertia of a pile group

Effect of torque can be eliminated if the centre of the pile group is made to coincide with the point of action of the eccentric horizontal shear. Similarly, effect of moments M_x and M_y can be eliminated if the centre of gravity of a pile group is made to coincide with the point of action of the biaxially eccentric load.

Circular Layout

Piles are arranged in concentric circles with pile spacing equal in each circle (Fig. 6.4).

Load Cases

i. $$P + M_x + M_y = M = \sqrt{Mx^2 + My^2} \quad \ldots (6.5)$$

$$\text{Pile load } = \frac{P}{n} = -\frac{M}{I}\gamma_n \quad \text{where } n \text{ is total number of piles.}$$

where $$I = \sum_{n=1}^{n=n} M(\gamma_n)^3$$

ii. $$V + T = \text{shear per pile } = \frac{V}{n} \pm \frac{I}{I_z} \cdot \gamma_n$$

where $$I_z = \sum_{n=1}^{n=n} M(\gamma_n)^2$$

Both the vertical and the horizontal capacities of the pile must not be exceeded.

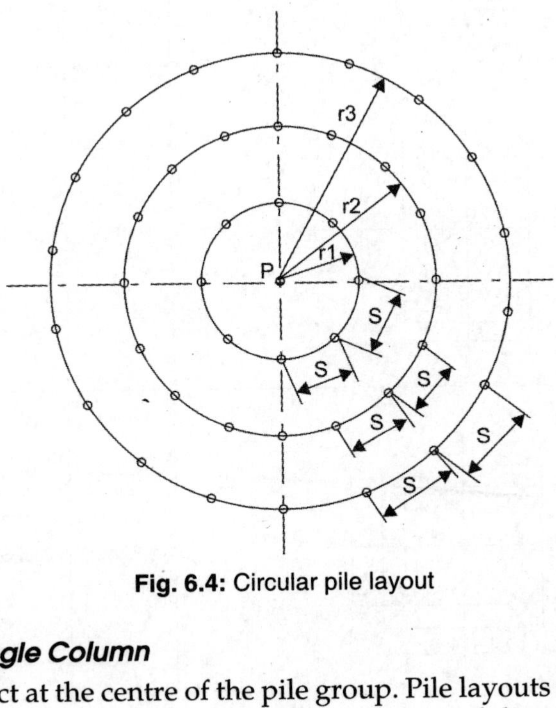

Fig. 6.4: Circular pile layout

Pile Layout for a Single Column

Column load is to act at the centre of the pile group. Pile layouts from 1 to 20 piles are given in Table 6.1, along with Z-values in the two principal directions, for ready use.

Table 6.1: Pile layout for a single column

No. of piles	*Pile layout*	*Zxx*	*Zyy*	*Remarks*
1	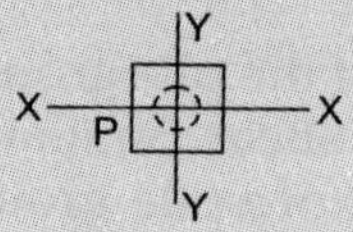	0	0	Incapable of taking moments either way
2	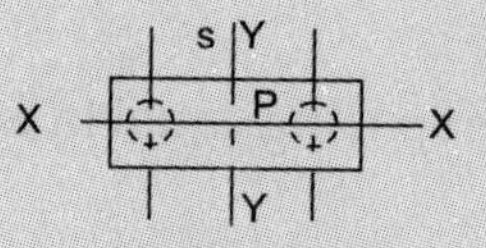	0	0	Incapable of taking moments about x-x axis
3	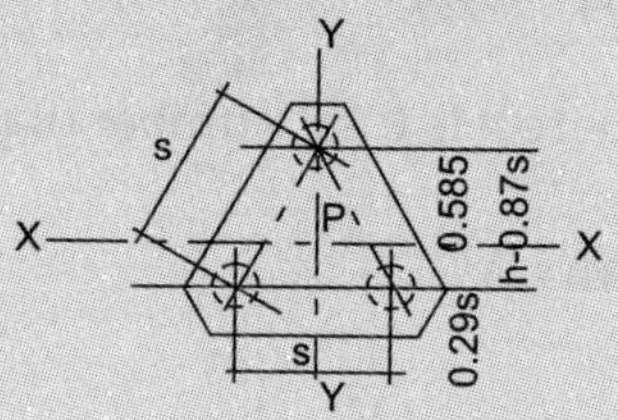	1.74	1.0 s	—
4	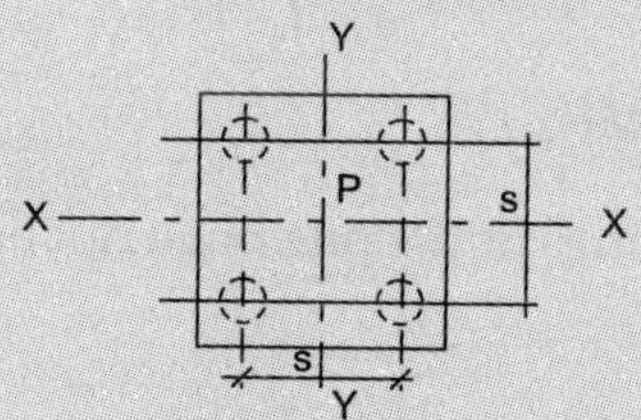	2.0 s	2.0 s	2 x 2 Rectangula grid
5	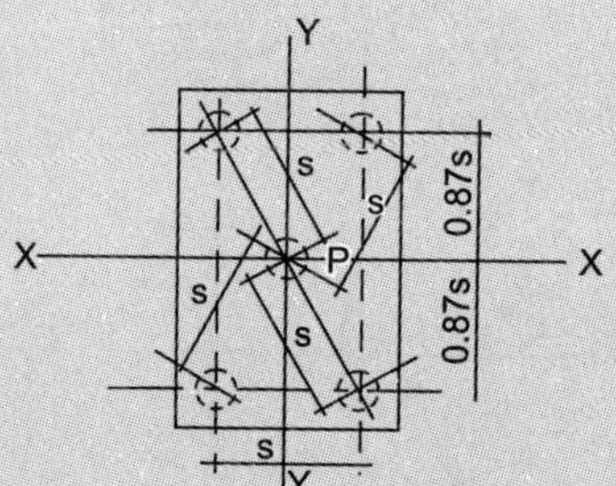	3.48 s	2.0 s	This is more close knit than the following arrangement
6	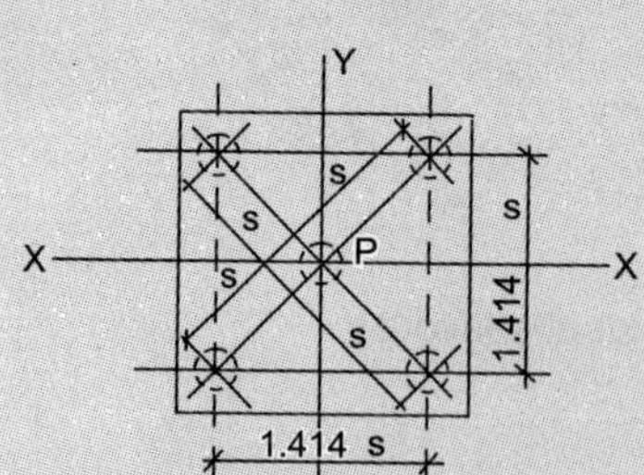	2.828 s	2.828 s	Alternate sysmmetrical arrangement

(Contd...)

Table 6.1: Pile layout for a single column *(Contd...)*

No. of piles	*Pile layout*	*Zxx*	*Zyy*	*Remarks*
7	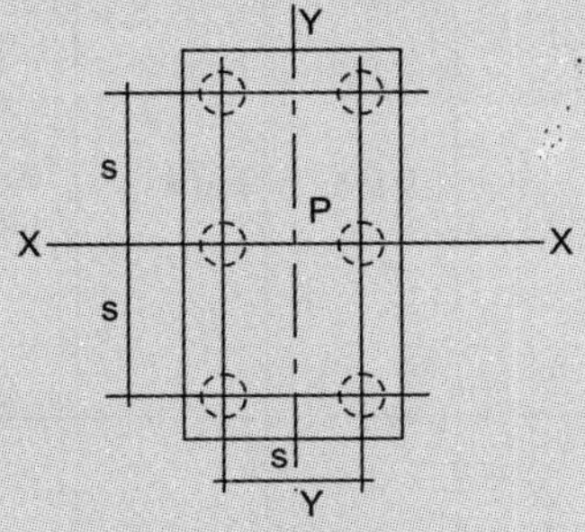	3.0 s	4.0 s	3 × 2 Rectangular grid
8	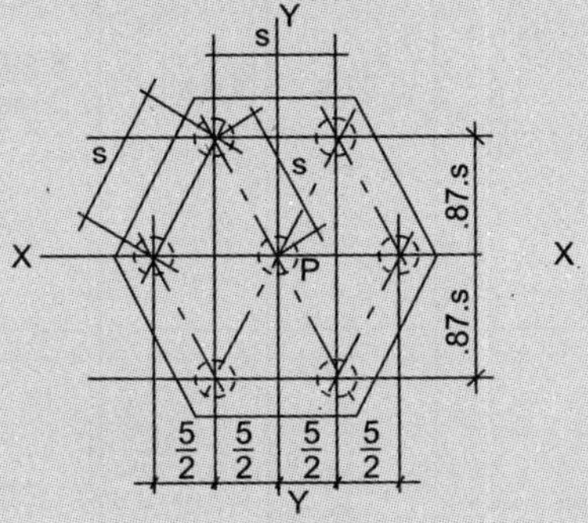	3.48 s	3.00 s	—
9	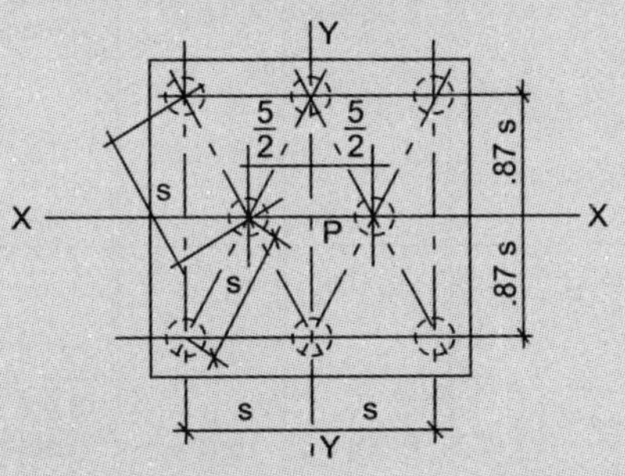	5.22 s	4.50 s	—
10	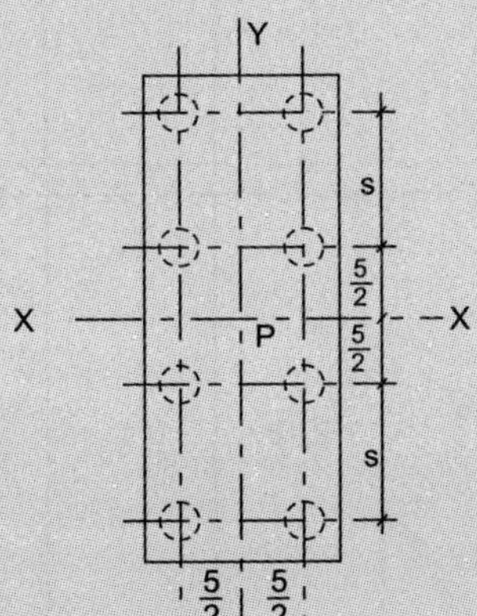	6.67 s	4.0 s	Alternate mon close-knit arrangement 4 × 2 grid

(Contd...)

Table 6.1: Pile layout for a single column *(Contd...)*

No. of piles	*Pile layout*	*Zxx*	*Zyy*	*Remarks*
11	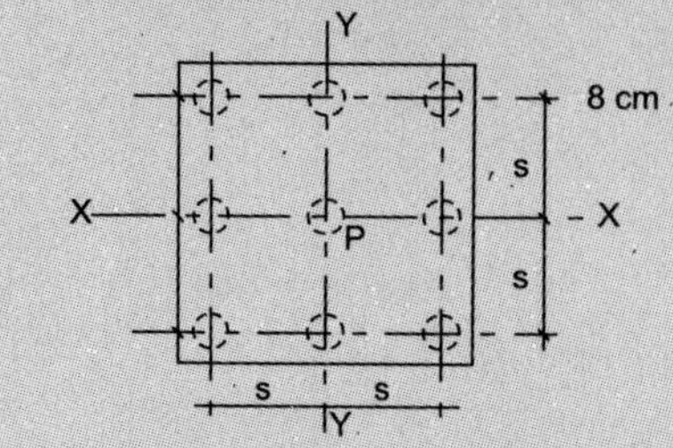	6.0 s	6.0 s	3 × 3 Rectangular grid
12	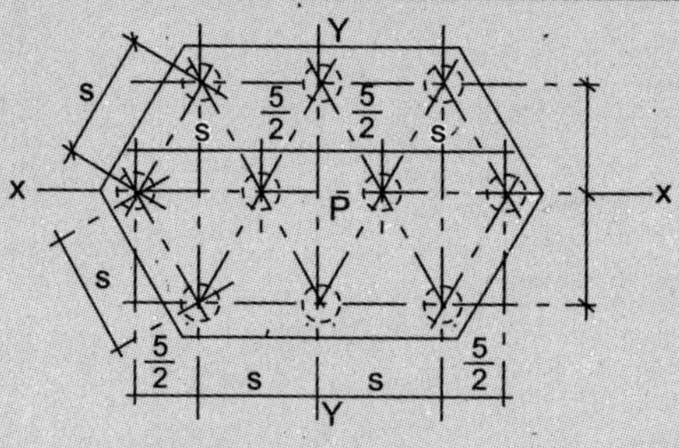	5.22 s	6.0 s	8 Piles * 2 more piles
13	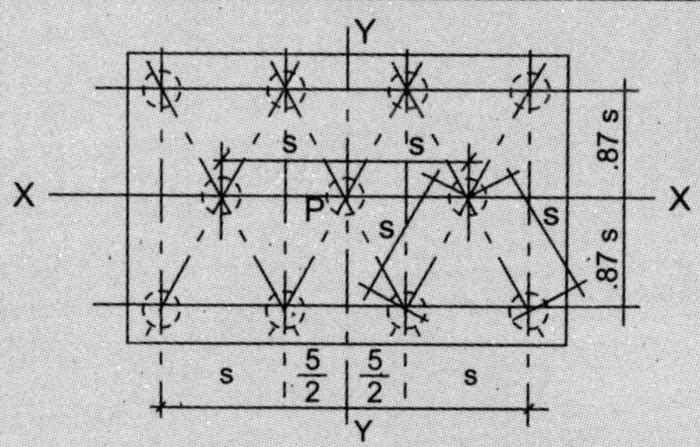	6.96 s	8.0 s	7 Piles * 6 more piles
14	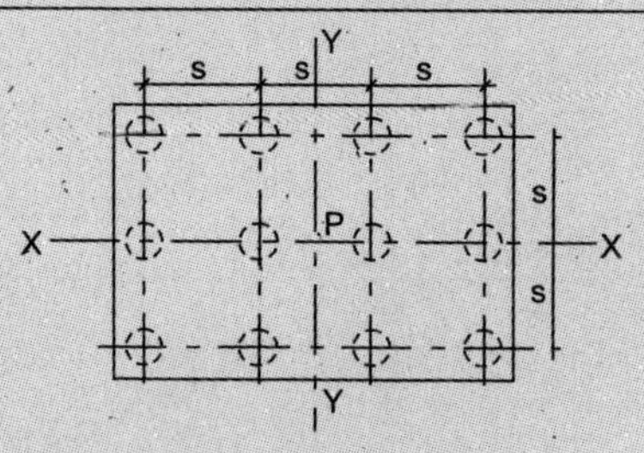	8.0 s	10.0 s	3 × 6 Rectangulat grid
15	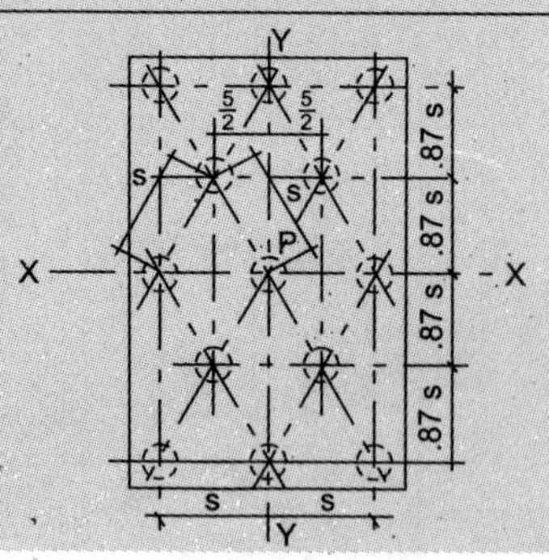	12.18 s	7.0 s	11 Piles * 2 more piles

(Contd...)

Table 6.1: Pile layout for a single column *(Contd...)*

No. of piles	*Pile layout*	*Zxx*	*Zyy*	*Remarks*
16	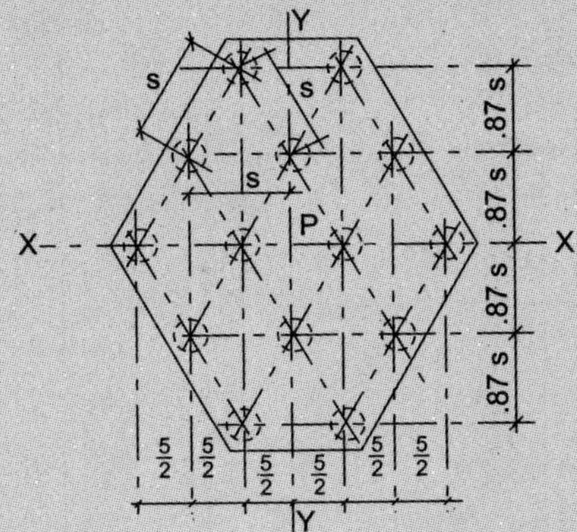	12.18 s	6.67 s	8 Piles * 6 more piles
17	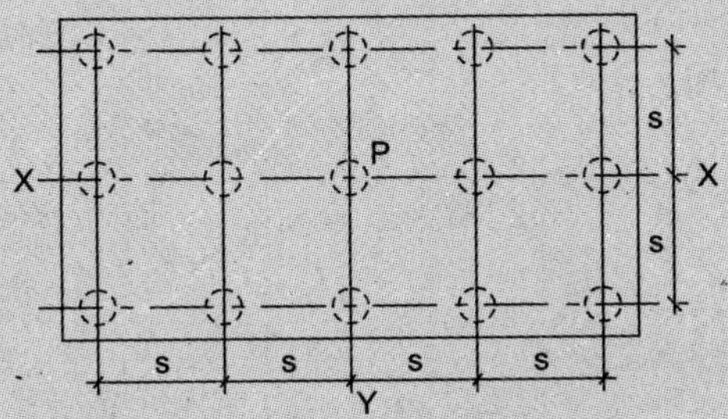	10.0 s	15.0 s	3 × 5 Rectangular grid
18	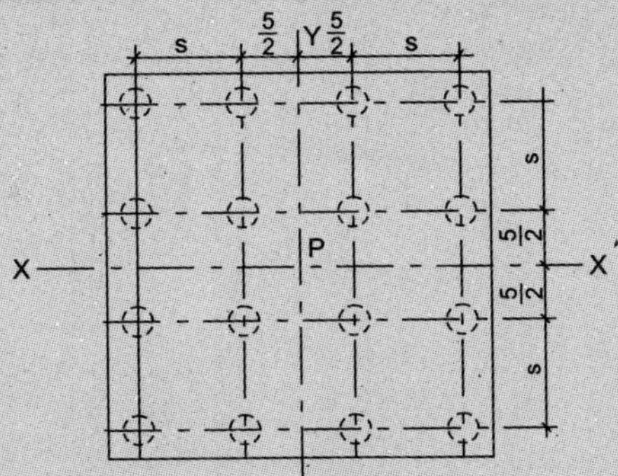	13.33 s	13.33 s	4 × 4 rectangular grid
19	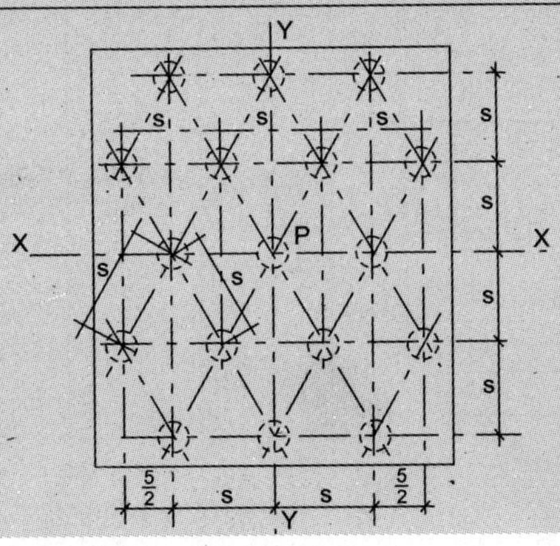	13.92 s	10.67 s	13 Piles * 4 more piles

(Contd...)

Table 6.1: Pile layout for a single column *(Contd...)*

No. of piles	*Pile layout*	*Zxx*	*Zyy*	*Remarks*
20	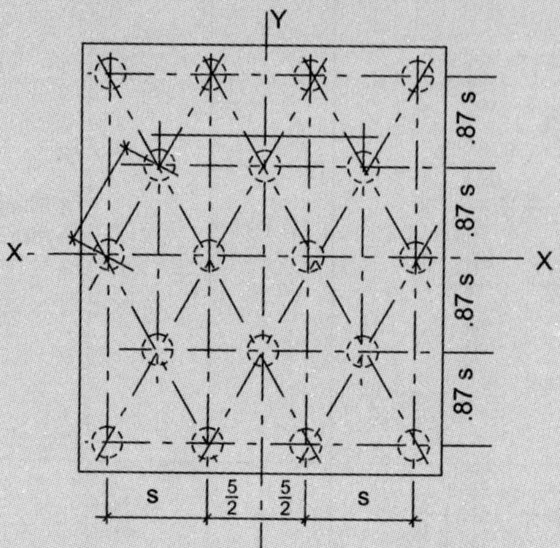	16.53 s	12.67 s	14 Piles * 4 more piles
21	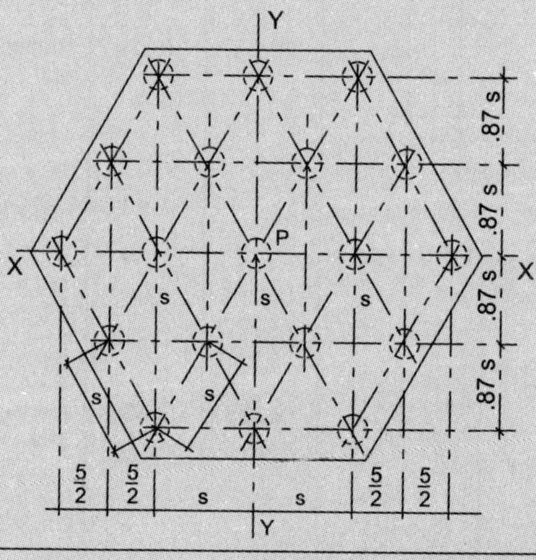	13.92 s	12.0 s	17 Piles * 2 more piles
22	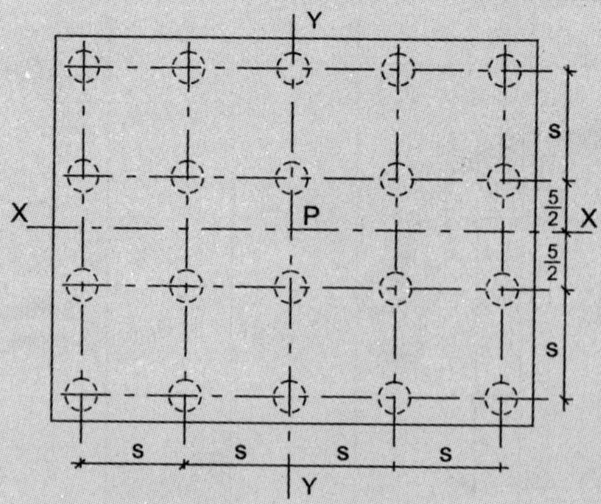	16.67 s	20.0 s	4 × 5 Rectangular grid

Pile Layout for Multiple Columns

This situation arises for two or more columns at the location of expansion joints or for a large number of columns provided in a lift core. Then a combined pile cap is to be designed which is to be supported on a common pile group. The general principle is to coincide the centre of the pile group with the centre of gravity of all column loads so that under the vertical load (ΣP), all the piles are equally loaded. In each principal direction, column base moments are to be added and the pile group is to be checked for the maximum pile load to be within the vertical load capacity of pile. The same procedure has to be adopted for earthquake loading in each principal direction separately. Horizontal shear per pile is the result of earthquake or wind loading only and it has to be checked to be less than the horizontal load capacity of the pile multiplied by 1.25 as 25% excess capacity is allowed for earthquake or wind loading.

6.3 DESIGN OF PILE CAP

6.3.1 Isolated Pile Cap

As suggested by IS: 456–1978, a pile cap is a footing on piles. The action of a pile cap is similar to the action of an isolated footing, the only difference being that instead of a uniform soil pressure in the case of isolated footings, a pile cap has concentrated pile reactions acting from below. Clause 33 of IS: 456–1978 gives the necessary guidance for design of footings on piles for moment, perimeter shear, beam shear and development length of bars. The critical section for moment is the face of column or pedestal, while for beam shear and perimeter shear, the critical section lies at $\left(\frac{d}{2}\right)$ away from column or pedestal face, *d* being the effective depth of pile cap. IS 2911 requires adequate depth of pile cap for anchoring column and pile reinforcements and also for providing adequate rigidity to distribute column load equally to the supporting piles. For pile caps to be rigid, pile cap has to be quite deep, with 60 cm as the minimum depth. For perimeter shear, the calculated shear stress should not exceed $^{0.25}\sqrt{f\,ck}$ in the limit state method and $^{0.16}\sqrt{f\,ck}$ in the working stress method of design as per clause 30.6.3.1 of the code.

The main problem in design of pile caps is posed by the beam shear. The design shear strength of concrete τ_c given in Table 13 of the code should be multiplied by a factor δ, (clause 39.2.2 of the code) where

$$\delta = 1 + \frac{3P_u}{Afgck} (> 1.5) \qquad \ldots (6.6)$$

Table 13 of the code is based on a formula which is related to *pt* in such a way that when $pt = 0$, $P_c = 0$. This appears to be absurd as plain concrete is known to resist shear without any aid from steel. ACI code gives for plain concrete $\tau_c = {}^{2}\sqrt{fc'}$ where fc' is cylinder strength of concrete in psi.

For usual concrete mixes,

$$\text{M15}\ fck = 150\ \text{kg/cm}^2 \text{ or } 2130\ \text{psi};$$

$$fc' = 0.8\, fck = 1704\ \text{psi};$$

$$\tau_c = {}^{2}\sqrt{1704} = 82.6\ \text{psi or } 5.8\ \text{kg/cm}^2$$

$$\text{M20}\ fck = 200\ \text{kg/cm}^2 \text{ or } 2840\ \text{psi};$$

$$fc' = 2272 \text{ psi};$$

$$= \sqrt[2]{2272} = 95.3 \text{ psi or } 6.7 \text{ kg/cm}^2$$

$$\text{M25 } fck = 250 \text{ kg/cm}^2 \text{ or } 3550 \text{ psi};$$

$$fc' = 2840 \text{ psi};$$

$$\tau_c = \sqrt[2]{2840} = 106.6 \text{ psi or } 7.5 \text{ kg/cm}^2$$

This values of τ_c are quite high. A minimum value of $\tau_c = 0.35$ N/mm^2 or 3.5 kg/cm^2 for $pt = 0_7$ to 0.25 is suggested. When $pt > 0.25$, Table 13 of the Code[7] should be followed. Multiplying factor will be very useful in design of pile caps for beam shear.

Some engineers, however wish to follow the code by the letter. For pile caps, minimum steel is to be provided at 0.12% of the cross-sectional area.

$$A_{st}(\text{min}) = 0.12bD$$

$$pt = \frac{Ast}{bd} = \left(\frac{Ast}{bd}\right).\frac{D}{d} = 0.12\frac{D}{d} = \frac{0.12}{\frac{d}{D}} = \frac{0.12}{0.95\ (say)} = 0.126$$

For M15, formula

$$\tau_c = 0.85\sqrt{\frac{0.8fck}{6\beta}}\left(\sqrt{1+5\beta}-1\right),$$

$$\beta = \frac{0.8fck}{6.89\,pt}(\not< 1.0)$$

on which Table 13 of the code is based, gives,

$$\tau_c = 0.26 \text{ N/mm}^2 \times (\delta \text{ factor})$$

Now, this value is quite small and beam shear requires more depth for pile caps. For an economical design of pile caps. For an economical design of pile caps, two steps are suggested; namely:

i. Provide a column pedestal, say, 30 cm all around the column. Provision of a pedestal is of considerable benefit as it shifts critical section for beam shear 30 cm outwards and gives a reasonable depth of pile cap. A pedestal will be designed for bearing stresses which shall not exceed 0.25 *fck* in the working stress method and 0.45 *fck* in the limit state method of design (clause 33.4 of the code). A minimum of 0.15% steel be provided in pedestal (clause 25.5.3.1 of the code).

ii. Provide depth of pile cap in such a way that the critical section for beam shear coincides with the outer edge of the outer-most pile of the pile group. In this way, design shear works out to be small and shear stress will be within

$$\tau_c = 0.26\ (\times\ \delta)\ \text{N/mm}^2$$

For working out shear force on account of pile loads, clause 33.2.4 of the code is to be followed. Let the critical section for shear be 1′ – 1′ (Fig. 6.5). Step (ii) given above, gives,

$$\frac{a}{2} + 30\frac{d}{2} = s + \frac{D_P}{2}$$

when a, S, D_P known, d can be easily calculated.

$$D = d + 10 \text{ cm (say)}$$

with this arrangement,

Design shear = Shear at pile centre = 0.

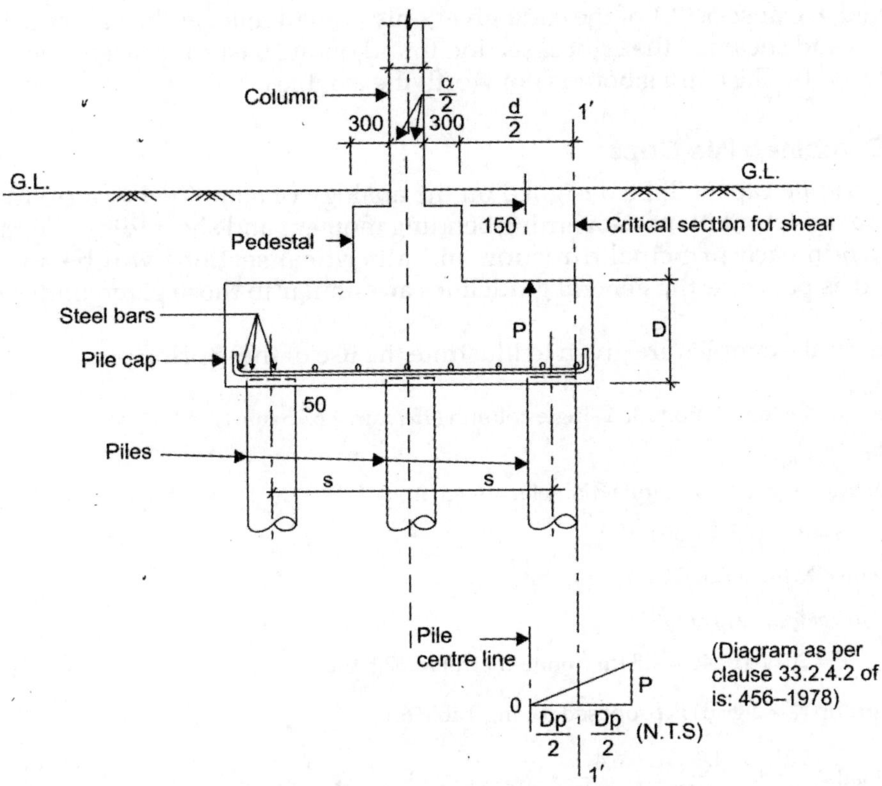

Fig. 6.5: Design of pile cap for beam shear

In practical cases, design shear works out to be either zero or a small value and the calculated shear stress will be well within $\tau_c = 26\delta$ N/mm². IS 2911 (Part I/sec 1) - 1979, clause 5.12 gives also guidelines for design of pile caps. The dispersion method (Fig. 6.6)

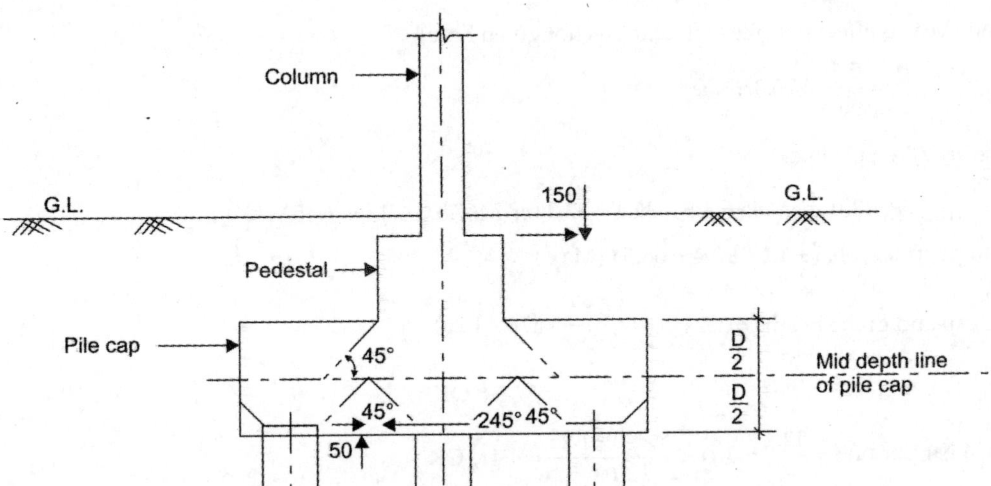

Fig. 6.6: Dispersion method of pile cap design

suggested in clause 5.12.1 of the code gives only a small relief in the values of bending moments and shears at the critical section found on the basis of concentrated column and pile loads. The extra labour is not worth the trouble.

6.3.2 Combined Pile Caps

Combined pile caps will be designed on the analogy of combined or strip footing or slab-type raft foundations. Governing bending moment and shear force diagrams will be drawn in each principal direction and all critical sections will be adequately designed as per code the general guidelines are similar to those given under isolated pile caps.

Numerical examples are given to illustrate the use of these principles.

Example 6.4: Pile foundation for a single column (450 mm × 67.5 mm)

Pile layout

Solution: Pile capacity = 70 *t*, pile diameter = 0.45 m

$$S = 3 \times 0.45 = 1.35 \text{ m}$$

Concrete mix M15, hence *fck* = 150 kg/m²

Case (i) Vertical load (VL) only

$$P = 201\ t\phi\ \text{M } 45 = 4.8 \text{ tm (tonne-metre) M } 67.5 \text{ tm}$$

A 4-pile group (2 × 2 grid) is proposed. Using Table 6.1,

$$\text{Pile load} = \frac{201}{4} \pm \frac{4.8}{2 \times 1.35} \pm \frac{6.4}{2 \times 1.35}$$

$$= 50.3 \pm 1.8 \pm 2.4 = 54.5 \text{ t (Max.)}$$

Pile cap (Fig. 6.7) self-weight for pile

$$= 2.1 \times 2.1 \times \frac{1.2(say) \times 2.5}{4} = 3.3\ t$$

Grade beams effect per pile (calculations not given here)

$$= \frac{5.5t}{63.3} < 70.0t, \textit{ safe}$$

Case (ii) VL + EQ (45 ↔)

$$P = 201\ t\phi, M\,45 = 4.8 + 32.4 = 37.2\ tm, M67.5 = 6.4\ tm, H45 = 12.4\ t$$

Load per pile = 50.3 ± 13.8 ± 2.4 = 66.5 *t* (Max.)

$$\text{Pile cap and grade beams} = 3.3 + 5.5 = \frac{8.8t}{75.3t} < (70 \times 1.25)$$

$$< 87.5 \text{ t. OK}$$

$$\text{EQ. shear per pile } = \frac{12.4}{4} = 3.1t < \frac{5 \times 70 \times 1.25}{100} = 4.4t, \text{ OK}$$

Case (iii) VL + EQ (67.5 ↔)

$$P = 201\ t\phi, M\,45 = 4.8, M\,67.5 = 6.4 + 40.5 = 46.9\ tm, H\,67.5 = 11.4\ t$$

Load per pile = 50.3 ± 1.8 ± 17.4 = 69.5 t (Max.)

Pile cap and grade beams $= \frac{8.8}{78.3} < 87.5t$, OK

EQ. Shear per pile = 11.4/4 = 2.4 t < 4.4 t, OK.

Thus a 4-pile group is adequate.

Check on pedestal stresses

Case (i) VL only

Column size: 45 cm × 67.5 cm

Pedestal size: 105 cm × 127.5 cm

$$A = 105 \times 127.5 = 13387.5 \text{ cm}^2$$

$$Z\ 45 - 45 = 127.5 \times (105)^2 = 234281.25 \text{ cm}^3$$

$$Z\ 67.5 - 67.5 = 105 \times \frac{(127.5)^2}{6} = 284484.375 \text{ cm}^3$$

$$fc = \frac{P}{A} \pm \frac{M45}{Z45-45} \pm \frac{M67.5}{Z67.5-67.5}$$

$$= \frac{201 \times 1000}{13387.5} + \frac{4.8 \times 10^5}{234281.25} \pm \frac{6.4 \times 10^5}{284484.375}$$

$$= 15.0 \pm 2.1 \pm 2.2 = 19.3,\ 10.7 \text{ kg/cm}^2$$

$$fc, p = 0.25\, fck\ \frac{\sqrt{A_1}}{A_2} \qquad \text{(Fig 6.7)}$$

$$= 0.25 \times 150\ \frac{\sqrt{13387.5}}{45 \times 67.5} = 37.5 \times 2.0 = 75 \text{ kg/cm}^2$$

Plain concrete pedestal 30 cm wide on all sides is safe in compression.

Case (ii) VL + EQ (45)

$$fc = 15.0 \pm 16.3 \pm 2.2 = 33.5,\ -3.5 \text{ kg/m}^2$$

$$fc, p = 75 \times 1.33 = 100 \text{ kg/cm}^2 \text{ in compression, OK.}$$

As per clause 5.2.2 of the IS code, allowable tension in plain concrete,

$$ft,\ p = \frac{1}{1.8m} \times 0.7\sqrt{fck} = \frac{1}{1.8} \times 7\sqrt{15} = 1.5 \text{ N/mm}^2$$

or 15.0 kg/cm², ft, p (for BQ)

$$= 15.0 \times 1.33 = 20 \text{ kg/cm}^2 \qquad \textbf{OK}$$

Nominal steel provided in the pedastal.

Case (iii) VL + EQ (67.5 ↔)

$$fc = 15.0 \pm 2.1 \pm 16.1 = 33.2, -3.2 \text{ kg/cm}^2 \qquad \textbf{OK}$$

Pile Cap Design

Coinciding the critical section for shear with the outer face of the pile gives (Fig. 6.7) in each direction,

$$= \frac{0.675}{2} + 0.30 + \frac{d}{2} = 0.675 + 0.225,\ d = 0.52\ m,\ D = 0.62\ m$$

or,

$$= \frac{0.45}{2} + 0.30 + \frac{d}{2} = 0.675 + 0.225,\ d = 0.75\ m,\ D = 0.85\ m$$

Provide $D = 0.90$ m, $d = 80$ cm, $\frac{d}{2} = 40$ cm

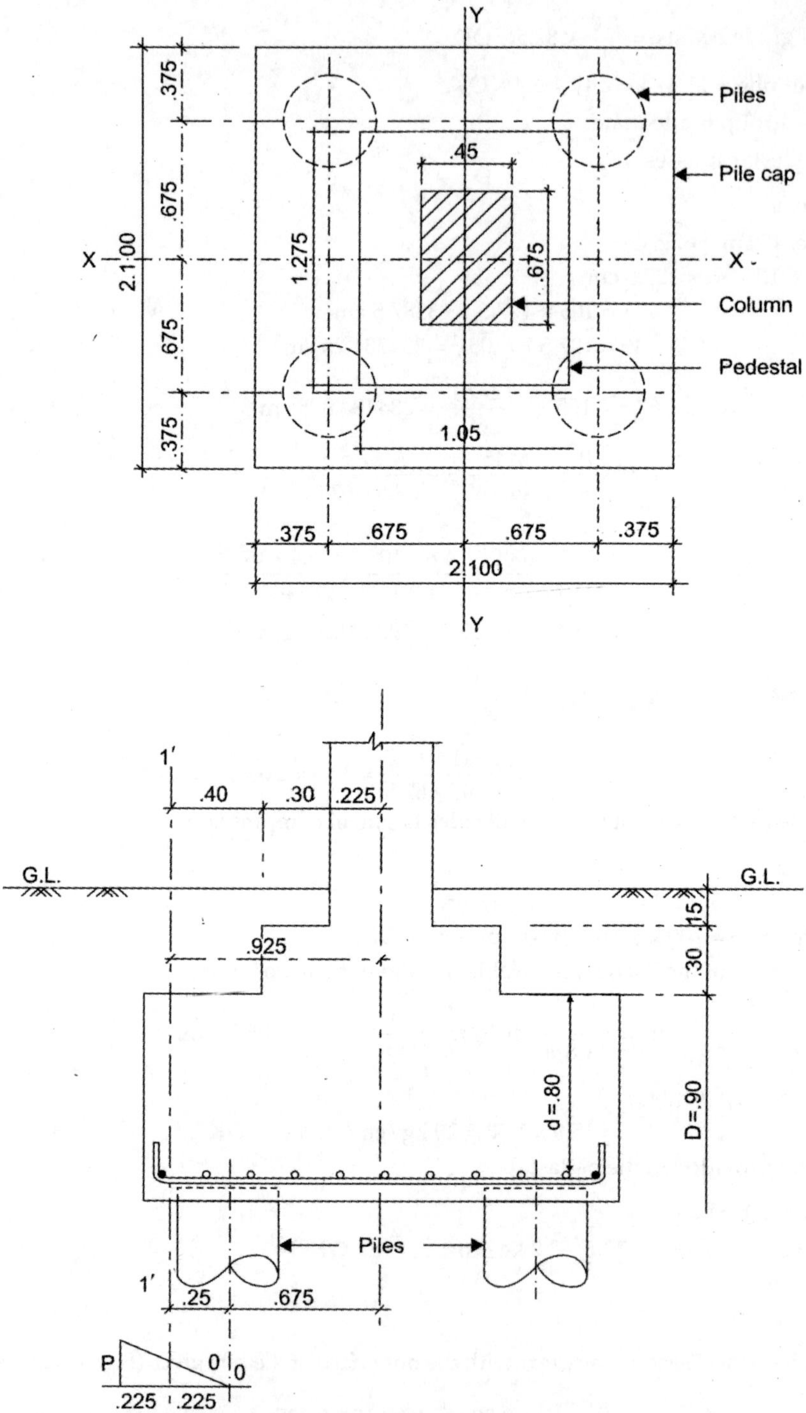

Fig. 6.7: Plan and section of pile footing of Example 1

Case (i) VL only

$P = 63.3t$ less self-weight of pile cap ($3.3t$)

$= 63.3 - 3.3 = 60\,t$

$M\,1\text{–}1 = 2P\,(0.675 - 0.525) = 18$ tm

$$K = \frac{1.5 \times 18.0 \times 105}{210 \times (80)^2 \times 10} = 0.20 \text{ N/mm}^2$$

Minimum $A_{St} = 0.12 \times 90 = 10.8$ cm²/m

Tor steel bar 12/100 = 11.31 cm²/m both ways at bottom

$S\,1' - 1' = 0$ (Fig 6.7)

$\tau_v = 0$

Case (ii) VL + EQ (67.5 ↔)

$P = 78.3 - 3.3 = 75.0$ t

$$\text{M1} - 1 = \frac{75}{60} \times 18 = 22.5$$

$$K = \frac{1.2 \times 22.5 \times 10^5}{210 \times (80)^2 \times 10} = 0.20 \text{ N/mm}^2$$

Min $A_{st} = 0.12 \times 90 = 10.8$ as before.

$D = 90$ cm with tor steel bar 12/100 both ways at bottom adequate.

Example 6.5: Pile foundation for a group of columns Fig 6.8 gives the plan of pile foundation of six columns (25 × 90) of a lift well. Pile capacity = 70.0*t*, pile diameter = 0.45 *m*, *S* = 1.35 *m*.

Solution: *Pile layout*

Case (i) VL only

$\Sigma P = 895\,t$ at centre of gravity of loads as given in Fig 6.8.

$\Sigma M\,25 = 69.6$ tm

$\Sigma M\,90 = 42.9$ tm

Let a 4 × 5 rectangular grid of piles be provided.

Using Table 1 in above example,

$$\text{Pile load} = \frac{845}{20} + \frac{69.6}{20 \times 1.35} \pm \frac{42.9}{16.67 \times 1.35}$$

$$= 44.8 \pm 2.6 \pm 1.9 = 49.3t$$

$$\text{Pile cap} = \frac{6.7 \times 4.8 \times 2.0 \times 2.5}{20} = 8.0$$

$$\text{Weight of walls per pile (say)} = \frac{2.0t}{59.3t} < 70.0t, \qquad \textbf{OK}$$

Case (ii) VL + EQ (↔)

$\Sigma P = 895t$

$\Sigma M25 = 100.0\ tm,\ \Sigma H\,25 = 12.0\ t$

$\Sigma M\,90 = 42.5\ tm$

Pile load = 44.8 ± 4.0 ± 1.9 = 50.7

$$\text{Pile cap plus walls} = \frac{10.1}{60.8} < (70.0 \times 1.25 = 87.5t)$$

$$\text{EQ shear per pile} = \frac{12.0}{20} = 0.6t < 4.4t, \qquad \textbf{OK}$$

Case (iii) VL + EQ (↕)

$\Sigma P = 895\,t$

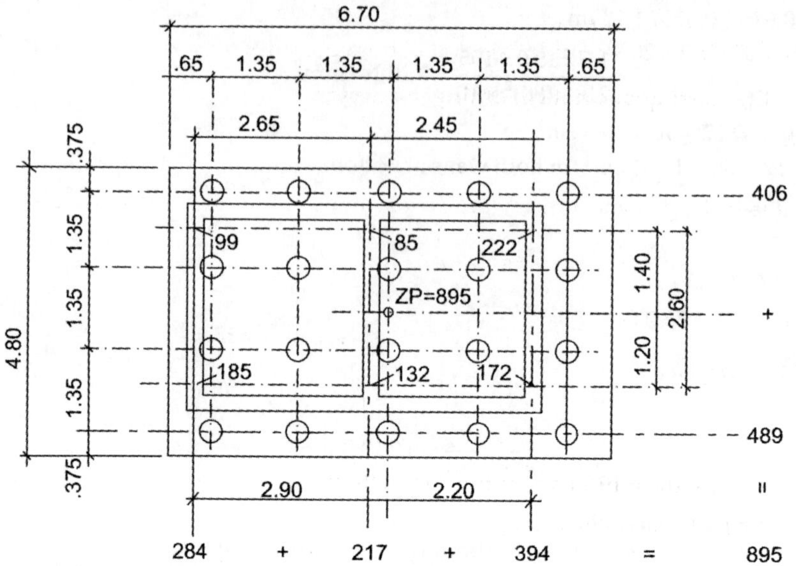

Fig. 6.8: Plan of pile foundation of lift-well columns—Example 6.5

$\Sigma M 25 = 69.6\ tm$
$\Sigma M 90 = 318\ tm,\ \Sigma H 90 = 83.5\ t$
Pile load = 44.8 ± 2.6 ± 14.0 = 61.4

Pile cap plus walls $= \dfrac{10.1}{71.5}$ $<(87.5t)$, **OK**

EQ shear per pile $= \dfrac{83.5}{20} = 4.21 < 4.4t$ **OK**

It is seen that a 20 pile group is required mainly for countering earthquake shear.

Pedestal stresses

No pedestals are provided for these columns, as 250 mm thick reinforced concrete walls are provided for lift pit to keep the earth out.

Pile cap design

Case (i) VL only

Longitudinal direction (↔)

Fig 6.9 gives the bending moment and shear force diagrams. With $D = 160$, $d = 150$ cm, $d/2 = 72$ cm at critical section (Fig. 6.9) = 105 t

$$\tau_v = \frac{1.5 \times 105 \times 1000}{480 \times 150 \times 10} = 0.22\ \text{N/mm}^2$$

$$\tau_c = 0.26 \times \delta \text{ for min steel of } 0.12\% \text{ of gross area}$$

$$\delta = 1 + \frac{3P_u}{fct\ Ag} = 1 + \frac{3 \times 394 \times 1.5 \times 1000}{150 \times (25+150)(90+150)} = 1.28$$

$$\tau_c = 0.26 \times 1.28 = 0.33\ \text{N/mm}^2\text{, safe}$$

Maximum $M_t = -103\,tm,\ K = \dfrac{1.5 \times 103 \times 10^5}{480 \times (150)^2 \times 10} = 0.14$

Minimum $A_{st} = 0.12 \times 160 = 19.2 \text{ cm}^2/\text{m}$

Tor steel bar 20/150 = 20.94 cm²/m top and bottom provided.

Case (ii) VL + EQ: Earthquake not governing the design.

Transverse direction **(ꝗ)**

Case (i) VL only:

Fig. 6.10 gives the bending moment and shear force diagrams. Maximum shear at critical section (Fig. 6.10) = 41 *t*

$$\tau = \frac{1.5 \times 41 \times 1000}{670 \times 150 \times 10} = 0.06 \text{ N/mm}^2$$

$$\tau_c = 0.6 \times 1.28 = 0.33 \text{ N/mm}^2, \textbf{ OK}$$

Maximum M = –218 *tm*

$$K = \frac{1.5 \times 218 \times 10^5}{670 \times (150)^2 \times 10} = 0.22$$

Minimum $A_{st} = 0.12 \times 160 = 19.2 \text{ cm}^2/\text{m}$ top and bottom

Case (ii) VL + EQ: Earthquake not governing.

Pile cap 6.7 × 4.8 × 1.6 with tor steel bar 20/150 both ways top and bottom provided.

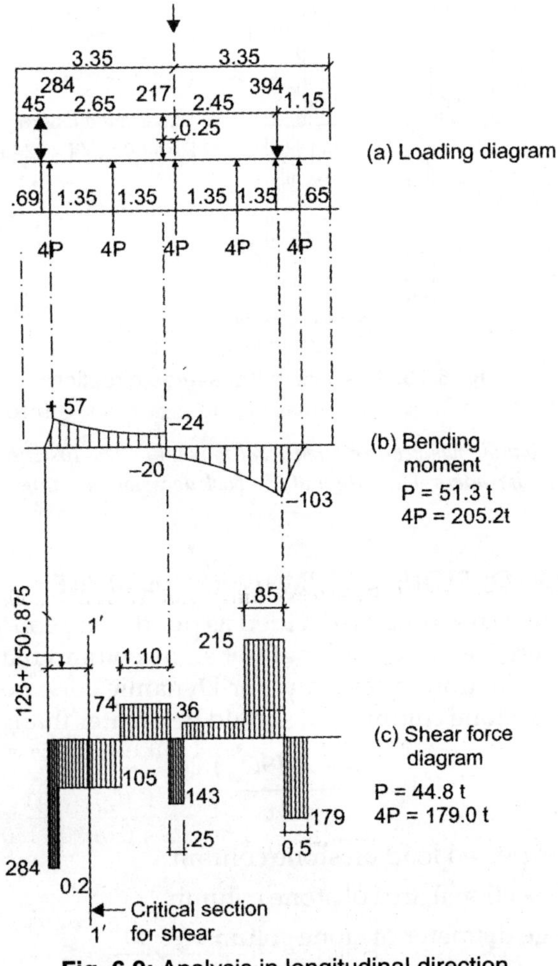

Fig. 6.9: Analysis in longitudinal direction

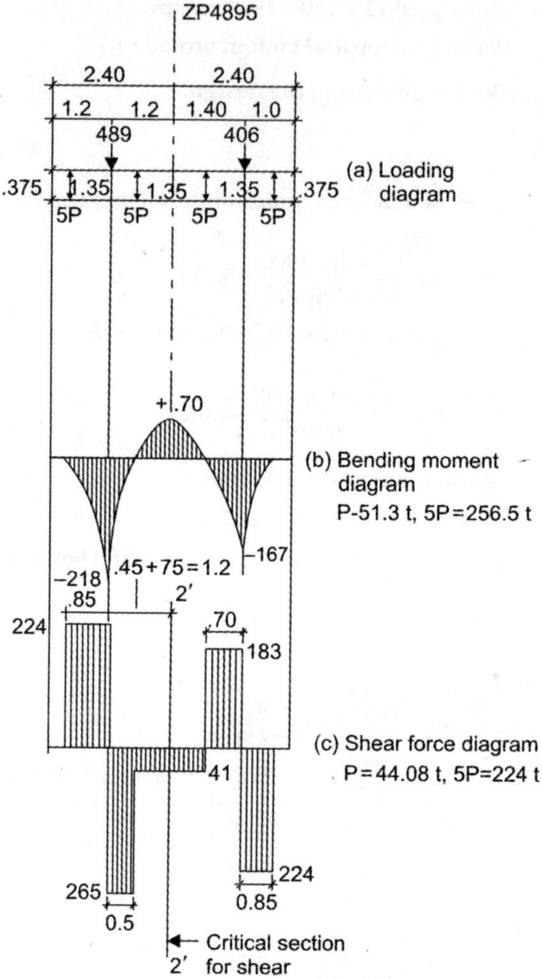

Fig. 6.10: Analysis in transverse direction

*(**Note:** The above design example has been referred from journal of The Institution of Engineers, Vol. 74, August 1993, for the benefit of readers. The author of this book does not own any responsibility, if anything is wrong in above example.)*

6.4 SIMPLIFIED DESIGN OF STONE COLUMN/GRANULAR PILE

The length of stone column should be decided as per the site condition and the load to be transferred by superstructure. The length of stone column may also be decided on the basis of standard penetration tests and/or Dynamic cone penetration test result. However, the length of stone column (L_c) should be greater than:

$$L_c \geq \frac{P - A_0(9C_{pt})}{\pi\, d.c}$$

where, P = Total expected load on stone column

A_0 = Cross–sectional area of stone column

d = Average diameter of stone column

c, C_{pt} = Side and point cohesion of soil respectively

Though the ground also takes some load, but for the conservative design it is assumed that the foundation loads are carried only by total no. of stone columns.

6.4.1 Load Capacity of Granular Pile/Stone Column

The load capacity of granular pile or stone column may be calculated by following formula:

$$q_a = \frac{K_p(4c + \sigma_r)}{FOS} \quad \ldots (6.8)$$

where K_p = Passive pressure co-efficient = $\tan^2(45 + \phi/2)$

ϕ = Angle of internal friction of stone ballast in drained condition

c = Drained cohesion

σ_r' = Effective radial stress as measured by a pressure metre (may be taken as 2c if pressure metre data is not given)

FOS = Factor of safety = 2 to 2.5

On the basis of theory proposed by Gibson and Anderson, the load supported by the stone column (granular pile) is given as

$$Q_d = K_p\,(8\,C_u + O_e + O_v)\,.\,A_p \quad \ldots (6.9)$$

where K_p = Passive pressure co-efficient = $\tan^2(45 + \phi/2)$

(Here, for computation of k_p, ϕ is normally taken as angle of internal friction of material used. For stone ballast, ϕ is 44°, but for conservative design ϕ is taken as two-thirds of 44°, i.e. 29.5°)

C_u = Undrained shear strength = 2 × 0

0_e = Effective average stress at critical pile depth

= $d_c \times \gamma_b$ (γ_b = bulk density)

d_c = Critical pile depth (may be taken as 4 to 5 times of pile diameter)

0_v = Effective Normal pressure (kg/cm²) = $k_B . q$

$$k_B = 1 - \left\{\frac{1}{1 + \left[\frac{a}{z}\right]^2}\right\}^{3/2} \quad \ldots (6.10)$$

a = Diameter of circular load (if circular foundation is there. Otherwise the shorter dimension of footing size).

z = Critical depth of pile (d_c) = 4 to 5 times of pile diameter

$$q = \text{Load intensity} = \frac{\text{Total load of superstructure}}{\text{Bearing area}}$$

A_p = Cross-sectional area of pile stem

$$Q_{safe} = \frac{Q_d}{FOS} \quad \ldots (6.11)$$

Example 6.6: A granular pile foundation has to be designed for a 23.8 m diameter molasses storage steel tank proposed to be laid on ground resting on silty clayey strata. The tank should satisfy both shear and settlement criteria as applicable to liquid storage tanks. The dynamic cone test data is as shown in the Fig. 6.11. The other data are:

i. Submerged unit weight = 0.95 t/m³, liquid limit of soil = 29%
ii. Undrained shear strength (c) = 2.7 t/m² (determined from unconfined compressive strength test)
iii. Ground water table (G.W.T) at 0.5 m depth below natural ground level (NGL).

Solution: From the dynamic cone test results (Fig. 6.11), the N-values indicate up to 10 m depth the soft to medium compact strata and thereafter it tends to become stiffer. This chart can give fairly good idea of deciding depth of stone column (granular pile).

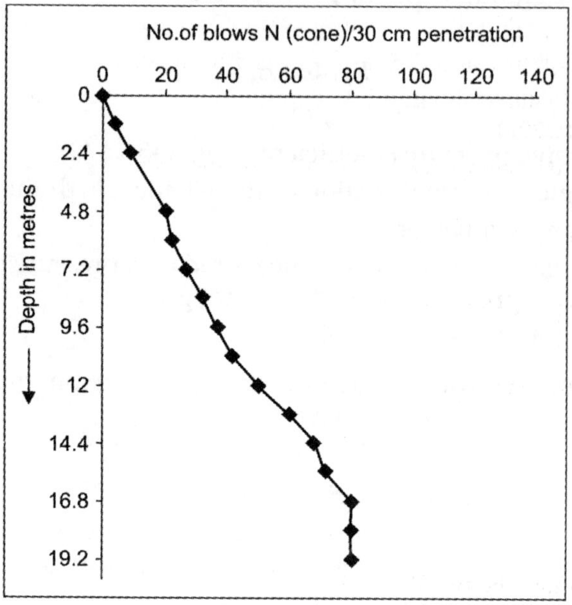

Fig. 6.11: Dynamic cone resistance curve

i. Computation of Safe bearing Capacity of Soil

a. Net ultimate bearing capacity from shear criteria (IS: 6403–1981, Reaffirmed 1997)

The net ultimate bearing capacity for foundations on fairly saturated cohesive soil is calculated by using following relationship (IS: 6403–1981)

$$q_d = CN_C S_C d_C i_c \qquad \dots (6.12)$$

where,

q_d = Net ultimate bearing capacity (t/m²)

c = Cohesion of the soil (2.7 t/m²)

N_c = 5.14 (IS: 6403–1981)

S_C = Shape factor = 1.3 (for circular foundations)

d_c = Depth factor = 1 (assumed)

i_c = Inclination factor = 1 (assumed)

Thus, $q_d = 2.7 \times 5.14 \times 1.3 \times 1 \times 1 = 18.04$ t/m²

Applying a factor of safety as 2.5.

$$Q_{safe} = \frac{18.04}{2.5} = 7.21 \text{ t/m}^2 \qquad \dots (6.13)$$

ii. Computation of Settlement

Firstly, the influence zone below foundation (equivalent to 1.5 times the foundation width) is divided into different layers (Fig. 6.12). Lesser the thickness of layer, more accurate will be the computation of settlement. Then settlement (S) in each layer is computed. The load dispersion may be taken as 2:1, i.e. 2 vertical to 1 horizontal. This becomes nearly 26° which has been taken for computations of settlement. For settlement computations in this example the influence zone below foundation has been taken as B, i.e. diameter of the tank = 23.8 m ≈ 25 m.

The settlement in each layer is given by (IS: 8009 Pt – 1, 1976. reaffirmed 1993)

$$S = \frac{C_c \times H}{1+e_0}\log_{10}\frac{p_0+\Delta P}{P_0} \quad \text{... (6.14)}$$

where, C_C = Compression index = 0.009 (w_L–10) after Skempton, 1944 ... (6.14a)

w_L = Being the liquid limit of the soil in percentage (= 29%)

H = Thickness of the compressible layer (read from Fig. 6.12)

e_o = Initial void ratio (computed from e-log Þ curve, Ref. Soil Testing for Engineers by Mittal and Shukla, 2014)

Alternatively, c_c can also be computed by any of following relationships:

a. $C_c = 0.2343\left[\frac{LL(\%)}{100}\right] \times G_s$ (After Nagraj and Murty, 1985)

b. $C_c = \frac{n_o}{371.747 - 4.275n_o}$ (After Park and Koumoto, 2004) ... (6.14b)

c. $C_c = \frac{PI(\%)}{74}$ (After Kulhawy and Mayne, 1990)

d. $C_c = 0.5G_s\left[\frac{PI(\%)}{100}\right]$ (After Wroth and Wood, 1978)

where, G_s = *Specific gravity*, n_o = *in situ* porosity of soil, PI = Plasticity index, LL = Liquid limit.

If soil is swelling in nature, the swelling index is $\frac{1}{4}$ to $\frac{1}{5}$ of c_c.

The swelling index is also referred to as the re-compression index.

p_o = Initial stress at centre of the layer = γZ

Z = Depth of the centre of the layer below natural ground

γ = Submerged density of the strata (0.95 t/m^3)

ΔP = Pressure increment (due to total liquid and structural load in the tank) at centre of each layer

= Total load including that of the structure (Q) multiplied by the original area and divided by the area of spread at the centre of layer.

$$\Delta P = \frac{\frac{\pi}{4}B^2}{\frac{\pi}{4}(B+H\tan\theta)^2} \times \frac{Q}{\frac{\pi}{4}\times B^2} \quad \text{...(6.15)}$$

B being width of footing (≈ diameter of tank = 23.8 m)

The total settlement (S) = $S_1 + S_2 + S_3 + S_4 + S_5$ ...(6.16)

where = S_1, S_2, S_3, S_4, S_5 are settlements in different layers of influence zone.

Note: $\theta = 26°$ (before piles) and $\theta = 50°$ or even more (after granular piles)

The soil samples are collected at various depths in the field and they are tested in the laboratory to determine their Atterberg limits. From the various values of liquid limits at various depths, from equation 6.4, the C_C is calculated from e-log Þ curves (computed from consolidation tests). C_c can also be computed using liquid limit test values as given in equations 6.14(a) and (b). The value of C_c can also be computed from e-log Þ curve. Similarly, void ratio is calculated. Substituting all the values, the settlement is calculated at various layers. Finally the total settlement is obtained by adding the various values of settlement at different layers. Referring the Fig. 6.12 the thickness of compressible layers within influence zone work out to 4 m, 6 m, 5 m, 5 m and 5 m respectively. From the laboratory test conducted at the sample collected from 2 m depth, the liquid limit, i.e.

$$w_L = 29\%, e_o = 0.59$$

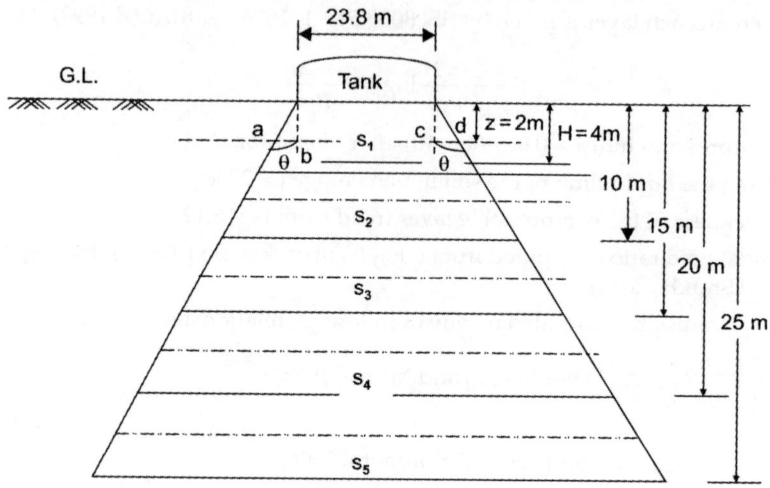

Fig. 6.12: Influence zone divided into different layers

Therefore, $C_c = 0.009\ (29–10)$
= 0.17 (incidentally, if Eqn. 'a' of 6.24 (B) is used for $G_s = 2.6$, the value of c_c is obtained as 0.17 only)

i. Here Q = total load of the liquid in tank and that of the structure = 5336 t

ii. q = intensity of loading at the bottom of tank $= Q/A = \dfrac{Q}{\frac{\pi}{4}\times D^2} = \dfrac{5336t}{\frac{\pi}{4}\times 23.8^2} = 11.994 = 12t/m^2$

iii. Settlement Computations

Ist Layer

H = 4 m

$p_0 = \gamma z_1$ (where z is depth at centre of each layer below foundation)

$= 0.95 \times 2 = 1.90\ t/m^2$

Here $z_1 = H/2 = 4/2 = 2$ m

$$\Delta p = \frac{B^2}{(B + H\tan\theta)^2} \times q$$

$$\Delta P = \left(\frac{D}{D_2}\right)^2 \times q$$

where,

D = diameter of tank

D_1 = diameter of centre of the layer

Here $D_1 = ab + bc + cd$ (Ref. Fig 6.12)

$= z_1 \tan\phi + 23.8 + z_1 \tan\phi$

$= 2\tan 26° + 23.8 + 2\tan 26°$

$= 23.8 + 4\tan 26° = 25.75$ m

Taking q as 12 t/m²

Settlement in Ist layer

$$\Delta P_1 = \left(\frac{23.8}{25.75}\right)^2 \times 12 = 10.25\ t/m^2$$

$$Þ_o = \gamma z_1 = 0.95 \times 2 = 1.90 \text{ t/m}^2$$

$$S_1 = \frac{0.17 \times 4}{1 + 0.59} \log \frac{1.90 + 10.25}{1.90} = 0.32 \text{ m}$$

IInd Layer

$$H = 6 \text{ m}, \ \Delta Þ_2 = \left(\frac{D}{D^2}\right)^2 \times 12, \ z_2 = 4 + 3 = 7 \text{ m}$$

$$Þ_o = 0.95 \times 2 = 6.65 \text{ t/m}^2$$

$$D_2 = (23.8 + 14 \tan 26°) = 23.8 + 6.83 = 30.63 \text{ m}$$

$$\Delta Þ_2 = \frac{23.8}{30.63}^2 \times 12 = 7.25 \text{ t/m}^2$$

$$S_2 = \frac{0.17 \times 6}{1.59} \log \frac{6.65 + 7.25}{6.65} = 0.19 \text{ m}$$

IIIrd Layer

$$\text{H} = 5 \text{ m, and } z_3 = 4 + 6 + 2.5 = 12.5 \text{ m}$$

$$Þo = 0.95 \times 12.5 = 11.875 \text{ t/m}^2$$

$$D_3 = (23.8 + 2 \times 12.5 \tan 26) = (23.8 + 12.19) = 35.99 \text{ m}$$

$$\Delta Þ_3 = \left(\frac{D}{D_3}\right)^2 \times 12$$

$$\Delta Þ_3 = \left(\frac{23.8}{35.99}\right)^2 \times 12 = 5.25 \text{ t/m}^2$$

$$S_3 = \frac{0.17 \times 5}{1.59} \log \frac{11.875 + 5.25}{11.875} = 0.08 \text{ m}$$

IVth Layer

$$H = 5 \text{ m}, z_4 = 4 + 6 + 5 + 2.5 = 17\ 5 \text{ m}$$

$$Þ_o = 0.95 \times 17.5 = 16.625 \text{ t/m}^2$$

$$\Delta Þ_4 = \left(\frac{D}{D_4}\right)^2 \times 12$$

$$D_4 = (23.8 + 2 \times 17.5 \tan 26) = 23.8 + 17.07 = 40.87 \text{ t/m}^2$$

$$\Delta Þ_4 = \frac{(23.8)^2}{(40.8)} \times 12 t/\text{m}^2$$

$$S_4 = \frac{0.17 \times 5}{1.59} \log \frac{16.625 + 4.07}{16.625} = 0.046 \text{ m}$$

Vth Layer

$$H = 5 \text{ m}, z_5 = 4 + 6 + 5 + 5 + 2.5 = 22.5 \text{ m}$$

$$Þ_o = 0.95 \times 22.5 = 21.375 \text{ t/m}^2$$

$$\Delta Þ_5 = \left(\frac{D}{D_5}\right)^2 \times 12$$

$$D_5 = (23.8 + 2 \times 22.5 \tan 26) = (23.8 + 21.95) = 45.75 \text{ m}$$

$$\Delta Þ_5 = \left(\frac{23.8}{45.75}\right)^2 \times 12 = 3.25 \text{ t/m}^2$$

$$S_5 = \frac{0.17 \times 5}{1.59} \log_{10} \frac{21.375 + 3.25}{21.375} = 0.03 \text{ m}$$

Total settlement = $S_1 + S_2 + S_3 + S_4 + S_5 = 0.32 + 0.19 + 0.08 + 0.046 + 0.03 = 0.66 \text{ m} = 660 \text{ mm}$

This settlement is on higher side hence to reduce this, the granular piles have to be provided. It is evident from the above results that settlement in the upper 10.5 m depth is around 0.46 m.

Thus, more than 75% of the settlement is expected within top 10.5 m depth of strata. The allowable soil pressure is 7.2 t/m^2 against the design load of about also the total settlement is on higher side. Hence it is essential to share the load on piles up to 10.5 m depth.

(This also gives a fairly good assessment of deciding the pile length. Since up to 10.5 m depth, the virgin ground settlement is of the order of 0.46 m, hence pile length should be more than 10.5 m depth so that the influence zone may be made strong and intact).

After strengthening the ground by stone column:

Y_{sub} = 1.25 t/m^3 and θ° = 50° (as stated earlier, it can even be as high as 62°, as per Giroud and Noiray's approach).

Here, value of '*e*' shall also be changed because clayey soil shall be replaced by gravel material.

Hence, taking *e* = 0.72 (as per literature available) and value of c_c can also be reduced to half, i.e. 0.08 (Say).

Settlement in Ist layer *(Refer Fig. 6.13)*

$$H = 4 \text{ m}, z = 2 \text{ m}$$

$$Þ_o = 1.25 \times 2 = 2.5 \text{ t/m}^2$$

$$\Delta Þ_1 = \left(\frac{D}{D_1}\right)^2 \times 12 D_1 = (23.8 + 4 \tan 50^\circ) = (23.8 + 4.76) = 28.6 \text{ m}$$

$$\Delta Þ_1 = \left(\frac{23.8}{28.6}\right)^2 \times 12 = 8.31 \text{ t/m}^2$$

$$S_1 = \frac{0.08 \times 4}{1 + 0.70} \log \frac{2.5 + 8.31}{2 \times 1.25} = 0.12 \text{ m}$$

IInd layer

$$z = 4 + 3 = 7 \text{ m}$$

$$Þ_o = 1.25 \times 7 = 8.75 \text{ t/m}^2$$

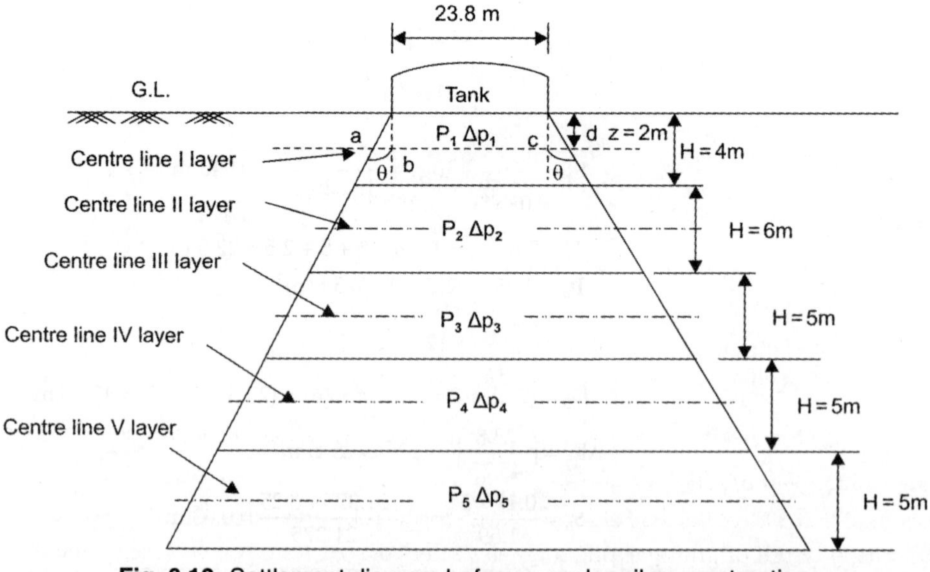

Fig. 6.13: Settlement diagram before granular piles construction

$$\Delta \text{Þ}_2 = \left(\frac{D}{D_2}\right)^2 \times 12, D_2 = (23.8 + 14 \tan 50) = 40.48 \text{ m}$$

$$\Delta \text{Þ}_2 = \left(\frac{23.8}{40.48}\right)^2 \times 12 = 4.14 \text{ t/m}^2$$

$$S_2 = \frac{0.08 \times 6}{1.7} \log \frac{8.75 + 4.14}{8.75} = 0.05 \text{ m}$$

IIIrd layer

$$\text{H} = 5 \text{ m}, z = 4 + 6 + 2.5 = 12.5 \text{ m}$$

$$\text{Þ}_o = 1.25 \times 12.5 = 15.625 \text{ t/m}^2$$

$$\Delta \text{Þ}_3 = \left(\frac{D}{D_3}\right)^2 \times 12, D_3 = (23.8 + 2 \times 12.5 \tan 50) = 53.6 \text{ m}$$

$$\Delta \text{Þ}_3 = \left(\frac{23.8}{53.6}\right)^2 \times 12 = 2.36 \text{ t/m}^2$$

$$S_3 = \frac{0.08 \times 5}{1.7} \log \frac{15.625 + 2.36}{15.625} = 0.015 \text{ m}$$

IVth layer

$$\text{H} = 5, z = 4 + 6 + 5 + 2.5 = 17.5 \text{ m}$$

$$\text{Þ}_o = 1.25 \times 17.5 = 21.875 \text{ t/m}^2$$

$$D_4 = (23.8 + 2 \times 17.5 \times \tan 50) = 65.5 \text{ m}$$

$$\Delta \text{Þ}_4 = \left(\frac{23.8}{65.5}\right)^2 \times 12 = 1.58 \text{ t/m}^2$$

$$S_4 = \frac{0.08 \times 5}{1.7} \log \frac{21.875 + 1.58}{21.875} = 0.007 \text{ m}$$

Vth layer

$$H = 5, z = 4 + 6 + 5 + 5 + 2.5 = 22.5 \text{ t/m}^2$$

$$\text{Þ}_o = 1.25 \times 22.5 = 28.125 \text{ t/m}^2$$

$$D_5 = (23.8 + 2 \times 22.5 \times \tan 50) = 77.43 \text{ m}$$

$$\Delta \text{Þ}_5 = \left(\frac{D}{D_5}\right)^2 \times 12$$

$$\Delta \text{Þ}_5 = \left(\frac{23.8}{77.43}\right)^2 \times 12 = 1.13 \text{ t/m}^2$$

$$S_5 = \frac{0.08 \times 5}{1 + 0.7} \log \frac{28.125 + 1.13}{28.125} = 0.004 \text{ m}$$

$$S = S_1 + S_2 + S_3 + S_4 + S_5 = 0.12 + 0.05 + 0.015 + 0.007 + 0.004 = 0.19 \text{ m} = 190 \text{ mm}$$

Settlement in natural strata of the influence zone = 660 mm

Settlement after strengthening the ground using stone column = 190 mm. Thus, it is noticed that the foundation on settlement after introduction of stone column reduces to about 29%.

Note: ($\theta = 26°$) before using stone column and $\theta = 50°$ after strengthening of ground by stone column.

i. Design and layout of piles

It is assumed that 80% of the load shall be shared by the piles and balance by the soil.

The critical depth of granular pile diameter after construction. If the average pile diameter is taken as 0.5 m (say), this depth is taken as 0.5 × 4 = 2.0 m

Now,

$$o_e = d \times \gamma_{sub}$$

$$d = 2.0 \text{ m}$$

$$\gamma_{sub} = 0.95 \text{ t/m}^3$$

$$O_e = 2 \times 0.95 = 1.9$$

$$C_u = 2 \times o_e = 2 \times 1.9 = 3.8$$

$$0_v = K_B \,.\, q$$

$$O_v = \left[1 - \left\{\frac{1}{1+\left(\frac{a}{z}\right)^2}\right\}^{3/2}\right] . q \qquad \left\{\begin{array}{l}\textit{Here, } z = \textit{critical depth } (d_c = 2 \text{ m}) \\ a = \textit{tank diameter} \\ q = \textit{lading intensity}\end{array}\right\}$$

$$O_v = \left[1 - \left\{\frac{1}{1+\left(\frac{23.8}{2.0}\right)^2}\right\}^{3/2}\right] \times 12$$

$$= 0.999 \times 12 = 11.95 \approx 12$$

Passive pressure co-efficient $K_p = \tan^2 (45 + \phi/2)$

Taking ϕ as $\frac{2}{3}$ of 44° = 29.3°, i.e. 29.5° (say)

$$K_p = \tan^2\left(45 + \frac{29.5}{2}\right)$$

$$= 2.95$$

Substituting the values in equation.

$$Q_d = K_p\,(8\,Cu + Oe + Ov).\,A_p$$

$$Q_d = 2.95\,(8 \times 3.8 + 1.4 + 12) \times \frac{\pi}{4}\,(0.5)^2 = 25.37 \text{ t}$$

Applying the factor of safety 2.0

$$Q_{safe} = \frac{25.37}{2.0} = 12.68 \text{ t}$$

Load shared by the piles = 0.80 × 5336 = 4269 t
Load shared by the soil = 0.20 × 5336 = 1067 t

No. of piles = $\frac{4269}{12.68}$ = 336.7, say 340

These piles shall be constructed in a zig-zag pattern at the corners of equilateral triangles. The layout is shown in the Fig. 6.14. The spacing of the pile is decided by hit and trial method. One pile is made centrally beneath the tank.

The intensity of load to be shared by soil = $\dfrac{1067}{\frac{\pi}{4}(23.8)^2} = 2.39 \approx 2.4 \text{ t/m}^2$

This is less than 7.21 t/m². Hence safe.

Note: *Only one quarter has been shown with piles. In other part also the layout shall be same.*

This is the complete design of the granular pile foundation for tanks. Similarly, the foundation may be designed for other structures also like transmission towers, multi-storeyed buildings, etc.

A RCC skirt wall can also be provided circumscribing all the piles. This wall reduces differential settlement and provides a positive confinement to a greater extent. The design of this skirt wall is beyond the scope of this book. This wall is designed like a cantilever. Besides this wall, a ring beam is also provided as shown in the Fig. 6.15. The whole tank rests on a soil padding made of local material adequately compacted. The Fig. 6.15 illustrates the cross-section of the skirted granular pile foundations.

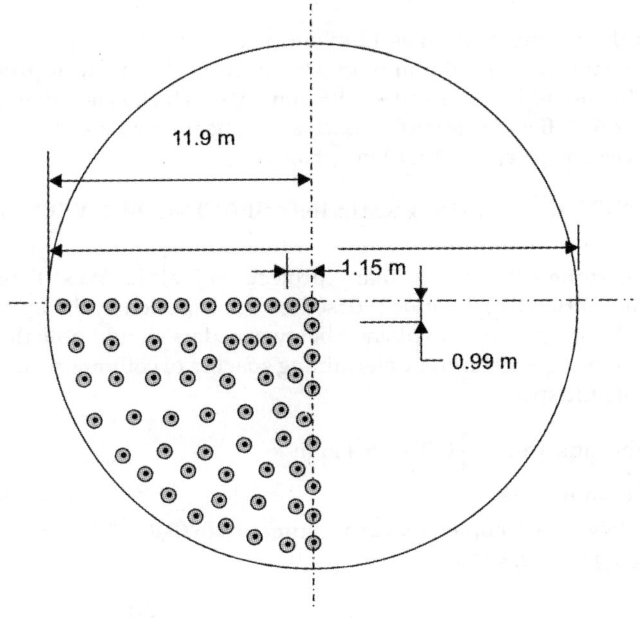

Fig. 6.14: Layout of piles below tank

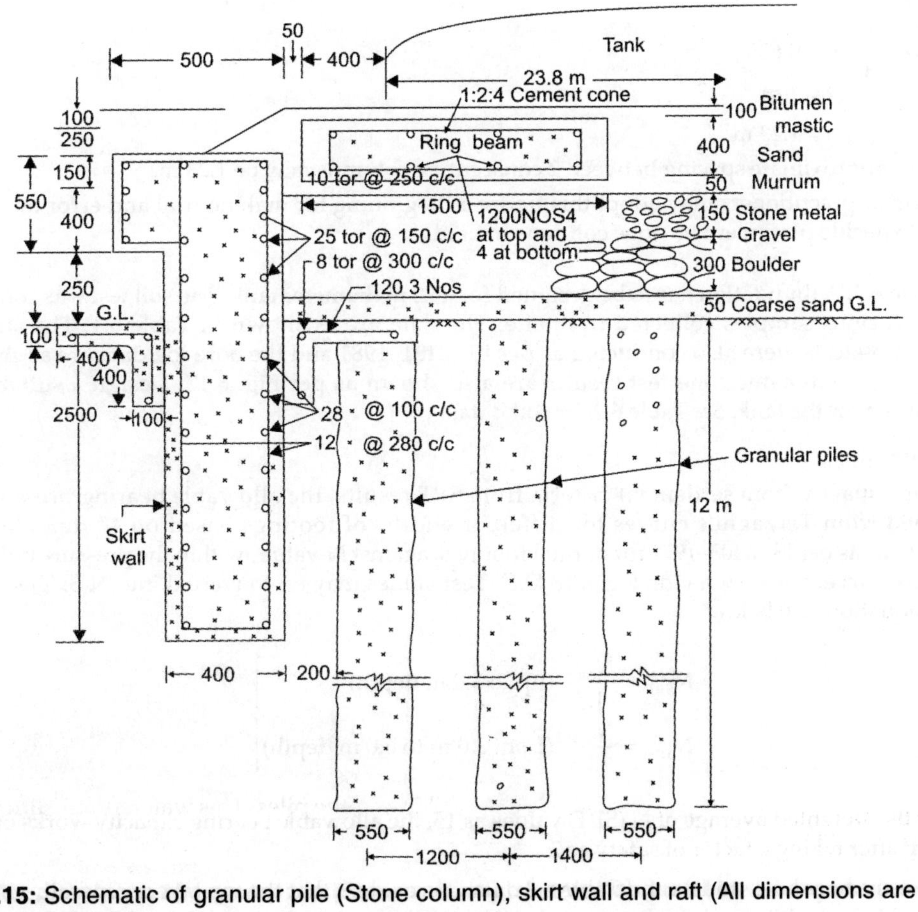

Fig. 6.15: Schematic of granular pile (Stone column), skirt wall and raft (All dimensions are in mm unless specified)

Settlement after Piles

Considering the depth of stone column as 12 m and load distribution as 50 p the total settlement calculated as above comes out to 190 mm only. Due to RCC skirt wall (if provided), the author's experience is that it further reduces to half say 95 mm only. Which is less than permissible 125 mm settlement in clayey soils. Besides the RCC skirt, a ring beam may also be provided as shown in Fig 6.15, which reduces the differential settlement also.

APPROXIMATE METHOD FOR DETERMINING SPACING BETWEEN STONE COLUMNS (Mittal, 2013)

Author has designed stone columns for many projects in Haldia (West Bengal) refinery, sugar factories, silos, etc. for various applications. Based on his experience, the approximate method of determining stone column spacing is explained below (readers should note that it has no technical basis. Rather it gives only a good start for determining spacing of columns to accommodate required no. of columns in available space).

In this example, the tank area = $\frac{\pi}{4}(23.8)^2 \approx 446 \text{ m}^2$

The no. of stone columns = 340

Hence approximately area occupied by each column = 446/340 = 1.31 m^2

Half of this area = 1.31/2 = 0.655 m^2

The columns should be placed at equilateral triangle pattern, the area of which is $\frac{\sqrt{3}}{4}y^2$ (y, being the length of each arm).

Take $\frac{\sqrt{3}}{4}y^2 = 0.655$

or, $y^2 = 1.51$

Hence $y = 1.23$ m

Thus, approximate spacing between 2 consecutive columns may be 1.23 m.

Users or practitioners may adopt this figure as beginning for making trial and error on a graph paper to decide placement of stone columns in field.

Example 6.7: A foundation has to be designed for a 22 m diameter tank. The soil test was conducted on the various samples collected from site. The data are as shown in Table 6.1. The standard Penetration tests were also conducted as per IS : 2131–1981 and the bore log data are as shown is Fig. 6.16. The dynamc cone test results are also shown as per Fig. 6.17. Design a suitable pile foundation for the tank. *See* Table 6.2 for soil data.

Solution:

Bearing capacity from settlement criteria from SPT results, the allowable bearing pressures are obtained from Terzaghi's curves for different widths of footings based on 40 mm allowable settlements as per IS : 6403–1971 for normal footings, against N-values within the pressure bulb zone.

These curve, are shown vide Fig. 6.18 D.C. Test values may be converted into N(SPT) value, by using equation 6.30 below:

$$\left.\begin{aligned} N_{SPT} &= \frac{N_{cone}}{1.5} \text{ (up to 3.0 m depth)} \\ N_{SPT} &= \frac{N_{cone}}{1.75} \text{ (from 3.0 m to 6.0 m depth)} \end{aligned}\right\} \quad \ldots (6.17)$$

For the weighted average of N (SPT) values as 15, the allowable bearing capacity works out to as 11 t/m^2 after taking a factor of safety as 2.

From the bore hole studies and Table 6.2 data, it is evident that the top layers generally comprise of gravel mixed with some and fine particles (GC or GM) up to a depth of about 2.5 m below ground

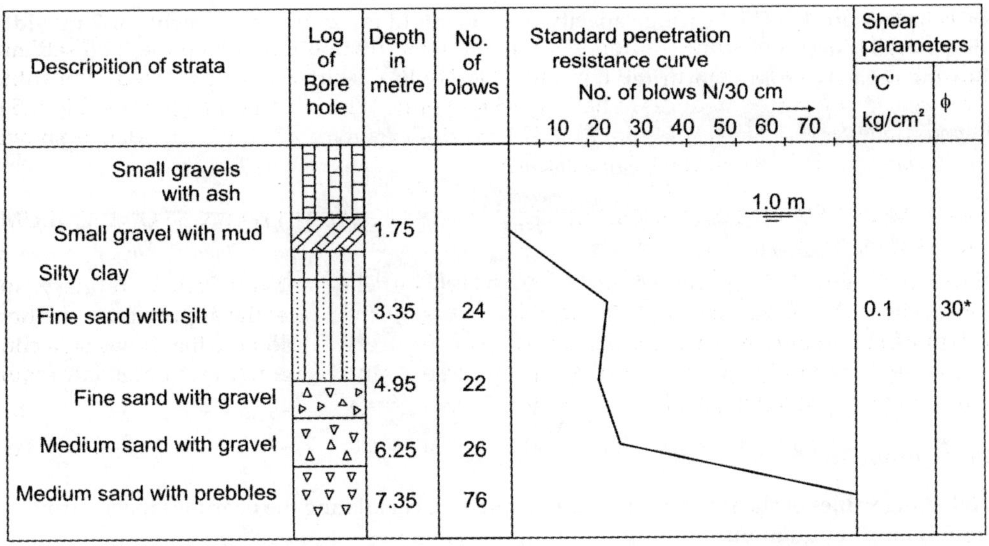

Fig. 6.16: Log of bore hole—1, SPT resistance curve and shear parameters

level followed by layers of silts mixed with sand and clay particles (CI or ML) or low to medium plasticity in about 2–2.5 m thickness.

Bearing capacity using DCPT test results:

The test results reveal the minimum N(cone) value as 21 at depth below GL and their converted

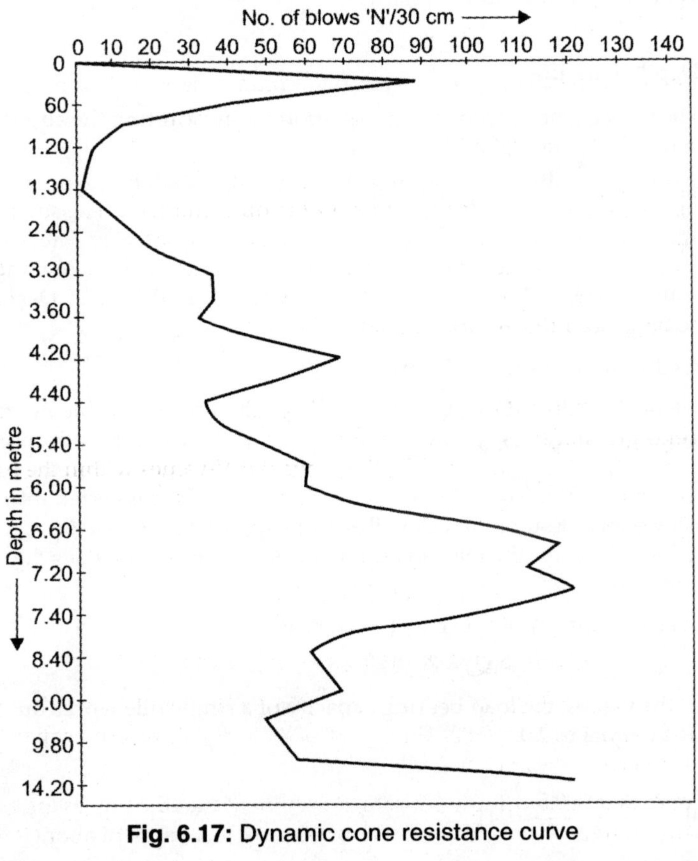

Fig. 6.17: Dynamic cone resistance curve

N (SPT) values are 12. The bearing capacity from Fig. 6.18 for 25 mm settlement for 2 m width of footing are 8.3 t/m^2.

Bearing capacity from shear criteria:

Table 6.2: Soil data

S. No.	*Depth (m)*	*Mechanical analyses (Sieve and hydrometer)*						*Atterberg limits*			*ISI*
		Gravel/ Cobble/ boulders > 4.75 mm	*Coarse Sand 4.75– 2.0 mm*	*Medium Sand 2– 0.425 mm*	*Fine Sand 0.425 – 0.075*	*Silt 0.075– 0.002*	*Clay < 0.002 mm*	*Liquid Limit %*	*Plastic Limit %*	*Plasticity Index %*	*Classifica-tions as per IS: 1498–1970*
1	1.50–2.20	49.0	6.0	7.0	5.0	23.0	10.0	40	24	16	GC
2	2.20–2.80	1.0	4.0	5.0	11.0	59.0	10.0	38	18	20	CI
3	2.80–4.85	0.0	1.0	3.0	36.0	55.0	5.0	–	–	–	ML
4	4.85–5.60	4.0	4.0	3.0	61.0	26.0	2.0	–	–	–	SM

Following values of shear parameters are adopted for calculating the bearing capacity from shear criteria. Using the equation:

$$Q_{dx} = \frac{1}{F}\left[1.3\ C.N + \gamma D_f(Nq - 1) + 0.4\gamma BN_r\right] + \gamma D_f \qquad \text{... (6.18)}$$

here F = Factor of safety

$C = 0.1$ kg/cm^2

$\phi = 30°$

$\gamma = 1.0$ gm/cm^3 (Submerged unit weight)

$B = 2.0$

$D_f = 2.5$ m (Depth of footing)

The allowable bearing capacity from above equation by substituting the above values works out to 18 t/m^2 using a factor of safety of 2.5.

For design purposes, the allowable bearing capacity value shall be east of the three values, i.e. 8.3 t/m^2.

Design of piles

i. Diameter of tank = 22 m

ii. Total weight = 4000 tonnes

iii. Unit submerged weight on soil = 0.7 t/m^3

iv. Un-drained shear strength = 0.3 t/m^2

v. Safe bearing capacity of soil = 8.3 t/m^2

vi. Intensity of loading I = 11 t/m^2

In this example, we may assume that the piles will support 50% of the design load and balance shall be shared by the soil. Thus, the load shared by the soil alone is $11 \times 0.5 = 5.5$ t/m^2 which is less than 8.3 t/m^3. Hence safe.

The ultimate bearing capacity of pile may be taken as

$$Q_d = K_p\,(8\,C_u + O_e + O_v)\,A_p \qquad \text{... (6.19)}$$

Substituting all the values, the load bearing capacity of a single pile works out to 8.3 tonnes only with a factor of safety equal to 2.0.

Thus, no. of piles $= \dfrac{4000 \times 0.5}{8.3} = 241$

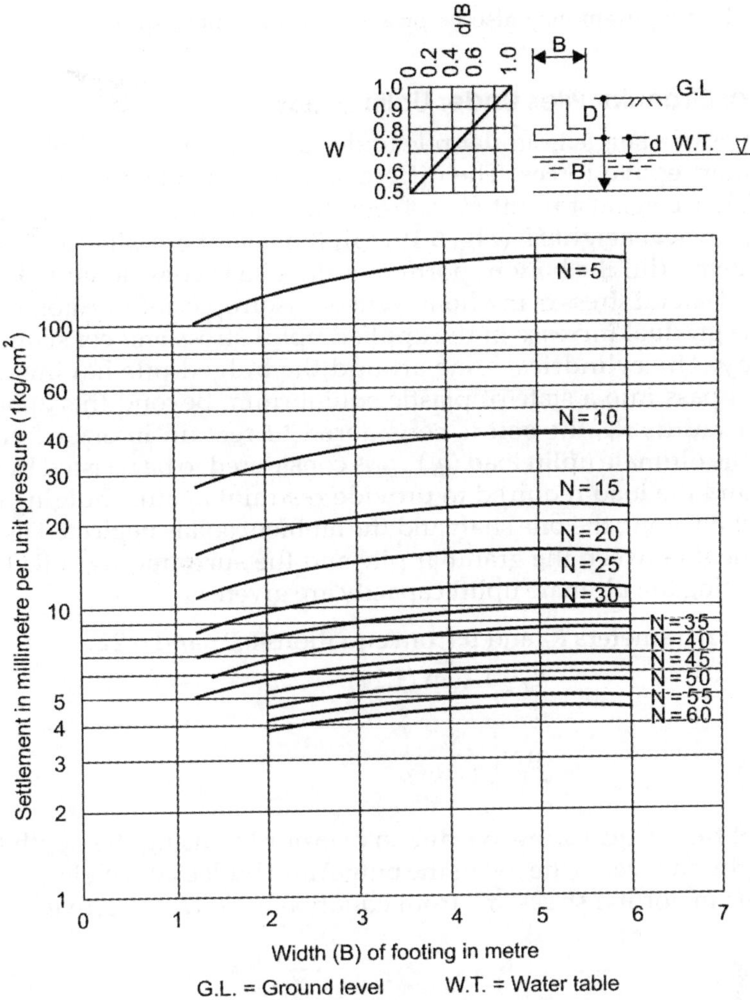

Fig. 6.18: Settlement per unit pressure from standard penetration resistance (taken from "Soil testing for Engineers" by Mittal and Shukla)

The granular piles of 30 cm diameter (after compaction 35 cm approximately) and of depths 3.5 m to 4.0 m may be provided. The spacing and layout of piles shall be computed same as in Example 6.6.

Computation of settlement before Piles

$$S = \frac{C_c \times H}{1 + e_o} \log_{10} \frac{Po + \Delta P}{P_o} \qquad \ldots (6.20)$$

Feeding the relevant data in above equation, the total settlement works out to be 57.1 cm, after considering 26° load dispersion.

Settlement after piles: Load dispersion may be assumed as 50°, and value of c_c can also be reduced to half, besides increasing the value of void ratio to a little less than double. The total settlement works out to be 46.5 cm, while only 26.8 cm shall be at the top layer.

It is assumed that due to piles and compaction of neighbouring ground due to pile driving, this settlement shall be reduced to 20%, i.e.

$$26.8 \times 0.2 = 5.4 \text{ cm}$$

If we add the balance settlement to it, It will be = 19.7 cm + 5.4 = 25.1 cm

As proposed in example 6.6, a skirt wall may be provided in this case also which will reduce this settlement to half i.e. 12.5 cm only.

A soil pad and a ring beam may also be provided to give more strength to the foundation (Fig. 6.15).

6.4.2 Design of Granular Piles Under Uplift Loads

Design principles of a single granular pile under uplift load are basically the same as those under compressive forces. The differences are that the pile is assumed (i) to derive the resistance against uplift force from the restraint provided by the ambient soil around the critical height H_c (Fig. 6.19), (ii) the height of bulging H_c is limited to 4 to 5 pile diameter, (iii) the bottom portion of the pile is considered to bulge due to uniform induced lateral stress σ_r in a homogeneous isotropic, infinite non-cohesive soil mass due to the gradual increase in the applied uplift stress and consequently the σ_r in the pile body. The cylindrical zone around the bulged pile having a radius R_u (Fig. 6.19a) will pass into a state of plastic equilibrium. Beyond this zone of plastic equilibrium of radius R_p, the soil is considered to remain in elastic equilibrium condition, (iv) the ultimate uplift load $(q_u)_{uplift}$ is considered to be resisted by the weight of the pile W_g and the load required to provide restraint against bulging of the pile, (v) the unit friction along the pile shaft and the ambient soil is neglected as there is no relative movement between the granular pile and the surrounding soil. The various steps for computing the ultimate uplift capacity are given as:

Step 1: Calculate parameters K_0 and μ from equations 6.21 and 6.22.

$$k_0 = 1 - \sin(1.2\phi) \quad \dots (6.21)$$

$$\mu = \left(\frac{k_o}{1-k_o}\right) \quad \dots (6.22)$$

Step 2: Find out the vertical stress σ'_v due to overburden using the depth (measured from average ground level to the top of the bulge) and bulk unit weight γ_{bulk} (Fig. 6.19). Calculate the mean normal stress, σ'_m from equation 6.23. The effective mean normal

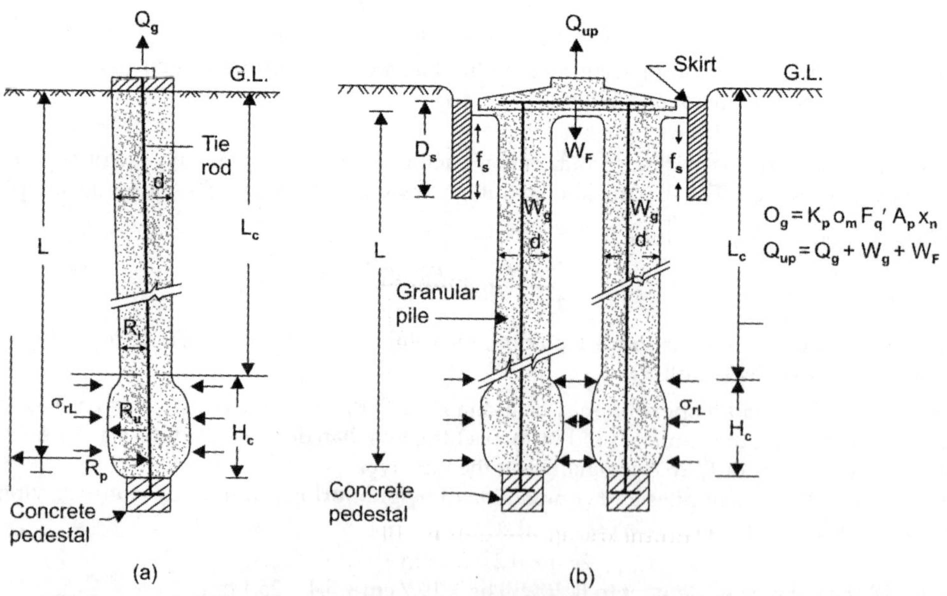

(a) Plain granular pile

(b) Skirted granular pile

Fig. 6.19a and b: Bulging failure mode of granular piles under uplift stress and effective vertical stress σ'_v shall be used in Eqn. 6.23, in case of water table is found at ground level and for intermediate levels, as explained earlier in step 2.

$$\sigma_m = \frac{1}{3}(1-2K_o)\sigma_v \qquad \ldots (6.23)$$

Step 3: Calculate the corrected soil modulus E'_s and the rigidity index I_r from Equations 6.24 and 6.25 respectively, where $\Delta\sigma'_m$ is the increase in mean normal stress due to additional load. Now, assigning σ'_v = 100 kN/m², the cavity expansion factor F'_q may be found out from Fig. 6.20.

$$E'_s = E\left[\frac{\sigma_m + \Delta\sigma_m}{\sigma_1}\right]^{0.5} \qquad \ldots (6.24)$$

$$I_r = \frac{E'_s}{2(1+\mu)(\sigma_m + \Delta\sigma_m)\tan\phi} \qquad \ldots (6.25)$$

Step 4: Calculate the uplift resistance, $(Q_u)_{uplift}$ of a single granular pile from equation 6.26,

$$(Q_u)_{uplift} = kp \cdot \sigma_m \cdot F'_q A_p \qquad \ldots (6.26)$$

and the uplift resistance of the pile group having, many piles, may be obtained from equation 6.27.

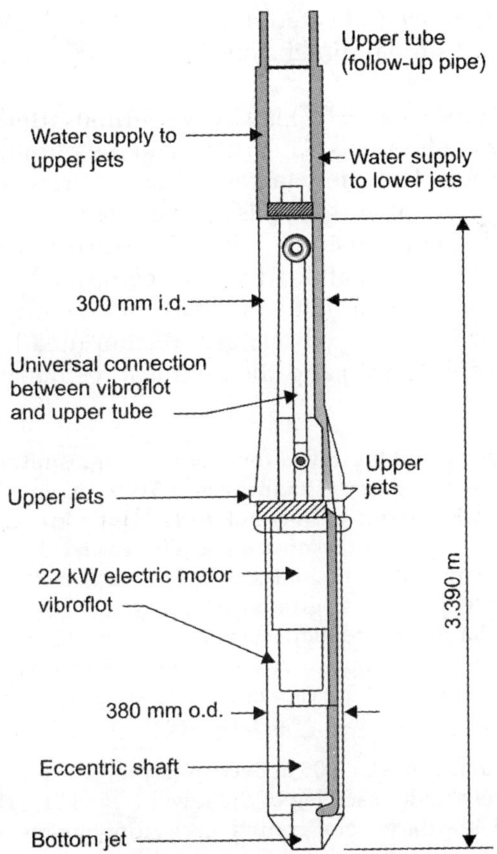

Fig. 6.20: Diagram of vibrofloat

$$(Q_u)_{uplift} = n \cdot kp \cdot \sigma_m \cdot F'_q A_p \qquad \ldots (6.27)$$

Find the weight of the pile group ($n.W_g$) and the footing W_F, and then calculate the increase in pile uplift capacity due to friction between soil plug and the skirt interface from equation 6.28.

$$Q_s = \left(0.5\gamma_{sub} . D_s + \frac{W_F}{A}\right) K_u \tan(1.2\phi) A_{sk} \qquad \ldots (6.28)$$

Step 5: Calculate the total uplift capacity of the skirted granular pile group $(Q_{skg})_{uplift}$ as

$$(Q_{skg})_{uplift} = (Q_{ug})_{uplift} + n \cdot W_g + Q_s + W_f \qquad \ldots (6.29)$$

6.4.3 Work Done in IIT Roorkee on Granular Pile Under uplift Loads (Kumar, 2002)

In situ tests on granular piles have earlier been carried out in cohesionless soil deposit (Rao and Ranjan 1983) having 250 mm diameter and 3.5 m deep as a research work. A similar study (Kumar 2002) has also been carried out in varying soil conditions in both type of soils having predominantly soft clayey silt deposit (Rajpura soil) and loose to medium cohesionless soil (CBRI, Roorkee). The study was initiated in the laboratory before it was applied to actual field conditions. Model granular piles (GAP systems) for two diameters (d) as 50 and 100 mm with varying L/d and S/d ratios were studied in the Laboratory and then in the Field in two different subsoil conditions namely—loose to medium dense cohesionless soil (Amanatgarh - Roorkee) and the soft cohesive soil (Rajpura–Punjab). The economy of the GAP as compared to that of a concrete pile has also been evaluated, which indicated that cost of GAP has been estimated as half the cost of concrete piles of similar dimensions (Kumar 2002) and hence is one amongst the most economical solutions.

Soil block method (Tomlinson, 1977) is also sometimes used to compute the uplift capacity of piles in cohesionless soils. The frustum of the pyramid/cone is assumed to be lifted with the pile group. The safety factor against uplift is taken as unity since skin friction around the periphery of the group is ignored. The safe uplift capacity of single pile is taken as the weight of the frustum of the soil block divided by the number of the piles in the group. The ultimate uplift load may be computed using a F.S. equal to 1.5.

The study by Ranjan and Kumar (2000) reveals that the ultimate uplift capacity of the single pile in a collectively skirted pile group computed by the 'modified cavity expansion approach' (Ranjan 1989) and the field test results are noted to be in close agreement.

Example 6.8: A 220 kN transmission line tower foundations is transmitting loads through each leg under normal condition as down thrust (Compression) 319.7 kN and uplift 261.75 kN whereas corresponding values under broken wire conditions (BWC) are 428.25 kN and 369.75 kN respectively. In addition, the foundation is subjected to a transverse thrust of 150 kN/m^2. The soils at the site consisted of predominantly cohesionless soil. The water table at the time of investigation was found to be at 0.6 m below the average ground level.

Design the skirted granular pile foundation against safety for (i) normal loading conditions and (ii) Broken wire condition, for both compressive and uplifting forces. The factor of safety under all conditions of loading may be taken as 1.5 except in uplift loading for broken wire conditions where a value of 2.5 may be adapted.

Solution: Considering 450 mm diameter 7.05 m deep, four granular pile group at spacing of 3 pile diameters skirted collectively under each leg of the tower (Fig. 6.21). The RCC pile cap 600 mm thick is placed at 600 mm below the general ground level. Adopting the dimensions of the pile cap (Fig. 6.21) the inside and outside dimensions of the skirt are 2.1 × 2.1 m and 2.6 × 2.6 m with 2.25 m depth from the cut-off level.

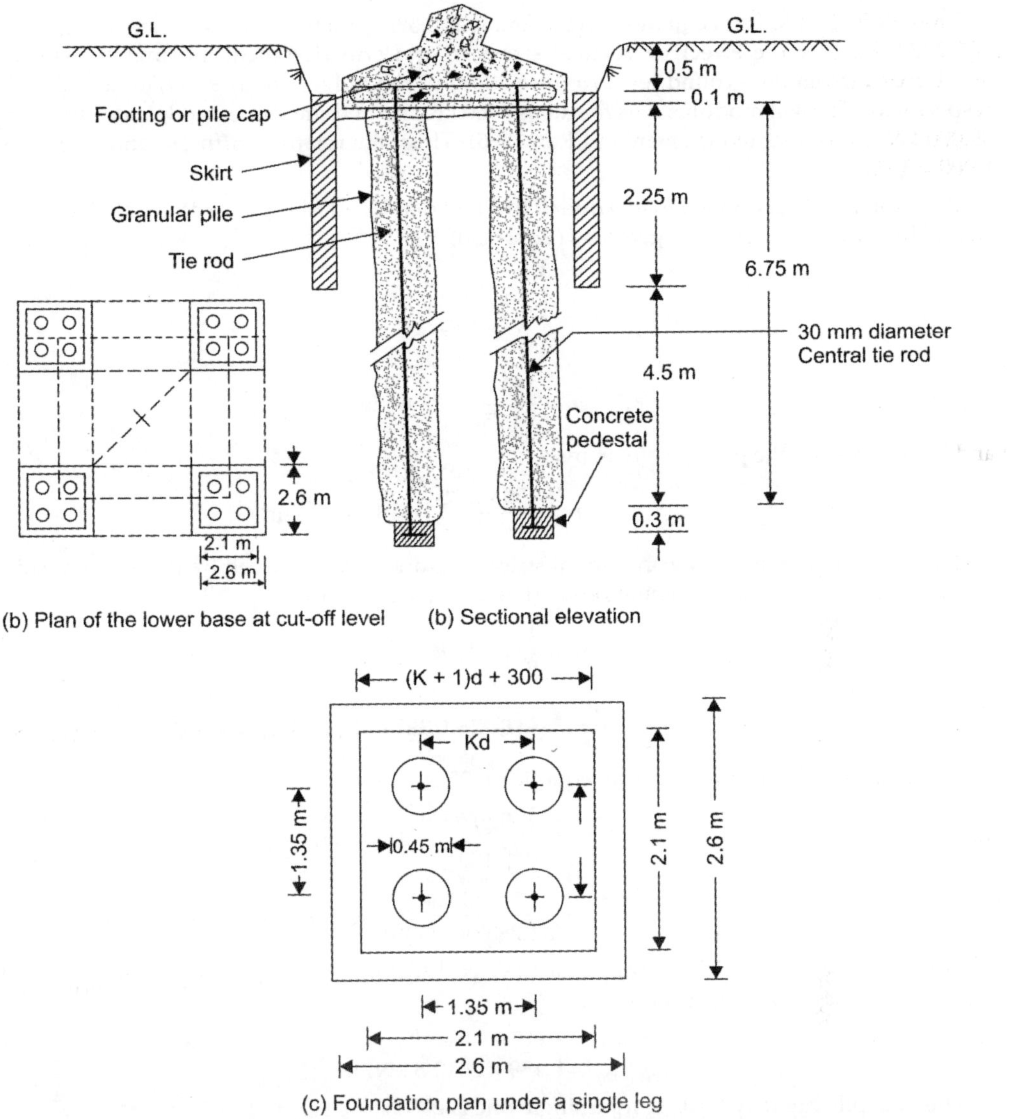

(b) Plan of the lower base at cut-off level (b) Sectional elevation

(c) Foundation plan under a single leg

Fig. 6.21: Granular pile foundation system for a 220 kV transmission line tower applicable for resisting compressive and uplifting forces (After Ranjan and Kumar 2000)

Computation of Bearing Capacity

Modified cavity expansion approach

Input sata: The effective angle of shearing resistance ϕ, and the effective cohesion 'c' are 31° and zero respectively. The bulk density and the submerged densities are 19.0 kN/m^3. The design thrust to be supported by the pile is 159 kN/m^2. The assumed pile length (Fig. 6.21) below cut off level is 6.75 m having initial and installed pile diameters as 0.37 m and 0.45 m (after compaction) respectively. The average standard penetration value N_{SPT} is found to be 8 while the static cone resistance is 3200 kN/m^2. The available safe bearing capacity is observed as 80 kN/m^2 and water table is at 0.6 m below ground level. Pile cap is placed at ground level on the top of the group of piles. The depth of the pile below cut off level is 6.75 m and overall depth below ground level is 7.65 m including the 0.30 deep concrete pedestal. The depth of the skirt from the cut off level is taken as 5 times pile dia which is 2.25 m (Fig. 6.21).

The co-efficient K_o from equation 6.11 is found as 0.3954 and the Poission's ratio μ (Equation 6.22) is 0.2833. Further the effective vertical stress σ_v and effective mean normal stress σ_m due to overburden (from the ground level to the bottom of the bulge) are 51.9 kN/m² and 30.9 KN/m² respectively. The soil modulus E_s is found as 8000 kN/m² and the value of pile modulus E_p equal to 30,000 kN/m² is adapted (Ranjan and Rao 1985). Thus, the relative stiffness ratio E_p/E_s = 30,000/8,000 = 3.75.

If the total vertical applied load q due to live and dead load (design load) is 150 kN/m², the load shared by the ambient soil q_s is given by (Eqn. 6.30).

$$q_s = q\left(\frac{E_s}{E_p + E_s}\right) \qquad \ldots (6.30)$$

$$q_s = \frac{150 \times 1}{4.75} = 31.6 \text{ kN/m}^2$$

and load shared by the pile q_p is given by,

$$q_p = \frac{150 \times 3.75}{4.75} = 118.46 \text{ kN/m}^2$$

The increase in effective mean normal stress $\Delta\sigma_m$ due to load shared by the ambient soil q_s and the corrected soil modulus E'_s equations 6.31 and 6.24 are given as

$$\Delta\sigma_m = \frac{1}{2}(1 + 2 \times k_o)_q \qquad \ldots (6.31)$$

$$\sigma'_m = \frac{1}{3}(1 + 2 \times 0.3954)\ 31.6 = 18.8 \text{ kN/m}^2$$

The rigidity index I_r as found from equation 6.25 is

and
$$E'_s = 8000\left[\frac{49.7}{100}\right]^{0.5} = 5639.86 \text{ kN/m}^2$$

$$I_r = \frac{5639.86}{2(1.2833)(49.7)(0.6)} = 73.7, \text{ say } (75)$$

Thus, for ϕ = 31°, I_r = 75, F'_q from Chart (Vesic 1972) may be found to be equal to 6.75 hence, ultimate capacity $(q_u)_{comp.}$ from equation 6.32 is,

$$(q_u)_{comp.} = K_p\,(\sigma_m + \Delta_{\sigma m})\,F'_q \qquad \ldots (6.32)$$

$$(q_u)_{comp.} = 0.159 \times 2012.8 \text{ kN/m}^2$$

The total pile capacity $(Q_u)_{comp.}$ for a single pile is given as,

$$(Q_u)_{comp.} = 0.159 \times 2012.8 = 320 \text{ kN}$$

The contribution of the friction between soil plug and the concrete skirt interface Q_s from equation 6.39 for a single pile is,

$$Q_s = \frac{1}{4}[(0.5 \times 9 \times 2.25 + 0.6 \times 24)\ 1.4 \times 0.759 \times (2.1 \times 4 \times 2.25)] = 123.13 \text{ kN}$$

Hence, the total, ultimate capacity of a single granular pile, when the skirt interface friction is also accounted for, is found as sum total of $(Q_u)_{comp.}$ and Q_s.

Therefore, $(Q_u)_{comp.} + Q_s = 320 + 123.13 = 443.13$ kN

Thus, the ultimate capacity of the 4 pile group $(Q_{ug})_{comp.}$ is given as,

$$(Q_{ug})_{comp.} = 4 \times 443.13 = 1772.5 \text{ kN}$$

Computation of uplift capacity

Modified cavity expansion approach

Taking the height of bulge as 5 times the installed diameter of pile the depth of the bulge from the ground level to the top of the bulge is 5.1 m. Thus utilising the input data provided earlier,

Effective normal stress $\sigma_v = 45.9$ kN/m^2

Effective mean normal stress $\sigma_m = 27.3$ kN/m^2

Corrected soil modulus $E'_s = 4179.9$ kN/m^2

Rigidity Index $I_r = 9942$ say (100)

Cavity expansion factor $F'q = 7$

Thus, ultimate uplift resistance of the single granular pile, $(Q_u)_{uplift}$ is found from equation 6.26.

$$(Q_u)_{uplift} = 6 \times 27.3 \times 7 \times 0.1590 = 182.3 \text{ kN},$$

and from equation 6.27

$$(Q_{ug})_{uplift} = 182.3 \times 4 = 729.2 \text{ kN}$$

Increase in uplift due to weight of 4 piles group.

$$n \cdot W_g = 4(0.1590 \times 7.05 \times 22) = 4 \times 24.7 = 98.8 \text{ kN}$$

Increase in upulift capacity of the pile due to soil-plug skirt interface friction (Q_s), will be same as in compression which is found to be as (123.13 × 4) and is equal to 492.5 kN, since there are 4 piles in the group, and increase in a pile capacity due to weight of footing W_F is,

$$W_F = (2.1 \times 2.1 \times 0.6 \times 24) = 63.5 \text{ kN}$$

Hence, the total uplift capacity of the skirted granular pile foundation having 4 piles in a group is given in equation 6.29.

$$(Q_{skg})_{uplift} = (Q_{ug})_{uplift} + \text{n. } W_g. + Q_s + W_F$$

$$= 729.2 + 98.8 + 492.5 + 63.2$$

$$= 1383.74 \text{ kN}$$

Therefore, the ultimate uplift capacity of a single pile is found as,

$$(Q_u)_{uplift} = \frac{1383.7}{4} = 345.6 \text{ kN}$$

A. Soil Block Method (Tomlinson 1977)

Volume of the frustum of the soil block ABCD, A'B'C'D' with the height KK' (Fig. 6.22) is given by

Volume = 1/3 (Area, A'B'C'D'× OK) = 1/3 (Area ABCD × OK)

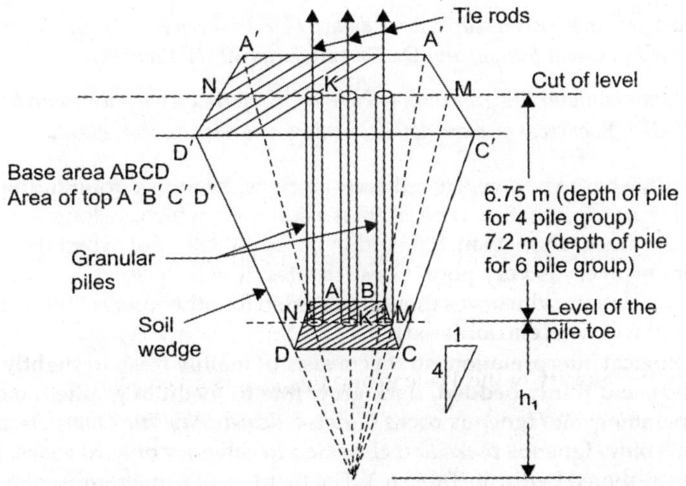

Fig. 6.22: Soil block method (After Tomlinson, 1917)

now h_1 = 3.59 m, weight of the block up to the cut-off level

$$= 9.0\,[1/3\,(5.18 \times 5.18 \times 10.34 - 1.8 \times 1.8 \times 3.59)] = 797.5 \text{ kN}$$

Weight of the pile cap = 63.5 kN

Thus, total weight = 797.5 + 63.5 = 861.0 kN

The total resistance against upulift is provided by the weight of the soil block the weight of the pile cap and the shearing resistance along the surface of the wedge of soil. Neglecting the shearing resistance, the total weight of the soil block and the pile cap may be taken as the safe uplift capacity of the pile group.

Safe uplift capacity of group = 861.0 kN

Taking factor of safety of 1.5 for the soil block, the ultimate capacity of the soil block is given as $(Q_u)_{Block}$ given as

$$(Q_u)_{Block} = 797.5 \times 1.5 + 63.5 = 1259 = 1259.8 \text{ kN}$$

Hence, ultimate uplift capacity for single pile = 1259.8/4 = 315 kN, whereas the proposed method by the authors (modified capacity expansion approach) gives ultimate uplift capacity of a single pile as 345.6 kN which is in good agreement.

a. ***Check for compressive force:***
   Normal condition: The force due to tower and the pile cap is 383.2 kN and the available capacity of the skirted granular pile is 1772.5 kN. Since the factor of safety is equal to 4.62 which is greater than 2.5, heance safe. Broken wire condition: The force due to tower and the pile cap is 491.7 kN hence factor of safety is 3.6 which is greater than 2.5. Hence safe.

b. ***Check for bearing capacity***
   Normal condition: The intensity of pressure under the cap (2.1 m × 2.1 m) is 86.8 kN/m^2 which is less than 160 kN/m^2 which is the safe bearing capacity of the granular pile group. Hence safe.

c. ***Check for up-lifting force:***
   Normal condition: The uplift force due to tower is 261.75 kN and the available ultimate uplift capacity is 1383.73 kN, thus the actual factor of safety is 5.2, which is greater than 2.5. Hence safe.
   Broken wire condition: The uplift force due to tower is 369.75 kN and the available ultimate uplift capacity is 138.73 kN. Hence the actual factor of safety is 3.74, which is greater than 1.5. Hence safe.

d. ***Design of the tie Bar***
   The ultimate uplift force for a single pile is 310 kN.
   Area of steel bar = 345.6/(4250 × 100) = 8.13 × 10^{-4} m^2
   Provide 35 mm bar in the centre as tie bar in accordance with Fig. 6.21.

(The above work from pare to pare is research work done by Dr. Pradeep Kumar, Sr. Scientist in CBRI Roorkee under the guidance of Dr. Gopal Ranjan and Dr. Swami Saram at IIT Roorkee).

Note: *The above discussion and design example have been referred from the research work of Dr. Pradeep Kumar, Scientist, CBRI, Roorkee. For more details, readers can refer Kumar, 2002.*

Example 6.9: A bridge has to be constructed across a river, for which foundation has to be designed where soil strata is varying in nature. The strata is a soft rock, which, so long is existing in its natural state, is hard (N_{SPT} values larger than 100, and even up to 200). But when its core is desired to be extracted, the core recovery is very poor. On such sites, it was decided to provide pile foundation (instead of open foundation, which was though provided for other piers at the same site) below some piers (wherever soft rock was encountered).

As per the geological interpretation, strata consists of mainly fresh to slightly weathered, highly fractured (sheared), and thinly bedded, dark grey, fine to medium grained, moderately strong to strong foliated metamorphic/Igneous rocks (Gneiss, *Schist, Phyllitic Quartzite and Granites*). These Rocks are metamorphic/Igneous rocks and classified in category of hard rocks. Point load strength Index test results presented below in Table 6.3. The pictures of some representative tests conducted on rock chunk pieces collected from this site are shown in Fig. 6.23.

Table 6.3: Results of point load strength index tests *(Tests done as per IS 8764: 1998 (reaffirmed 2003)*

Sl. No	Sample ID	Depth (m)	Souree of sample	Location	Specimen No.	Orientation of loading axis	Rock type	Width perpendicular to the loading direction (W) (mm) (D) (mm)	Distance between platen contact points (D) (mm)	Point load index strength $I_{s(50)}$ (MPa)	Condition of test
1	P-4/3.0–4.5/01	3.0–4.5	High level bridge over rive Mahanadi at nelson Mandela chowk to churpur road Sambalpur Odisha	P-4	1	NA	Streaky gneiss	52.84	37.50	4.00	All specimen to be done in axial mode and saturate condition
2	P-4/3.0–4.5/02				2	NA	Gneiss	52.98	26.32	2.09	
3	P-4/4.5–6.0/01	4.5–6.0			1	NA	Chloride schist	53.82	34.24	3.53	
4	P-4/4.5–6.0/02				2	NA		52.94	20.29	6.58	
5	P-4/6.0–7.50/01	6.0–7.50			1	NA	Granite	54.31	37.80	7.91	
6	P-4/6.0–7.50/02				2	NA	Gneiss	53.82	31.41	4.02	
7	P-4/7.5–9.0/01	7.50–9.0			1	NA		54.30	47.50	1.08	
8	P-4/7.5–9.0/02				2	NA	Chloride schist	54.40	36.99	4.40	
9	P-19/3.0–4.0/01	3.0–4.0		P-19	1	NA	Chloride schist	54.60	45.44	1.39	
10	P-19/4.0–5.0/01	4.0–5.0			1	NA		49.20	39.94	1.40	
11	P-19/4.0–5.0/02				2	NA		54.32	31.47	6.01	
12	P-19/5.0–6.0/01	5.0–6.0			1	NA		54.44	43.45	2.96	
13	P-19/5.0–6.0/02				2	NA		53.80	45.24	1.63	
14	P-19/6.0–7.0/03	6.0–7.0			3	NA	Basalt	54.20	44.89	12.23	
15	P-19/6.0–7.0/04				4	NA		54.28	44.68	13.50	

NA, not applicable

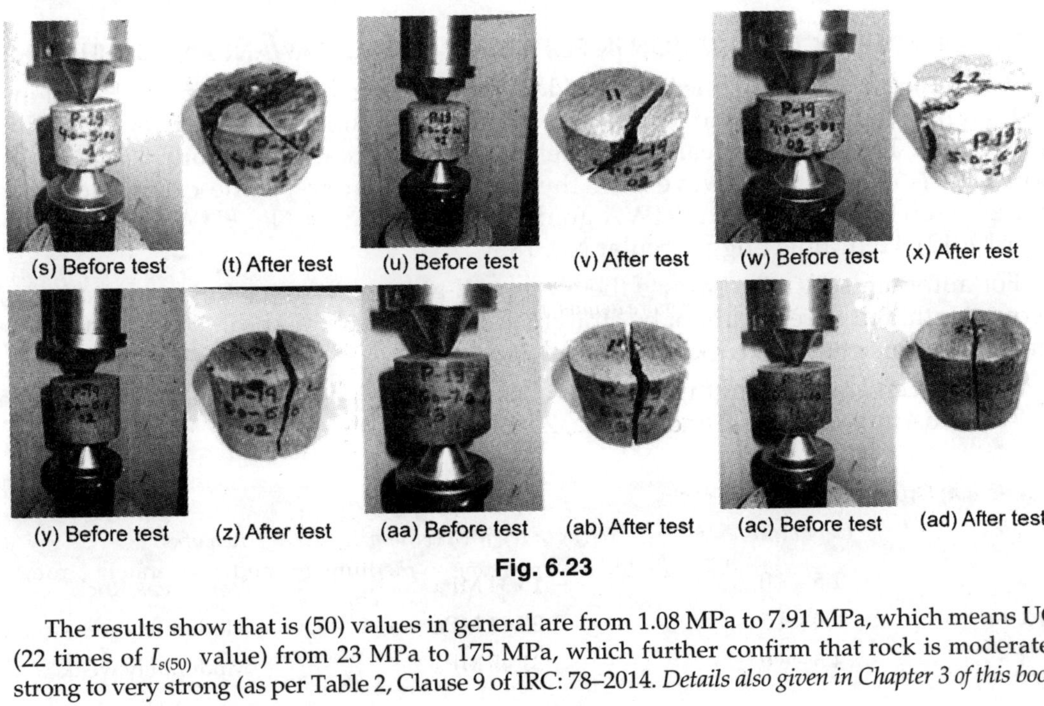

(s) Before test (t) After test (u) Before test (v) After test (w) Before test (x) After test

(y) Before test (z) After test (aa) Before test (ab) After test (ac) Before test (ad) After test

Fig. 6.23

The results show that is (50) values in general are from 1.08 MPa to 7.91 MPa, which means UCS (22 times of $I_{s(50)}$ value) from 23 MPa to 175 MPa, which further confirm that rock is moderately strong to very strong (as per Table 2, Clause 9 of IRC: 78–2014. *Details also given in Chapter 3 of this book*).

The river bedrock is observed from top of borehole/existing bed level, however during drilling rock cores could not be collected due to highly fractured/sheared nature of rock.

Refer to Method 2, Clause 9.1 Appendix-5 of IRC:78-2014 and considering the bedrock data, a pile of diameter 1.0 m with socket length of 3.3 m was proposed to be provided.

Ignoring first 0.3 m as recommended in this clause, the net socket length for design = 3.0 m

Axial Load Carrying Capacity

Extrapolated SPT values are more than 200, hence the shear strength as per table in method 2 of above said IRC code is 1.9 MPa.

$$R_e = \text{Ultimate end bearing} = C_{ub} \times N_c \times A_b \quad \ldots (6.33)$$

C_{ub} = 1.9 MPa= 190 t/m^2

$N_c = 9$

A_b = Area of pile at base = 0.785 m^2

$R_e = 190 \times 9 \times 0.7855 = 1343$ t

R_{af} = Ultimate side socket shear = $C_{us} \times A_s$

C_{us} = 190 t/m^2

L = 3.0 M

$A_s = 3.142 \times 1 \times 3 = 9.426$ sq m

$R_{af} = 190 \times 9.426 = 1791$ t

Q_{allow} = Allowable pile capacity

$= R_e/3 + R_{af}/6$

$= 1343/3 + 1791/6$

= 746 t

However, the same is restricted to **400 t** and pile will be load tested for the same.

Lateral load/moment capacity:

Moment carrying capacity of socketed piles shall be calculated by Clause 9.2 of Appendix-5 of IRC: 78–2014.

6.5 SCOUR DEPTH COMPUTATION IN SOFT ROCK (MITTAL, SAWANT AND SAHU, 2016)

Note: The above design of pile has been done by ignoring the scour depth calculations, although, it is being provided in a river bed where the discharge was also quite substantial. There are no clear cut guidelines for computation of scour depth where soft rock is available. However, for the benefit of readers, some calculations are being given below based on FHWA guidelines (Publication No. FHWA-HIF-12-003, April 2012), Hydraulic Engg. circular No. 18.

For alluvium soils, the Lacey's theory is well-accepted theory for computation of scour depth. But here, it cannot be straight away used as here the soil type is varying in nature, which has soft rock and hard rock also.

Based on crushing strength for given rocks, rock quality can be designated as given in Tables 6.4 and 6.5 (Mittal and Shukla, 2013).

Table 6.4: Crushing strength of rocks

S. No.	*Depth (m)*	*Strength reported (N/mm^2)*	*Remarks*
1	1.5 – 3.0	15.34 MPa	Very weak Rock
2	3 .0 – 4.5	25.74 MPa	Moderately weak
3	4.5 – 6.0	31.45 MPa	Moderately weak

According to FHWA, publication no. HIF-12-003, the erodibility index of rocks is given as

$$K = M_s \times K_b \times J_s \times K_d \qquad \text{...(6.34)}$$

where,

M_s = Intact rock mass strength = 0.87 for very weak rock

K_b = Block size parameter

Table 6.5: Values of rock mass strength parameter M_s

Hardness	*Identification in profile compressive*	*Unconfined strength (MPa)*	*Mass strength number (Ms)*
Very soft rock	Material crumbles under firm (moderate) blows with sharp end of geological pick and can be peeled off with a knife; is too hard to cut triaxial sample by hand	Less than 1.7 1.7–3.3	0.87 1.86
Soft rock	Can just be scraped and peeled with a knife; indentations 1 to 3 mm show in the specimen with firm (moderate) blows of the pick point	3.3–6.6 6.6–13.2	3.95 8.39
Hard rock	Cannot be scraped or peeled with a knife; hand-held specimen and be broken with hammer end of geological pick with a single firm (moderate) blow	13.2–26.4	17.70
Very hard rock	Hand-held specimen breaks with hammer end of pick under more than one blow	26.4–53.0 53.00–106.0	35.0 70.0
Extremely hard rock	Specimen requires many blows with geological pick to breakthrough intact material	Larger than 212.0	280.0

$$K_b = RQD/J_N \qquad \text{...(6.35)}$$

Presuming RQD = 1 (negligible) and rock joint set number J_n = 5 (for multiple/fissure joint sets; Table 6.6)

$$K_b = 1/5 = 0.2$$

Table 6.6: Rock joint set number J_n

Number of joint sets	*Joint set number (J_n)*
Intact, no or few joints/fissures	1.00
One joint/fissure set	1.22
One joint/fissure set plus random	1.50
Two joint/fissure sets	1.83
Two joint/fissure sets plus random	2.24
Three joint/fissure sets	2.73
three joint/fissure sets plus random	3.34
Four joint/fissure sets	4.09
Multiple joint/fissure sets	5.00

K_b = shear strength parameter $K_b = J_R/J_A$ ...(6.36)

where,

J_R = Joint roughness number = 1 (Table 6.7)

J_A = Joint alternation number = 10 (Table 6.8)

J_s = Relative Orientation parameter = 0.5 (Table 6.9)

Hence, $K = 0.87 \times 0.2 \times 0.5 \times 0.1 = 0.0087$

Table 6.7: Joint roughness number (J_r)

Condition of joint	*Joint roughness number* (J_R)
Stepped joints/fissure	4.0
Rough or irregular, undulating	3.0
Smooth undulating	2.0
Slickensided undulating	1.5
Rough or irregular, planar	1.5
Smooth planar	1.0
Slickensided planar	0.5
Joints/fissures either open or containing relatively soft gouge of sufficient thickness to prevent joint/fissure wall contact upon excavation	1.0
Shattered or micro-shattered clays	1.0

Table 6.8: Joint alteration number *Ja*

Description of gouge	*Joint alteration number* (J_A) *for joint separation (mm)*		
	1.0(1)	*1.0–5.0*(2)	*5.0*(3)
Tightly healed, hard, non-softening impermeable filling	0.75	–	–
Unaltered joint walls, surface staining only	1.0	–	–
Slightly altered, non-softening, non-cohesive rock mineral or crushed rock filling	2.0	2.0	4.0
Non-softening, slightly clayey non-cohesive filling	3.0	6.0	10.0
Non-softening, strongly over-consolidated clay mineral filling, with or without crushed rock	3.0	6.0**	10.0
Softening or low friction clay mineral coating and small quantities of swelling clays	4.0	8.0	13.0
Softening moderately over-consolidated clay mineral filling, with or without crushed rock	4.0	8.00**	13.0
Shattered or micro-shattered (swelling) clay gouge, with or without crushed rock	5.0	10.0**	18.0

Note:
1. Joint walls effectively in contact.
2. Joint walls come into contact after approximately 100 mm shear.
3. Joint walls do not come into contact at all upon shear.

** Also applies when crushed rock occurs in clay gouge without rock wall contact.

Table 6.9: Relative orientation parameter (J_s)

Dip direction of closer spaced joint set (degrees)	*Dip angle of closer spaced joint set (degrees)*	*Ratio of joint spacing, r*			
Dip direction	*Dip angle*	*Ratio 1:1*	*Ratio 1:2*	*Ratio 1:4*	*Ratio 1:8*
180/0	90	1.14	1.20	1.24	1.26
In direction of stream flow	89	0.78	0.71	0.65	0.61
In direction of stream flow	85	0.73	0.66	0.61	0.57
In direction of stream flow	80	0.67	0.60	0.55	0.52
In direction of stream flow	70	0.56	0.50	0.46	0.43
In direction of stream flow	60	0.50	0.46	0.42	0.40
In direction of stream flow	50	0.49	0.46	0.43	0.41
In direction of stream flow	40	0.53	0.49	0.46	0.45
In direction of stream flow	30	0.63	0.59	0.55	0.53
In direction of stream flow	20	0.84	0.77	0.71	0.67
In direction of stream flow	10	1.25	1.10	0.98	0.90
In direction of stream flow	5	1.39	1.23	1.09	1.01
In direction of stream flow	1	1.50	1.33	1.19	1.10
0/180	0	1.14	109	1.05	1.02
Against direction of stream flow	–1	0.78	0.85	0.90	0.94
Against direction of stream flow	–5	0.73	0.79	0.84	0.88
Against direction of stream flow	–10	0.67	0.72	0.78	0.81
Against direction of stream flow	–20	0.56	0.62	0.66	0.69
Against direction of stream flow	–30	0.50	0.55	0.58	0.60
Against direction of stream flow	–40	0.49	0.52	0.55	0.57
Against direction of stream flow	–50	0.53	0.56	0.59	0.61
Against direction of stream flow	–60	0.63	0.68	0.71	0.73
Against direction of stream flow	–70	0.84	0.91	0.97	1.01
Against direction of stream flow	–80	1.26	1.41	1.53	1.61
Against direction of stream flow	–85	1.39	1.55	1.69	1.77
Against direction of stream flow	–89	1.50	1.68	1.82	1.91
180/0	–90	1.14	1.20	1.24	1.26

Note:

1. For intact material take $J_s = 1.0$.
2. For values of r greater than 8 take J_s as for $r = 8$
3. If the flow direction *FD* is not in the direction of the true dip *TD*, the effective dip *ED* is determined by adding the ground slope to the apparent dip *AD*: $ED = AD + GS$

However, as a general rule, rock masses on which bridge piers are founded typically exhibit erodibility index *K* values ranging from 0.1 (very poor rock) up to 10000 or greater (very good rock).

Adopting $K = 0.1$, the critical stream power P_c for initiating quarrying and plucking is related to *K* as given by Annandale (1995, 2006).

$$P_c = K^{0.75} = (0.1)^{0.75} = 0.17783 \text{ kW/m}^2 \quad \dots (6.37)$$

As developed by Annandale, the stream power is calculated by considering the turbulence production near the bed of the stream:

$$P_a = 7.853\rho\left(\frac{\tau}{\rho}\right)^{1.5} \text{ W/m}^2 \quad \dots (6.38)$$

where,

τ = Bed shear stress of approach flow (N/m^2)

ρ = Mass density of water = 1000 kg/m^3

In Eq. (5), P_a is expressed in units of W/m^2, whereas the critical shear stress P_c (given by Eq. 6.37) is expressed in kW/m^2. To convert P_A to KW/m^2, the value from Eq. (6.38) shall be divided by 1000.

In the vicinity of a bridge pier, the downward flow at the upstream face of the pier creates additional local turbulence in the form of horse shoe vortex. As scour occurs, the stream power (P) at the bottom of the scour hole decreases as the scour hole becomes deeper.

Note: Tables 6.5 – 6.9 have been taken from FHWA publication number HIF-12–003.

Bottom of the scour hole becomes less than the critical stream power (P_c) at which point the scouring process can no longer be sustained. The relationship relating the relative depth of the scour hole to the stream power at the bottom of the hole for a variety of pier shapes (round, square and rectangular) can be expressed as:

$$\frac{P}{P_a} = 8.42^{e-0.712(y_s/b)} \text{ W/m}^2 \quad \dots (6.39)$$

where,

P = Stream power at the bottom of the scour hole (W/m^2)

P_a = Stream power of the approach flow near the stream bed (W/m^2)

Y_s = Depth of scour hole (m)

b = Pier width perpendicular to the flow direction (m)

Now, bed shear stress of approach flow

$$\tau = \gamma\, y\, S_f \quad \dots (6.40)$$

S_f = Slope of energy grade = 1 in 1000 (Say)

γ = Weight of water (N/m^3) = 9800

Therefore,

$$\tau = 9800 \times 12 \times \frac{1}{1000} = 117.6 \text{ N/m}^2$$

From Eq. (6.38),

$$P_a = 7.853 \times 1000 \times \left(\frac{117.6}{1000}\right)^{1.5} = 316.7 \text{ W/m}^2$$

or,

$$P_a = 0.3167 \text{ kW/m}^2$$

For critical condition as per FHWA guidelines, $P = P_C$

$$P_c = P_a \times 8.42\, e^{-0.712(y_S/b)}$$

$$0.17783 = 0.3167 \times 8.42\, e^{-0.712(y_S/b)}$$

or, $$e^{-0.712(y_s/b)} = \frac{0.17783}{0.3167 \times 8.42} = 0.066687$$

or, $$-0.712\,(y_s/b) = \ln(0.066687) = -2.707745$$

$$(y_s/b) = 3.803$$

$$y_s = 3.803 \times b = 3.803 \times 1.2 = 4.563 \text{ m}$$

This value is further increased to 5 m to account for the abrasion of bed rock.

The general cross-section showing various levels is shown in PLATE-1. The rock present at site is shown in Figs 6.24 and 6.25.

According to Ranjan and Rao (2000) the foundation depth below scour level should in no case be less than 2 m for piers and abutments with arches and 1.2 m for piers and abutments in other structures (IS : 3955, 1967 and IRC specification 1966). Adopting here 2 m as grip length.

Therefore foundation level = 132.0–2.0 = 130.0 for pier which has bed level 137.0 m.

Similarly levels can be decided for other piers also.

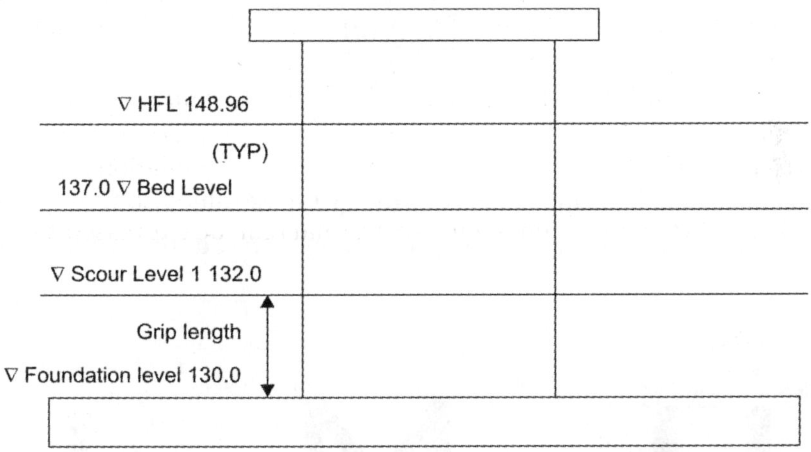

Plate 1: A general cross-section, showing various levels

Fig. 6.24: A view of constructed pier at site (Left bank)

Fig. 6.25: View of exposed rock near GL

6.6 PILES IN A GROUP *(After Braja M. Das, 2013)*

6.6.1 Group Efficiency

In most cases, piles are used in groups, as shown in Fig. 6.26, to transmit the structural load to the soil. A pile cap is constructed over group of piles. The cap can be in contact with the ground, as in most cases (see Fig. 6.26a), or well above the ground, as in the case of offshore platforms (see Fig. 6.26b).

Determining the load-bearing capacity of group of piles is extremely complicated and has not yet been fully resolved. When the piles are placed close to each other, a reasonable assumption is that the stresses transmitted by the piles to the soil will overlap (see Fig. 6.26c), reducing the load-bearing capacity of the piles. Ideally, the piles in a group should be spaced so that the load-bearing capacity of the group is not less than the sum of the bearing capacity of the individual piles. In practice, the minimum centre-to-centre pile spacing, d, is 2.5D and, in ordinary situations, is actually about 3 to 3.5D.

The efficiency of the load-bearing capacity of a group of pile may be defined as

$$n = \frac{Q_g(u)}{\Sigma Q_u} \quad \dots (6.41)$$

where

n = Group efficiency

$Q_{g(u)}$ = Ultimate load-bearing capacity of the group of pile

Q_u = Ultimate load-bearing capacity of each pile without the group effect

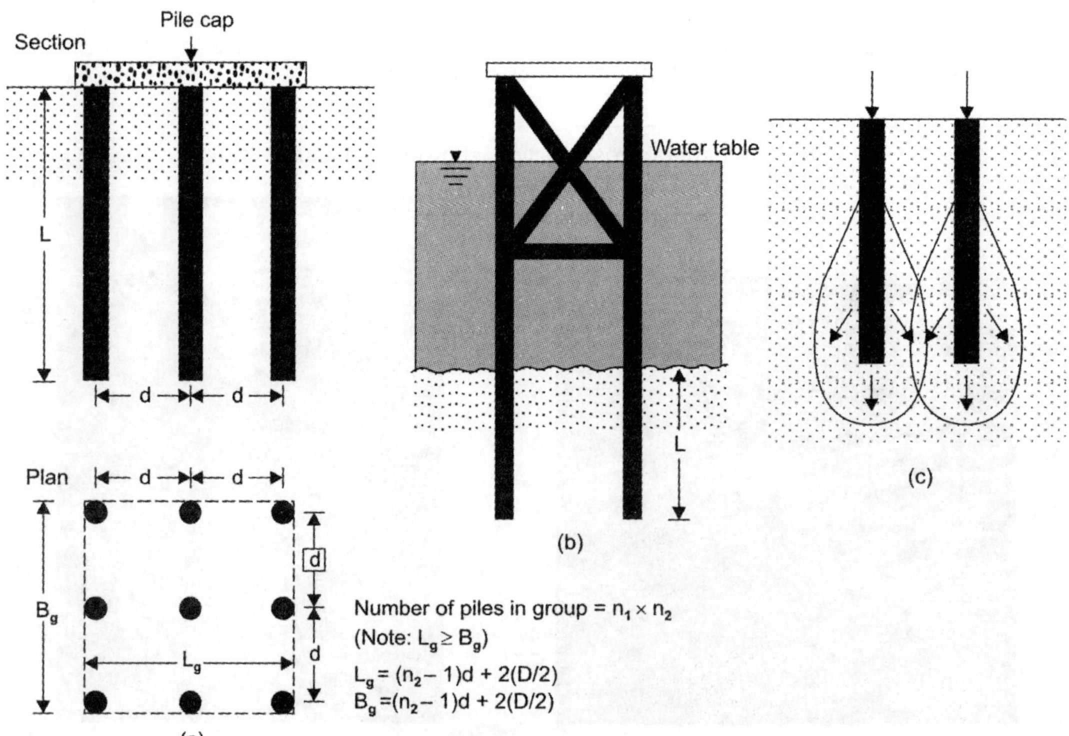

Fig. 6.26: Group of piles

Many structural engineers use a simplified analysis to obtain the group of efficiency for friction piles, particularly in sand. This type of analysis can be explained with the aid of Fig. 6.26a. Depending on their spacing within the group, the piles may act in one of two ways: (1) as a block, with dimensions $L_g \times B_g \times L$, or (2) as individual piles. If the piles act as a block, the frictional capacity is $f_{av}\, p_g\, L \approx Q_{g(u)}$. (Note: p_g = perimeter of the cross-section of block = $2(n_1 + n_2 - 2)\, d + 4D$, and f_{av} = average unit frictional resistance.) Similarly, for each pile acting individually, $Q_u \approx pLf_{av}$. (Note: p = perimeter of the cross-section of each pile).

Thus,

$$\eta = \frac{Q_g(u)}{\Sigma Q_u} = \frac{f_{av}[2(n_1+n_2)d+4D]L}{n_1 n_2 pLf_{av}} \quad \ldots (6.42)$$

$$= \frac{2(n_1+n_2-2)d+4D}{pn_1n_2}$$

Hence,

$$Q_{g(u)} = \left[\frac{2(n_1+n_2-2)d+4D}{pn_1n_2}\right]\Sigma Q_u \quad \ldots (6.43)$$

From Eq. 6.43 , if the centre-to-centre spacing d is large enough, $\eta > 1$. In that case, the piles will behave as individual piles. Thus, in practice, if $\eta < 1$, then

$$Q_{g(u)} = \eta\, \Sigma\, Q_u \quad \ldots (6.43a)$$

and if $n \geq 1$, then

$$Q_{g(u)} = \Sigma\, Q_u \quad \ldots (6.43b)$$

There are several other equations like Eq. 6.43 for calculating the group efficiency of friction piles. Some of these are given in Table 6.4.

It is important, however, to recognize that relationship such as Eq. 6.43 are simplistic and should not be used. In fact, in a group of pile, the magnitude of f_{av} depends on the location of the pile in the group (ex., Fig. 6.27).

Figure 6.28 shows the variation of the group efficiency η for a 3 × 3 group pile in sand (Kishida and Meyerhof, 1965). It can be seen that, for loose and medium sands, the magnitude of the group efficiency can be larger than unity. This is primarily due to the densification of sand surrounding the pile. More details are given in Table 6.10.

Table 6.10: Equation for group efficiency of friction piles

Name	*Equation*
Converse-Labarre equation	$\eta = 1 - \left[\frac{(n_1-1)n_2+(n_2-1)n_1}{90n_1n_2}\right]\theta$ where θ (deg) = $\tan^{-1}(D/d)$
Los Angeles group action equation	$\eta = 1 - \frac{D}{\pi d n_1 n_2}[n_1(n_2-1)+n_2(n_1-1)+\sqrt{2}(n_1-1)(n_2-1)]$
Seiler-Keeney equation (Seiler and Keeney, 1944)	$\eta = \left\{1-\left[\frac{11d}{7(d^2-1)}\right]\left[\frac{n_1+n_2-2}{n_1+n_2-1}\right]\right\}+\frac{0.3}{n_1+n_2}$ where d is in ft.

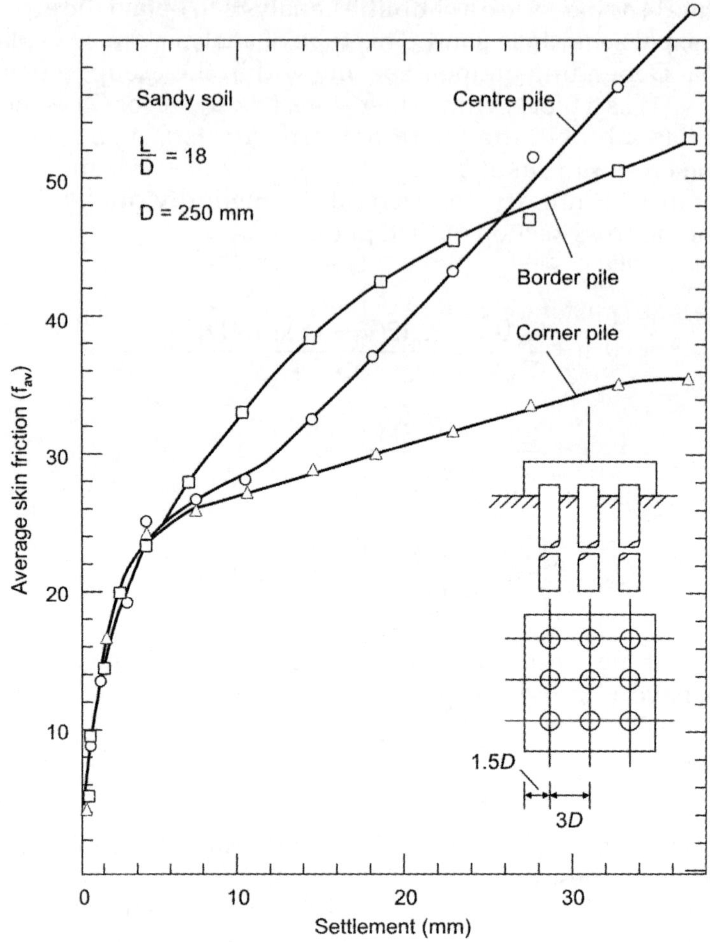

Fig. 6.27: Average skin friction (f_{av}) based on pile location (Based on Liu, et al. 1985)

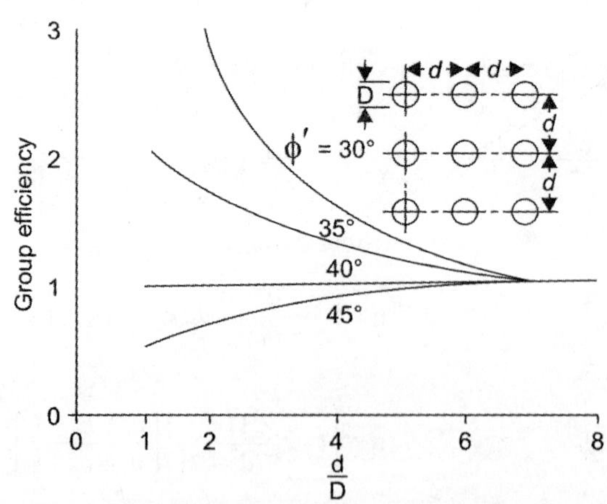

Fig. 6.28: Variation of efficiency of pile groups in sand (Based on Kishida and Meyerhof, 1965)

6.6.2 Ultimate Capacity of Group Piles in Saturated Clay

Figure 6.29 shows a group of pile in saturated clay. Using the figure, ultimate load-bearing capacity of group piles can be estimated in the following manner:

Step 1: Determine $\Sigma Q_u = n_1 n_2 (Q_p + Q_s)$.

Since, $$Q_p \approx N_c^* c_u A_p = 9c_u A_p \quad \ldots(6.44)$$

Here, $$Q_p = A_p [9A_p c_{u(p)}]$$

where $c_{u(p)}$ = undrained cohesion of the clay at the pile tip.

The ultimate side resistance of pile can be given as

and $$Q_s = \Sigma\, \alpha p_{cu} \Delta L \quad \ldots(6.45)$$

So,

$$\Sigma Q_u = n_1 n_2 [9A_p c_{u(p)} + \Sigma\, \alpha p c_u \Delta L] \quad \ldots(6.46)$$

Step 2: Determine the ultimate capacity by assuming that the piles in the group act as a block with dimensions $L_g \times B_g \times L$. The skin resistance of the block is:

$$\Sigma P_g\, c_u \Delta L = \Sigma\, 2\,(L_g + B_g)\, c_u \Delta L \quad \ldots(6.47)$$

Calculate the point bearing capacity:

$$A_p q_p = A_p c_{u(p)} N_c^* = (L_g B_g) c_{u(p)} N_c^* \quad \ldots(6.48)$$

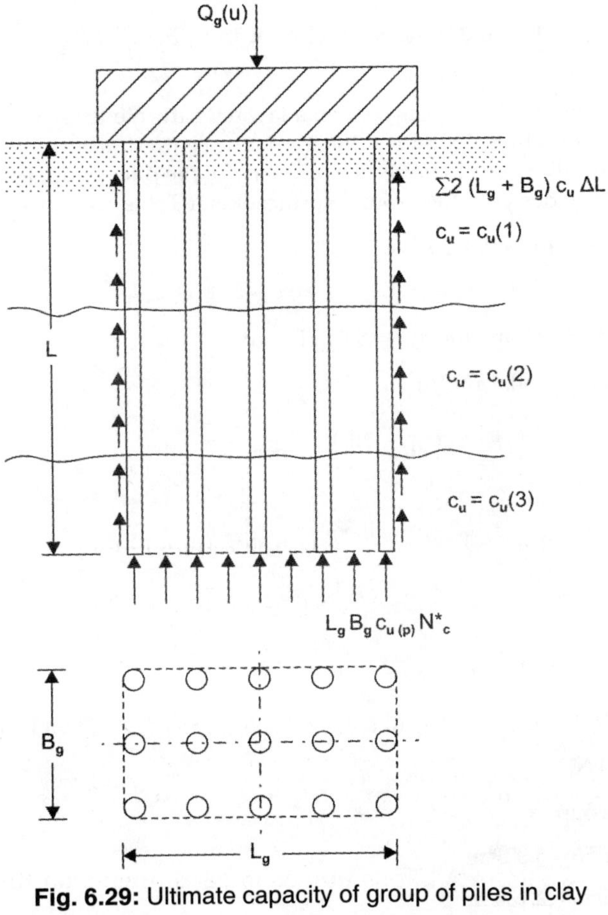

Fig. 6.29: Ultimate capacity of group of piles in clay

Obtain the value of the bearing capacity factor N_c^* from Fig. 6.30. Thus, the ultimate load is:

$$\Sigma Q_u = L_g B_g c_{u(p)} N_c^* + \Sigma\, 2\,(L_g + B_g) c_u \Delta L \quad \ldots (6.49)$$

Step 3: Compare the values obtained from Eqns. (6.46) and (6.49). The lower of the two values is $Q_{g(u)}$.

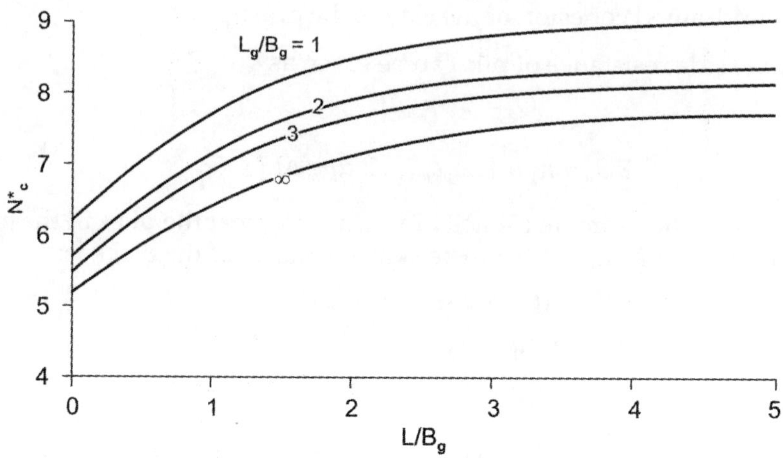

Fig. 6.30: Variation of N_c^* with L_g/B_g and L/B_g

Example 6.9: The section of a 3 × 4 group pile in a layered saturated clay is shown in Fig. 6.31. The piles are square in cross-section (356 mm × 356 mm). The centre-to-centre spacing, *d*, of the piles is 889 mm. Determine the allowable load-bearing capacity of the pile group. Use $FS = 4$. Note that the ground water table coincides with the ground surface. Refer Table 6.11 for some data.

Solution: From Eqns 6.46 to 6.49, we have

$$\Sigma Q_u = n_1\, n_2 [9 A_p c_{u(p)} + \alpha_1 p c_{u(2)} L_1 + \alpha_2 p c_{u(2)} L_2] \quad \ldots (6.50)$$

From Fig. 6.31, $c_{u(1)} = 50.3\ \text{kN/m}^2$ and $c_{u(2)} = 85.1\ \text{kN/m}^2$.

For the top layer with $c_{u(1)} = 50.3\ \text{kN/m}^2$,

$$\frac{c_{u(1)}}{p_a} = \frac{50.3}{100} = 0.503$$

$$\alpha_1 \approx 0.68.$$

Similarly,

$$\frac{c_{u(2)}}{p_a} = \frac{85.1}{100} \approx 0.85$$

$$\alpha_2 \approx 0.51$$

$$\Sigma Q_u = (3)(4)\,[(9)(0.356)^2(85.1) + (0.68)(4 \times 0.356)(503)(4.57) + (0.5)(4 \times 0.356)(8.51)(13.72)]$$

$$= 14011\ \text{kN}$$

For piles acting as a group

$L_g = (3)\,(0.889) + 0.356 = 3.023$ m

$B_g = (2)\,(0.889) + 0.356 = 2.134$ m

Table 6.11: Variation of α (interpolated values based on Terzaghi, Peck and Mesri, 1996)

$\frac{c_u}{p_a}$	α
≤ 0.1	1.00
0.2	0.92
0.3	0.82
0.4	1.74
0.6	0.62
0.8	0.54
1.0	0.48
1.2	0.42
1.4	0.40
1.6	0.38
1.8	0.36
2.0	0.35
2.4	0.34
2.8	0.34

Note: P_a = atmospheric pressure ≈ 100kN/m^2

$$\frac{L_g}{B_g} = \frac{3.023}{2.134} = 1.42$$

$$\frac{L}{B_g} = \frac{18.29}{2.134} = 8.57$$

From Fig. 6.30, $N_c^* = 8.75$. From Eqns 6.49,

$$\sum Q_u = L_g B_g c_{u(p)} N_c^* + \sum 2(L_g + B_g) c_u \Delta L$$

$$= (3.023)\,(2.134)\,(85.1)\,(8.75) + (2)\,(3.023 + 2.134)\,[(50.3)\,(4.57) + (85.1)\,(13.72)]$$

$$= 19217 \text{ kN}$$

Hence, $\Sigma Q_u = 14011$ kN.

$$\sum Q_{all} = \frac{14,011}{FS} = \frac{14,011}{4} \approx 3503\ kN$$

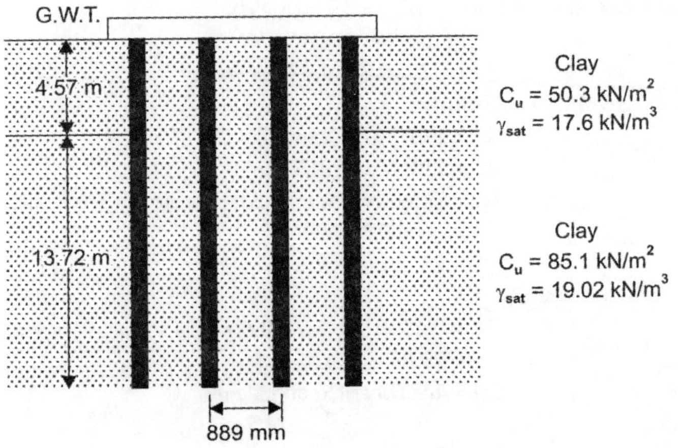

Fig. 6.31: Group pile of layered saturated clay

Example 6.10: Design of Piles for Patanjali IT Park

Solution: Author was involved for design of this pile foundation for a 8 storeyed building in Patanjali IT Park Haridwar.

The soil strata is as follows:

a. WT-2 m (which can rise upto 1m in monsoon month)
b. Soil report (Sep 2023)

Depth (m)	*N-Values*	*Soil Strata*
1.5	7	Sand 39%, Fines 61%, $c = 0$, $\Phi = 30°$ Non plastic
3.0	10	Sand 45%, Fines 55%, $c = 0$, $\Phi = 30°$ Non plastic
4.5	11	Sand 65%, Fines 35%, $c = 0$, $\Phi = 30°$ Non plastic
6.0	12	Sand 71%, Fines 29% Non plastic
7.5	12	Sand 66%, Fines 34% Non plastic

Beyond this depth, details of soil strata is not available.

Let's assume that the strata of 7.5 m depth continues upto 30 m depth or so.

Thus, design parameters for piles could be as follows:

$\Phi = 30°$, $c = 0$, Pile cut off level = 1.5 m from NGL because strata is purely Non-cohesive, also, space is required for construction of raft.

$$\text{Let pile dia} = 600 \text{ mm},$$

and $$\text{Length} = 20 \text{ m}$$

(The higher dia and length are adopted as the structure proposed is 8 storeyed. The ultimate load capacity (Q_u) of piles is given as follows (by IS:2911, Part 1/Sec 2:2010)

$$Q_u = A_p(0.5D\cdot\gamma N\gamma + P_D N_q) + \Sigma_{i=1}^{n} K_i\, P_{Di} \tan\delta_i \cdot A_{si} \qquad ...(1)$$

Here, $A_p = 0.786\ (0.6)^2 = 0.28 \text{ m}^2$, $A_{si} = \pi DL = 37.68 \text{ m}^2$, $D = 0.6$ m

γ = Effective unit wt. of soil at pile tip

$= 1.65 \times 1.10 - 1 = 0.8\ t/\text{m}^3$

$N\gamma$ & N_q = BC factors as per Φ value

Take, $\delta = \varnothing$

Effective overburden pressure at pile tip (P_D) = 15 × pile dia

$= 15 \times 06 = 9$ m

Thus, $P_D = \gamma Z = 0.8 \times 9 = 7.2\ t/\text{m}^2$

For conservative design, take Φ as follows:

$\Phi' = \tan^{-1}(0.67 \tan 30) = 21°$ (Say 25°)

For $\Phi' = 25°$, $N_q = 10$, $N\gamma = 10$

P_{Di} = Effective overburden pressure for i^{th} layer

$= P_D = 7.2\ t/\text{m}^2$

$Ki = 1$

Substituting these values in Eq (I),

$$Qu = 0.28\ [0.5 \times 0.6 \times 0.8 \times 10 + 7.2 \times 10] + 1 \times 7.2 \times \tan(25) \times 37.68$$
$$= 20.83 + 126.5$$

$\therefore$ $$Qu = 147.33\ t$$

Taking a FOS of 2.5,

$$(Q)_a = 58.9 \text{ Say } 60t$$

Notes:

i. The cut off level can be 1.5 m. At that depth, RCC raft can be laid.

ii. Bored cast insitu RCC pile of 600mm dia and 20m length is recommended to attain safe load capacity of 60T with a FOS of 2.5.

iii. It is recommended to carryout a pile load test (after 28 days of curing) under the supervision of geotechnical and structural experts. The test should be conducted by a competent agency, as per IS:2911 (Part-4):2013.

iv. The above design has been done on the basis of soil report (prepared by M/s SMEC, Roorkee), provided by Patanjali. This investigation was done only upto 10m depth. As advised by Patanjali officials, same strata has been assumed upto 20m depth for pile design.

v. After making the bore hole of design depth, PCC of about 8" thickness should be poured (through tremie) and it should be allowed to set. After that only, the reinforcement cage should be lowered and concreting be done (strictly through tremie pipe only)

vi. The conc mix for pile should be M35 or M30. The structure engineer may take final decision after load test also.

vii. Care should be taken that bore hole does not collapse at all. Therefore, either casing pipe or bentonite (or any other effective chemical) must be used for pile casting. Contractor may also decide if any dewatering system is to be used.

Modified Pile Capacity After Arrival of Refusal at 15.5 m Depth from NGL (21 May 2024), as Reported by Patanjali Officials

In Equation (1) the first term is for end bearing, i.e.

$$A_p(0.5\, D{\cdot}\gamma N\gamma + P_D N_q) \quad ...(2)$$

As reported by Mr. Shiva, V.P. (Constn), the refusal is obtained at about 15.50 m depth from NGL.

For refusal, $N(SPT)$ are > 30

Hence as per Fig. 6.32, for $(N)SPT$ values higher than 30, the Ø – value is 36° or more.

For $Ø = 36°$, $N_q \approx 40$ and $N\gamma = 50$ (Table 6.12)

Hence, from Eq (2)

For (15.5 – 3) = 12.5 m long pile, $P_D = 12.5 \times 0.6 = 7.5t/\text{sqm}$ (where, 3 m is cut off level)

$$= A_p(0.5\, D{\cdot}\gamma N\gamma + P_D N_q)$$
$$= 0.28\,[0.5 \times 0.6 \times 0.8 \times 50 + 7.5 \times 40]$$
$$= 87.4\, t \text{ say } 87$$

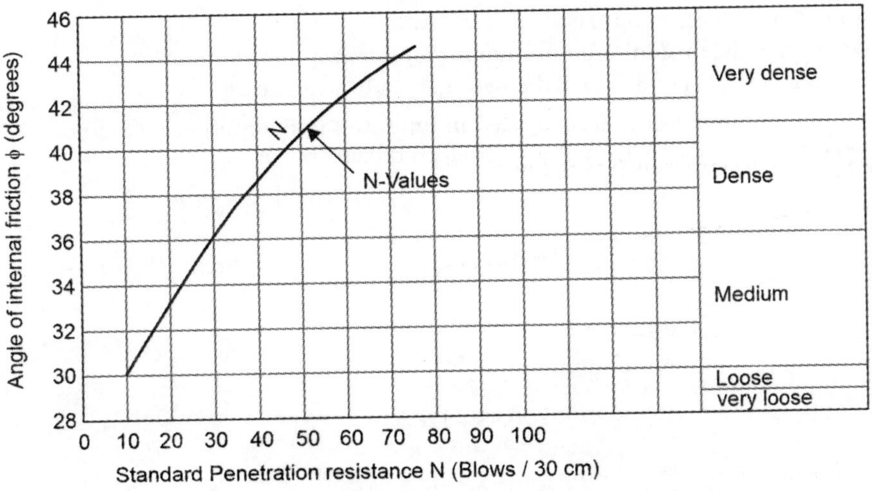

Fig. 6.32: Relationship between Ø and N

Table 6.12: Bearing Capacity Factors

ϕ(Degrees)	N_c	N_q	$N\gamma$
0	5.14	1.00	0.00
5	6.49	1.57	0.45
10	8.35	2.47	1.22
15	10.98	3.94	2.65
20	14.83	6.40	5.39
25	20.72	10.66	10.88
30	30.14	18.40	22.40
35	46.12	33.30	48.03
40	75.31	64.20	109.41
45	138.88	134.88	271.76
50	266.89	319.07	762.89

In the second term of eqn (1)

$$P_{Di} = 7.5, A_{si} = \pi \times 0.6 \times 12.5 = 23.55$$

Hence, 2nd term is computed as follows

$$K_i\, P_{Di} \tan \delta_i\, A_{si} \quad ...(3)$$

$$= 1 \times 7.5 \times \tan (0.67 \times 36) \times 23.55$$

$$= 82\ t$$

Total (Qu) with refusal $= 87.0 + 82 = 169\ t$

$$Q_a = 169/2.5 = 67.6\ t (\text{say } 68\ t)$$

Notes:

i. The above capacity of pile is with arrival of refusal at 15.5 m depth; 3 m cut off and soil Strata as provided by Patanjali officials.

ii. The pile load test should be conducted at 28 days of curing. If pile was casted on 18 May 2024 at IT Park, it should be tested on June 15, 2024.

iii. The loading Kentledge should be designed for $70 \times 3 \approx 210\ t$ capacity (minimum) say 225 t.

iv. The load test be conducted as per IS 2911 (Part 4):2013.

v. The safe vertical load on pile shall be as follows:
   - Two third of the final load at which total displacement is 12 mm.
   - 50% of final load at which total displacement is 10% of pile dia.

Important: The pile load test was conducted in June 2024 in the presence of author. The pile load capacity was worked out as 69t, exactly same as predicted above.

Chapter

7

Introduction of Spliced Piles in India

(Contribuited by Er AK Sharma, CBRI, Roorkee)

7.1 INTRODUCTION

Deep deposits of soft saturated clays are usually found in coastal areas, around lakes, ports and harbours. In such conditions there is a practice to provide foundation either of bored piles or driven cast-*in situ* piles up to the firm stratum. This is reported that in several situations bored concrete piles may suffer from several defects. These defects are not always on account of material and mixes or ignorance of quality control. It is also reported that the faults at time lies in driving piling system which may not be well-suited to highly compressible strata such as soft clays and this can be sorted out by providing pre-cast driven piles. Driving of long piles in a single length becomes inconvenient and expensive. Pre-cast piles in smaller segments driven and jointed together with a suitable joint may provide cost-effective and efficient alternative for foundation in deep deposit of marine clays. This pile foundation in soft deposits also generates negative drag on piles. This chapter deals with preliminary introduction of different types of splicing system to develop a pile foundation suitable for soft clay deposits. The results obtained from some studies carried out at CBRI Roorkee by using spliced piles are also reported in brief.

Geotechnical Engineers are often faced with the problem of designing foundations for difficult and problematic sub-soil conditions. These are often found at filled up sites, low-lying water-logged waste lands, saturated fine to medium silty sand deposits and creak lands having deep deposits of soft saturated marine clays with very low shear strength and high compressibility up to a great depth. The conventional practices of providing foundation in such deep deposits are either to use driven cast-*in situ* or bored cast-*in situ* piles. Many a times the deeper depth piles suffer from several defects (Mohan, et al. 1978, Bhandari, et al. 1982). A number of structures in the port areas are supported either on deep cast-*in situ* bored piles or driven cast-*in situ* piles. The unsatisfactory performance of these structures, to a large extent, is attributed to questionable integrity of these piles. The known defects thoroughly documented by Bhandari, et al. 1982, are honey-combing of concrete along pile shaft, excessive necking, unacceptable positional deviation and defective pile toe. These defects are not always on account of material and mixes or laxity in quality control, but at time fault lies in selecting piling system, which may not be well-suited to highly compressible soft marine clay deposits. In such situations precast driven piles provide an alternative solution. But driving of pre-cast piles of longer length (end bearing or friction piles) in one length becomes inconvenient and costly due to the fact that they need heavy rigs to drive piles. Hence, pre-cast piles having smaller length, driven in segment and jointing together with a suitable jointing system may provide efficient and cost-effective solution to such situations and use of spliced piles reported by other researchers (Bredberg, et al 1979, Bruce, RN, et al. 1974, Brenner, RP, et al. 1979) are excellent example.

This chapter describes in brief the findings of the study carried out at CBRI on development of different type of joints (Bhandari, RK. Jain, MP, and Sharma, AK-1989) for the use in pre-cast piles under load with respect to intact pile. The efficacy of spliced piles carried out at Creek land Bombay is also reported.

7.2 DEVELOPMENT OF PILE JOINTS

Four types of joints analogous to Hercules, ABB, Coupler and Square (Bruce and David 1974 and Bredenberg and Broms 1979) were selected for the study. These joints thereafter called as splice - A, *splice - B, Splice - C* and *Splice - D* respectively. The splice A, B and C are shown in Fig. 7.1 and splice C and D (line sketch) are shown in Figs 7.2 and 7.3 respectively. The male and female part of splice -B is identical. The male and female part of splice - C is circular in shape and has a groove along the periphery and the two halves are connected by two pieces of connector rings (Fig. 7.2. The connector ring in two halves projected at the ends having same dimensions equal to the depth of the grooves in two connecting plates. The detail of various components of spliced - D, male, female parts and locking key is shown in Fig. 7.3. The jointing of splice with reinforcement is shown in Fig. 7.4.

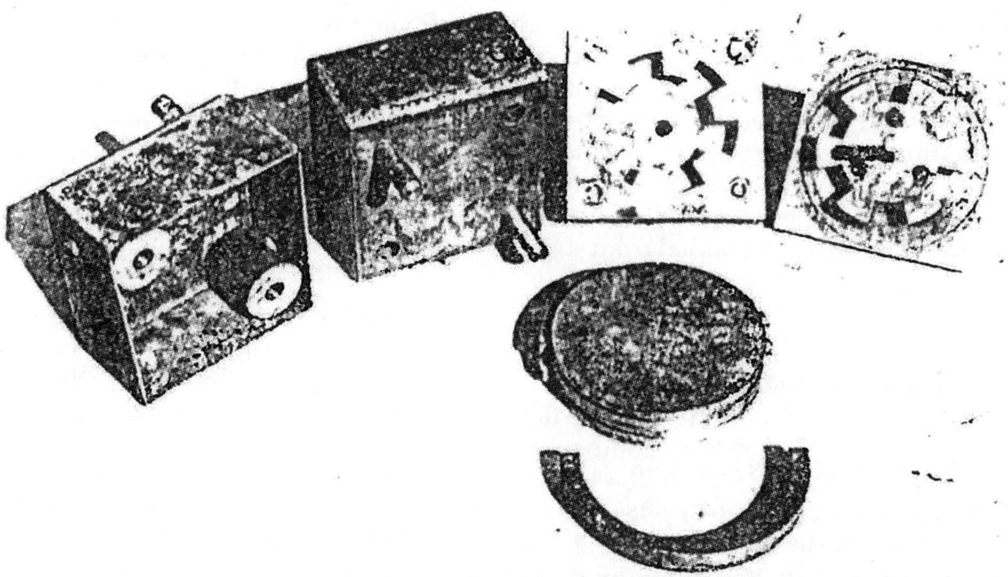

Fig. 7.1: Detail of splices developed at CBRI

7.3 PROCEDURE OF CASTING PILES

The pile segments were cast in 20 × 20 cm^2 square sizes having 1.5 m length each. After joining segments together the pile length become 3 m. Five pre-cast piles with each splice were cast. In addition to spliced piles, five intact piles having no joint same cross-section and length were also cast for comparative studies. Welded splice with reinforcement is shown in Fig. 7.5 for rechecking the proper fitting of joints and its alignment prior to casting of pile. The piles were then cast in wooden formwork, prepared from timber for the purpose of casting piles segments. The details of piles, length, and number of segments are provided in Tables 7.1 and 7.2 for test site I and site II respectively.

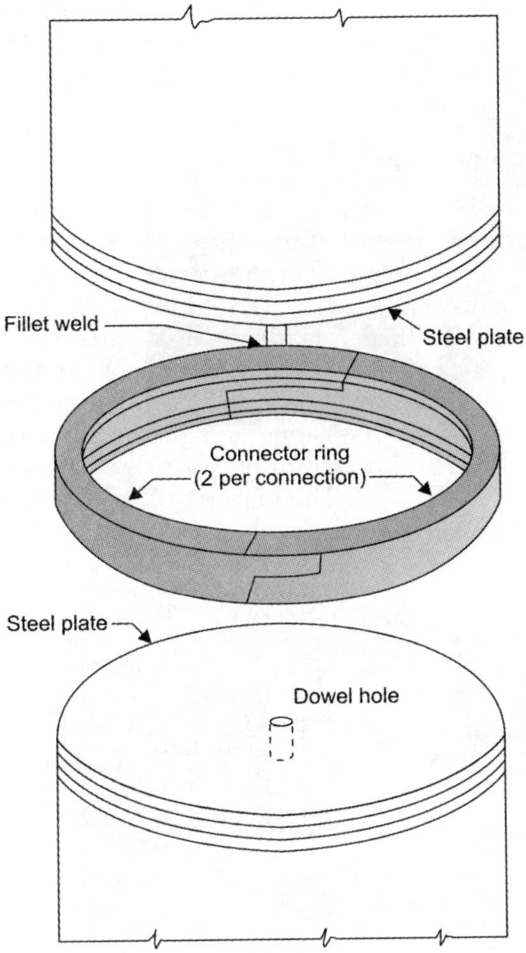

Fig. 7.2: Component of splice - C

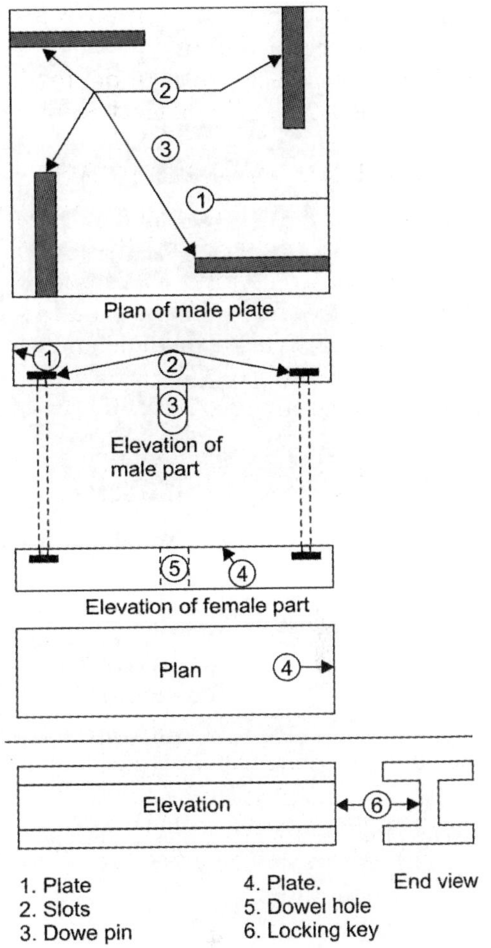

Fig. 7.3: Component of splice - D

7.4 BENDING TEST ON PRE-CAST SPLICED PILES

Bending test on spliced piles 3 m long having two segments jointed together with a joint in the center were conducted in laboratory. Apart from it the bending test on intact pile having same cross-section and length were also conducted (Fig. 7.6).

Table 7.1: Pile details at site I

Pile length (m)	*No. of segments*	*Splices used*
3	–	Intact pile
3	2	Splice - A
3	2	Splice - B
3	2	Splice - C
3	2	Splice - D

* Cross-section of piles used is 200 mm × 200 mm

* Length of each segment is 1.5 m.

Fig. 7.4: Splice joint with reinforcement cage

Table 7.2: Pile details at site II

Pile length (m)	*Length of each segment (m)*	*No. of segments*	*Remarks*
6	6.0	–	Intact pile
6	3.0	2	Splice pile
6	2.0	3	Splice pile
6	1.5	4	Splice pile
6	1.5	4	Splice pile in group of 4

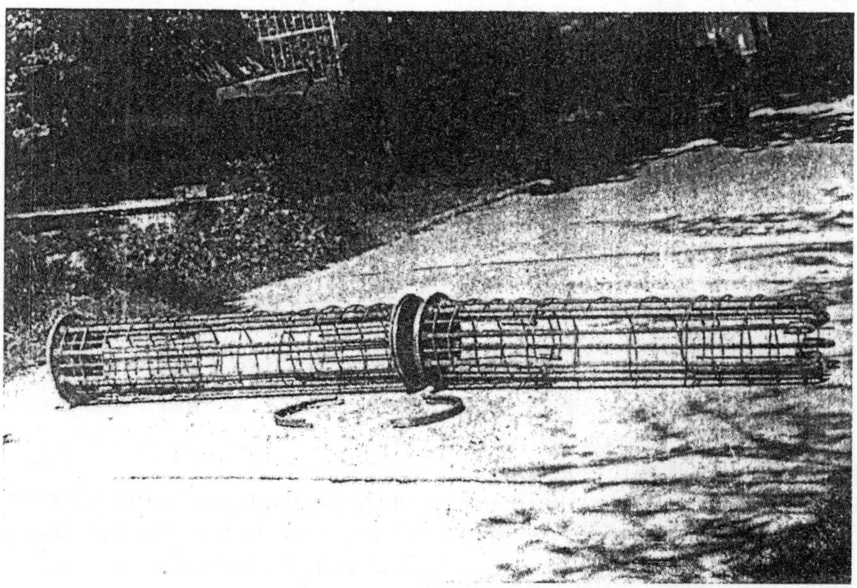

Fig. 7.5: Alignment check of splice-C before casting

Fig. 7.6: Casting of splice pile in wooden formwork

The load-deflection relationship lot along with theoretical predicted curve is presented in Fig. 7.7. The modulus of elasticity of the pile material was 2.8 × 105 kg/cm^2). The study of shows that the bending load for spliced piles varies between 53–89% of load taken by intact pile at a deflection varied from 8 to 20 mm. It was found that even the minimum bending load observed in splice - B that is 53–60% is sufficient to support actual bending stresses subjected to these piles during their service period under structure.

7.5 SUB-SOIL CONDITION

Field studies were conducted at two different sites *viz.* (i) Site I, having predominantly cohesionless soil deposits and (ii) Site - II having soft marine clay deposits. At both sites, detailed sub-soil investigations were carried out. It consisted of boring with standard penetration test including undisturbed soil sampling at an interval of 75 cm alternatively and dynamic cone penetration tests. These tests were carried out in accordance with IS: 213 1–1981 and IS: 4968 (Part - 1)-1976. The test results at site I are presented in Fig. 7.8.

The study of borehole indicated 4.5 m thick sandy silt layer (SM) deposit and further it revealed a layer of clay deposit with medium plasticity (CI) up to 6 m, the depth at which the borehole was terminated. The water table was found at 4 m in the month of November when the investigation was carried out.

The average N-value was observed 4 up to 4 m depth and 8 between 4–6 m depth increasing approximately at constant rate. The dynamic cone penetration test values also corroborate with N-value (Fig. 7.9). The sub-soil strata was found loose silty sand deposit (SM) having angle of shearing resistance (∅) 27°, obtained from triaxial test carried out on samples collected from bore-hole.

Similarly, boring, sampling, SPT, dynamic cone penetration test, vane shear test as per IS: 2131–1981, IS: 4968-(Pt. I-1976 and IS: 4434–1978 at site 11 were carried out. The test results are presenteed in Fig. 7.10. The study of borehole indicated the sub-soil from 1.0 to 5 m depth is soft marine clay underlain by a filled up layer of 1.0 m. The soft marine clay is followed by poorly graded sand deposit (SP) between 5 to 15 m depths.

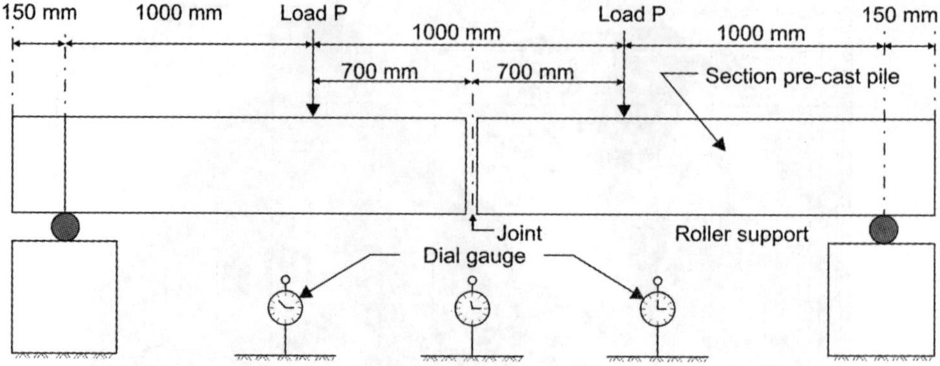

Set-up for bending test

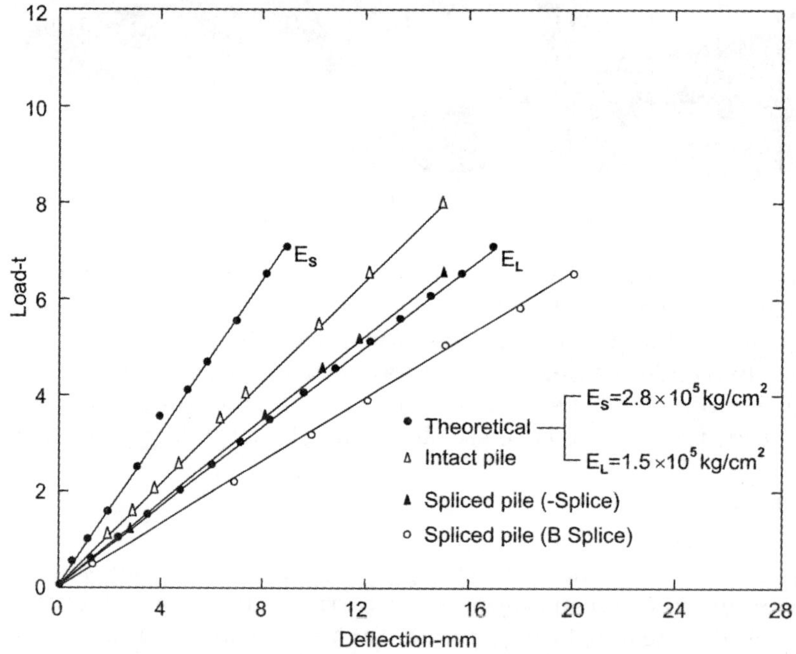

Fig. 7.7: Load deflection curve in bending test (mid-span)

The N_s value varies between 4 to 10 from 2 to 4 m depth beyond it shows a decreasing trend up to 6 m depth the value at this depth is 8 afterwards it again followed increasing trend up to 8 m depth beyond which it almost remained constant at 15 to 16.75 m depth of exploration. The natural moisture content varies from 49 to 64% and bulk density 1.34 to 1.36 t/cum. The liquid limit ranges from 94% to 96%, PI 60 to 61 %. The value of cohesion was found between 0.05 to 0.08 kg/cm^2 with angle of shearing resistance (ϕ) 2°. The high moisture content closer to liquid limit indicates the reason of low shearing strength.

7.6 DRIVING OF PILES

Precast piles were driven to compare the results of intact and spliced piles at both the sites. The driving of piles were carried out by dropping the hammer weighing 250 kg with free fall of 1 m height with the help of mechanical winch. During driving the precast spliced piles no damage was observed except minor spalling of surfacial

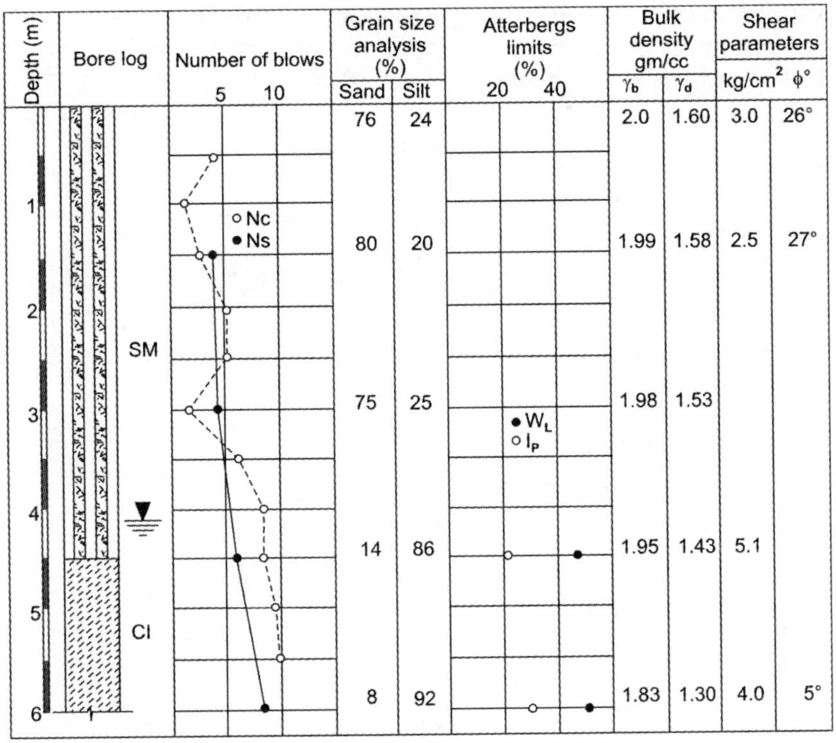

Nc Number of blows in dynamic cone penetration test
Ns Number of blows in standard cone penetration test
Test carried out in November 1988
as obtained from triaxial shear test

Fig. 7.8: Bore-log and other sub-soil characteristics (Site I)

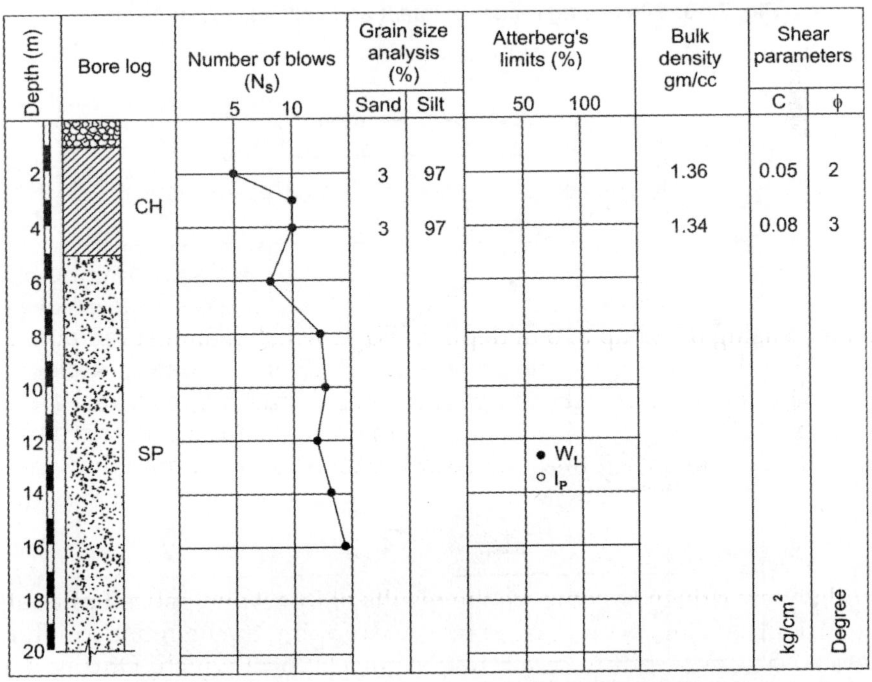

Fig. 7.9: Bore-log and other sub-soil characteristics (Site II)

concrete from the corner that may be due to excessive stress concentration at corners at the time of hammering. The pile penetration records of precast spliced piles installed in fields at site I and site II are shown in Figs 7.10 and 7.11.

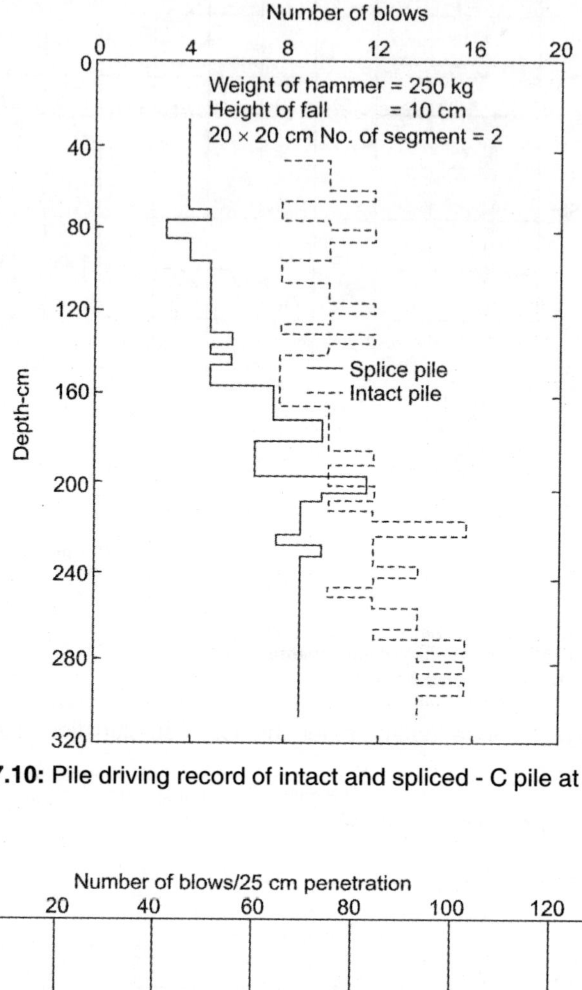

Fig. 7.10: Pile driving record of intact and spliced - C pile at site I

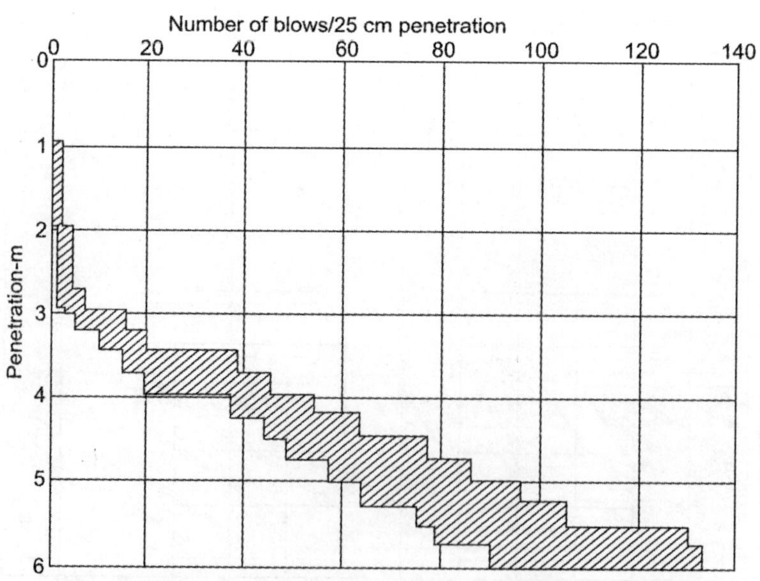

Fig. 7.11: Pile driving record of intact and spliced pile at site II

7.7 LOAD TEST ON PILES

7.7.1 Compressive Load Test

The load test on spliced and intact pile was conducted under vertical compressive loading and displacements were recorded using four dial gauges having 0.01 mm least count. The load-settlement plots have been presented in Figs 7.12 and 7.13 respectively for both the sites (site I and site II).

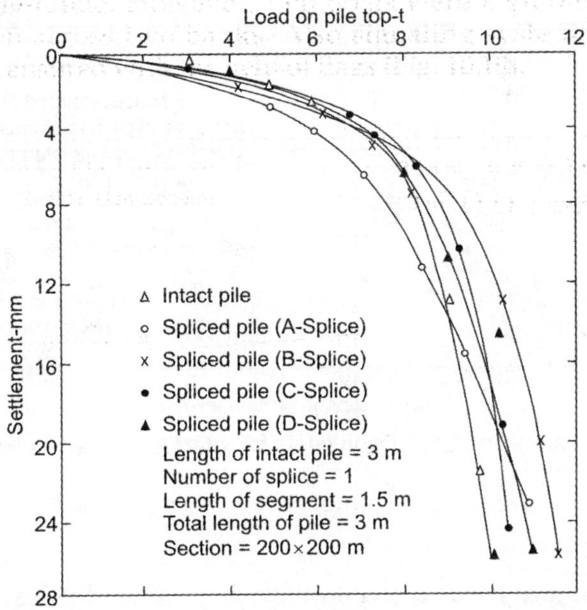

Fig. 7.12: Load settlement curve under compressive load at site I

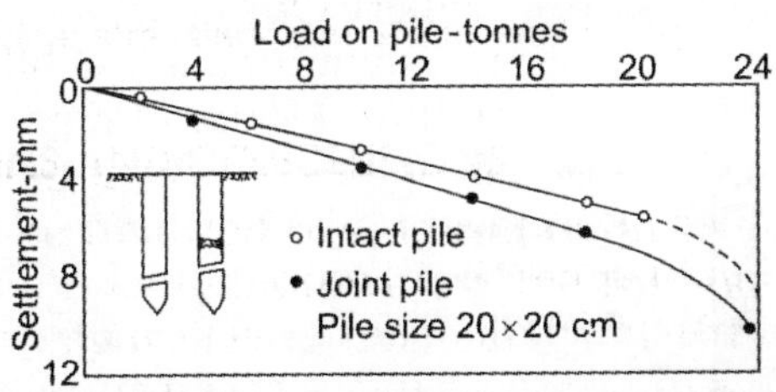

Load settlement curve of intact and spliced pile

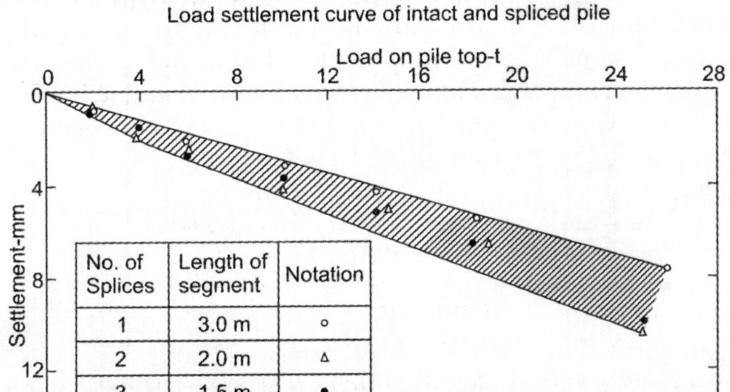

Fig. 7.13: Load settlement curve for intact and spliced piles using 1, 2 and 3 segments

7.7.2 Lateral Load Test

Lateral load tests were conducted on spliced piles having splice-A, splice-C and splice-D including intact piles. The plots of lateral load versus deflection curves are presented in Fig. 7.14.

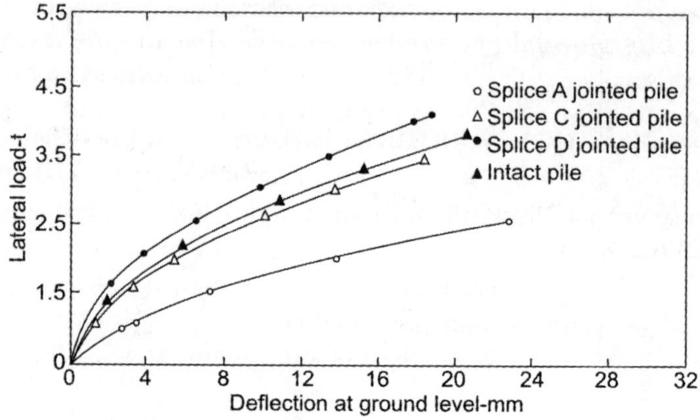

Fig. 7.14: Lateral load deflection curve for intact and spliced pile at site I

7.7.3 Pull Out Tests

Pullout load tests at site I was also conducted on spliced piles with splice-A and splice-C along with intact pile. The pullout load versus pullout movement curve is presented in Fig. 7.15.

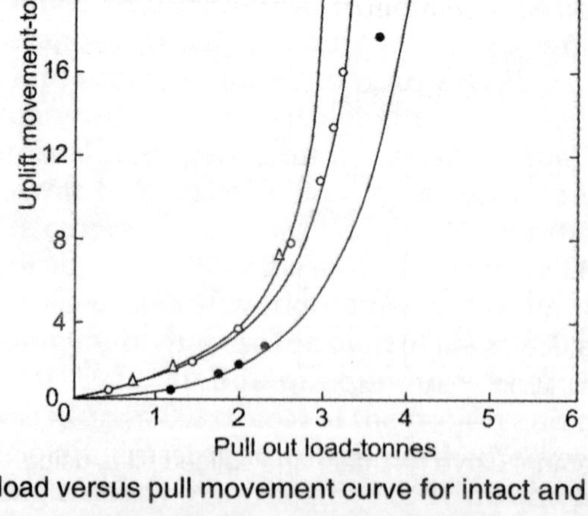

Fig. 7.15: Pull-out load versus pull movement curve for intact and spliced pile at site I

7.8 SUMMARY

Base on experimental results carried out at four different splices at two different sites in the form of prototype study the results are summarised as below:

- Study of load settlement curve (Fig. 7.12) of spliced and intact piles indicated the load of 8 to 9.3 tonnes corresponding to 12 mm settlement and 9.5 to 10.5 tonnes corresponding to 1/10th of pile diameter, i.e. 20 mm according to IS: 2911 (Pt. IV)-1985. The corresponding safe loads from both the criteria varied between 4.75 to 6.0 tonnes. The ultimate load of 7 tonnes was also computed considering $\phi = 27°$ on the basis of soil properties and static formula for cohesionless soils according to IS: 2911 (Pt. I/Sec 4)-1984. The corresponding settlement at 7.00 tonnes load was found varying from 4 to 7.5 mm. It is therefore concluded that the performance of spliced piles using joints developed at CBRI Roorkee with respect to intact pile were found practically same on the basis of the load test results.
- Study of load settlement curve (Fig. 7.12) of Site II of spliced and intact piles indicated the load of 24.7 to 26.2 tonnes corresponding to 12 mm settlement and 24.7 to 27.7 tonnes corresponding to settlement 1/10th of pile diameter, i.e. 20 mm according to IS: 2911 (Pt IV)-1985. The safe load from both the criteria varied between 16.8 to 18.8 tonnes. The ultimate load computed from soil properties using static formula for cohesive soils was obtained as 17.5 tonnes according to IS: 2911 (Part I/Sec. 4)-1984 and corresponding settlement 5 mm to 7.9 mm.
- Study of lateral load versus deflection curve Fig. 7.14 of spliced and intact piles varied between 1.25 to 1.75 tonnes respectively corresponding to 50% of load at 12 mm displacement and 1.6 to 2 tonnes corresponding to displacement of 5 mm according to Indian Standard Code of Practice. Hence, it is concluded that the behaviour of spliced piles in comparison with intact pile is practically same in lateral loading.
- Study of pullout load versus pullout movement (Fig. 7.15) of spliced and intact pile indicated 2.75 to 3.5 tonnes load at 12 mm movement and 2.5 to 4 tonnes at the load in movement curve shows the clear break. The corresponding safe loads vary between 1.75 to 2.3 tonnes from both the criteria. After pulling out of spliced piles completely the joint do not shown any adverse affect.
- The bending test results of spliced piles were found varying from 53–89% in comparison to intact pile. These tests were carried out in extreme bending load condition but in actual field conditions this situation may not be encountered. Hence, the bending load as obtained is sufficient to sustain bending stresses actually developed during driving of spliced piles. However, it is desirable to mention that the splice is to be placed at such a depth where bending stresses will be minimum.
- The development of spliced pile system at CBRI Roorkee is found cost-effective and easier to drive in soft clay deposits using light driving rigs of lower height in place of using driven cast *in situ* or bored cast *in situ* piles, to avoid defect in such piles. Hence, spliced piles may be a good solution for deep soft saturated marine clay deposit.

REFERENCES

1. Bhandari, RK; Prakash, C, and Sharma, A. (1988): 'Failures in Cast-in-Place-Piles', Commorative International Conference Maxican Soil Mech. Society, 25th Anniversary of its Foundation, Maxico August.
2. Bhandari, RK; Jain, MP and Shanna, AK (1989): 'An Improved Device for Joining Precast Piles in Segments', Indian Patent No. 165155 to 165158.

3. Bredenberg, H, and Broms, BB (1979): 'Joints Used in Sweden for Precast Concrete Piles, Recent Development in the Design and Construction of Piles', Proc Conference held at Institute of Electrical Engineers, 21–22 March. The Institution of Civil Engineers, London, pp. 11–22.
4. Bruce, RN, and David, CH (1974): 'Splicing of Precast Prestressed Concerete Piles', Part I- Review and Performance of Splices', Jr. of Prestressed Concrete Institute, Vol. 19, No.1, Sept.– Oct., pp. 70–97.
5. Bruce, RN, and David, CH (1974): 'Splicing of Precast Prestressed Concrete Piles' , Part II - Tests and Analysis of Cement–Dowel splice', Jr. of Prestressed Concrete Institute, Vol. 19, No. 6, Nov.- Dec., pp. 40–66.
6. Brenner, RP, et al. (1979): 'Proc. Pressure from pile driving in Bangkok clay'. Proc. 6th ARC, SMFE, Vol. 1, p. 133.
7. IS: 2131–1981: 'Indian Standard, Method for Standard Penetration Test for Soils', ISI, New Delhi.
8. IS: 4968 (Part-1) - 1976: 'Indian Standard Method for Sub-surface Sounding for Soils - Part - 1, Dynamic Method Using Cone and without Bentonite Slurry', ISI, New Delhi.
9. IS: 2911 (Part-IV) - 1979: 'Indian Standard Code of Practice for Design and Construction of Piles Foundations, Part - 12 Load Test on Piles', ISI, New Delhi.
10. IS: 2911 (Part-II Sec. 4) - 1989: 'Code of Practice for Design and Construction of Pile Foundations, Section 4, Bored Precast Concrete Piles', ISI, New Delhi.
11. Mohan, D; Jain, GRS, and Bhandari, RK. (1978): 'Remedial Underpinning of Steel Tank Foundation', Journal of Geotechnical Engineering Division, ASCE, Vol. 104, No. GTS, Proc. Paper 3756, Italy, pp. 639–655.

Chapter

8

Pile Construction Equipments and Construction Technology

8.1 INTRODUCTION

There are several equipments used for construction of pile foundations manually. With advent of new machines and equipments, some new equipments are also developed for speedy construction of piles. The important amongst them is vibrofloat. The latter is used where time is the main factor because it ensures the most quicker way to construct the piles. It also strengthens the surrounding weak natural soils up to appreciable depths. It is very much suitable for soils ranging from soft organic and alluvial clays to gravels and coarse rubble fills. The types of structure supported on foundation vary from single storey to several storeys and from traditional brick construction to industrial plant and oil tanks. This method is used in order to limit the settlement resulting from vertical loads by compaction of loosely packed granular soils by vibration.

The vibrofloat produces amplitudes and maintains maximum amplitude and lateral reaction at the tip for a given power. With regard to the damping effect of the soil. The larger the power and amplitude of the interactions, the larger is the radius of effect.

Nevertheless, vibrofloat is a quick way of piles construction and arrive at the solution but its running cost is high. It requires high capacity compressor, a crane, high density hydraulic oil, regular water supply, etc. The initial investments to meet out such features are high and hence mostly the mechanical means of construction of pile foundations are used.

The general equipments required for pile construction are:

a. The auger
b. Under-reamer
c. Boring guide
d. General tools like cutting tool extension rods and general tools and plants.

The above set of equipments is available for 20, 25, 30, 37.5, 40, 45 and 50 cm stem diameter piles with under-reamed diameter ratio varying between 2 and 3. However, the tools are also available for piles of 55 and 60 cm stem diameter. The extension rods are available in 1 m, 1.5 m or 2.0 m lengths.

8.2 TRIPOD

For the piles longer than 3.5 m and/or of the diameter larger than 37.5 cm stem diameter, a tripod is required. The tripod and its general components are shown in Fig. 8.1. The tripod is mounted with a winch operated manually or motorised.

The dimensions given in the Fig. 8.1 may vary depending upon the requirements on site. The tripod is a portable assembly and can be assembled or de-assembled at site.

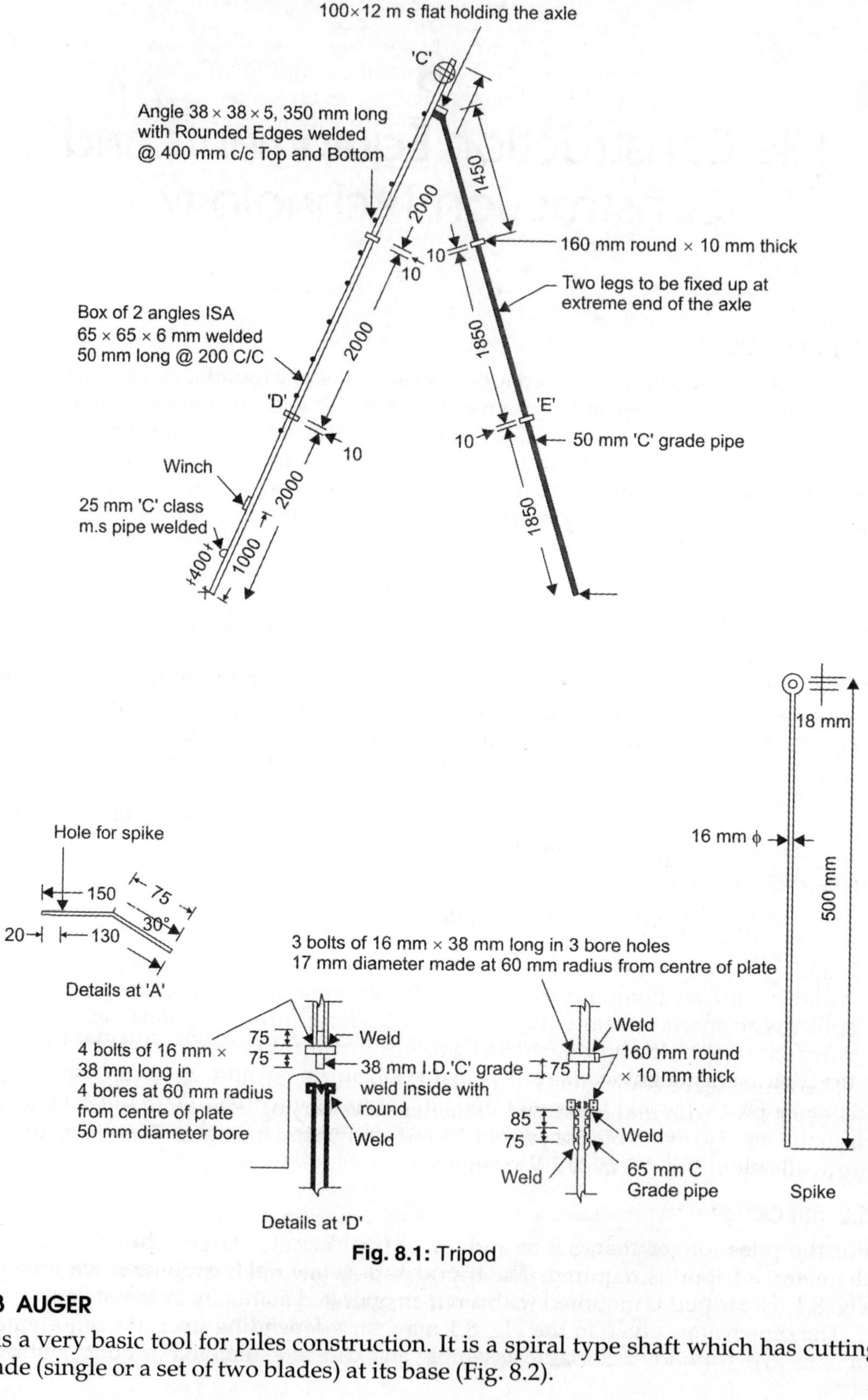

Fig. 8.1: Tripod

8.3 AUGER

It is a very basic tool for piles construction. It is a spiral type shaft which has cutting blade (single or a set of two blades) at its base (Fig. 8.2).

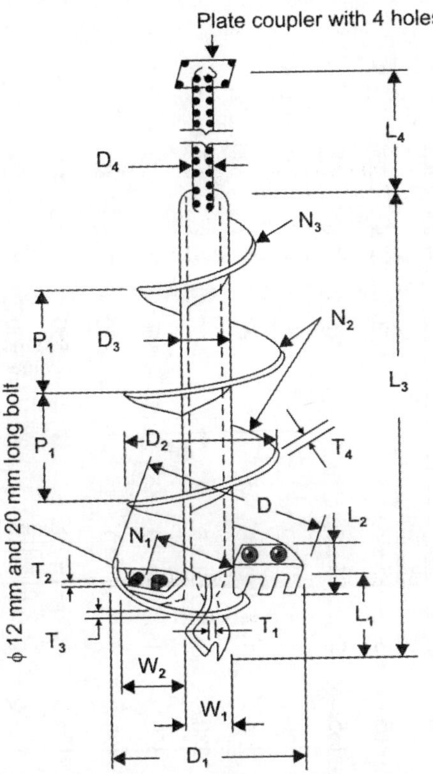

Fig. 8.2: Auger (Spiral type)

It has a flange on its top to be connected with extension rods. The various sizes of auger are available for making the bore holes. IS: 10442–1983 recomrnends the various sizes as given in the Table. 8.1.

The auger can be used for almost all types of soils except thickly gravelly soils. For such soils, a hard cutting blade is available which is toothed one. This blade is connected with the auger and thus work goes on. For large depths, the Auger has to be guided by a boring guide.

8.4 AUGER BORING GUIDE

It ensures an auger to remain straightened during its operation. It is very helpful tool while driving an auger. It has four legs which are supported and kept in position by spikes. The specially made flaps are used to allow the auger rod to pass through it. After the auger is lowered to a small depth, the boring guide is fixed. Figure 8.3 shows the typical boring guide.

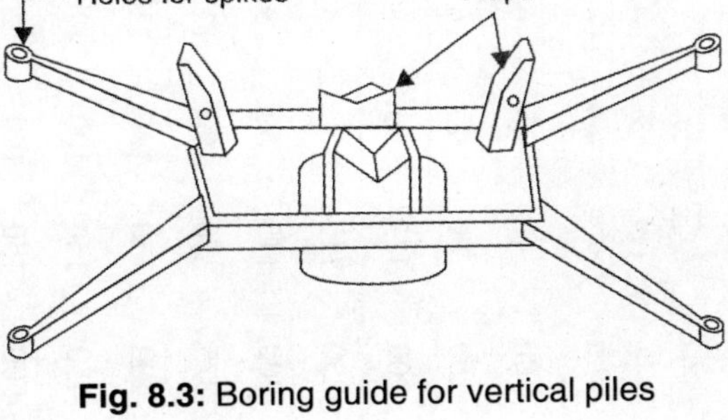

Fig. 8.3: Boring guide for vertical piles

Table 8.1: Sizes, dimensions and tolerances of anger (After IS: 10442–1983)

Nominal	*Pilot bits*			*Blades*				*Base plate*		*Spirals*					*Shaft*			
size D	$L_{1\,-3}^{\,+5}$	$W_{1\,-3}^{\,+5}$	$T_{1\,-1}^{\,+1}$	$L_{2\,-3}^{\,+5}$	$W_{2\,-3}^{\,+5}$	$T_{2\,-1}^{\,+1}$	N_1	$D_{1\,-4}^{\,+5}$	$T_{3\,-1}^{\,+1}$	$D_{2\,-4}^{\,+5}$	$T_{4\,-0.5}^{\,+1}$	P_1	N_2	N_3	$L_3 \pm 5$	$D_3 \pm 0.5$	$L_4 \pm 5$	$D_4 \pm 0.5$
(1)	(2)	(3)	(4)	(5)	(6)	(7)	(8)	(9)	(10)	(11)	(12)	(13)	(14)	(15)	(16)	(17)	(18)	(19)
mm	mm	mm	mm	mm	mm	mm	mm	mm	mm	mm	mm	mm	non	non	mm	mm	mm	mm
100	60	25	6	50	40.0	6	1	90	8	85	3	85 ± 5	3	1	250	26.9	750	26.9
150	90	40	6	85	57.5	6	1	140	8	135	3	85 ± 5	3	1	300	33.8	700	33.8
200	115	50	8	115	77.5	8	1	185	8	150	3	110 ± 5	3	1	400	42.5	600	42.5
250	115	50	8	115	102.5	8	1	235	8	230	3	110 ± 5	3	1	400	42.5	600	42.5
300	115	50	8	125	127.5	8	1	285	8	280	3	110 ± 5	3	1	400	42.5	600	42.5
375	150	100	12	150	147.5	10	2	360	10	350	4	165 ± 10	2	1	550	76.0	450	42.5
400	150	100	12	150	160.0	10	2	385	10	375	4	165 ± 10	2	1	550	76.0	450	42.5
450	150	100	12	150	185.0	10	2	435	12	425	4	165 ± 10	2	1	550	76.0	450	48.4
500	150	100	12	150	210.0	10	2	485	12	475	4	165 ± 10	2	1	550	76.0	450	48.4
550	150	100	12	150	235.0	10	2	535	12	520	4	200 ± 10	2	1	650	88.7	350	60.2
600	150	100	12	150	260.0	10	2	585	12	570	4	200 ± 10	2	1	650	88.7	350	60.2

For construction of batter piles, the design of boring guide is slightly changed. In such cases, the frame can be adjusted up to an angle of 30° with the horizontal Fig. 8.4 shows the boring guide for batter pile.

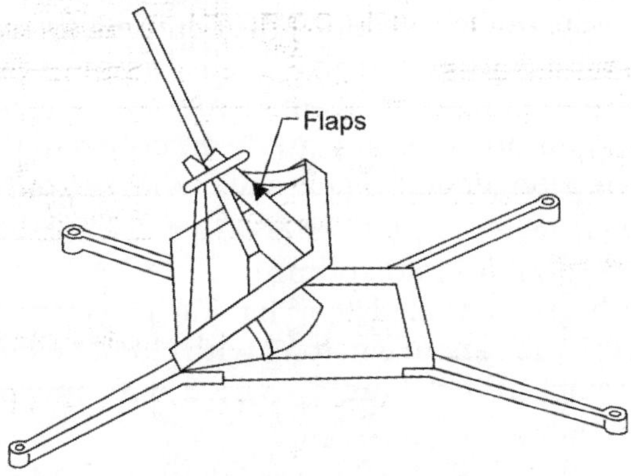

Fig. 8.4: Boring guide for battered piles

8.5 UNDER REAMER

It is the only tool available today to make the under-reams (bulbs) at various depths below ground level. After the hole is made by auger, the under-reamer is lowered into the hole and kept resting at desired depth where the bulb has to be made. It is rotated at its place. The blades of under-reamer gets widen through pressure and they cut the soil from sides due to rotations. A bucket is attached beforehand below the under-reamer (Fig. 8.5) which collects the soil due to cutting. When the bucket is filled, the under-reamer is pulled up and the bucket is cleaned. The under-reamer is lowered again and the process is repeated.

In case of multi under-reamed pile, the uppermost bulb is formed first and then by the same process other bulbs are made.

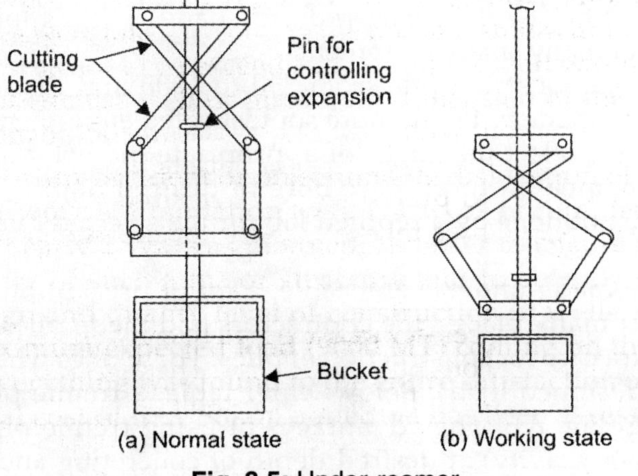

Fig. 8.5: Under-reamer

After making all the bulbs, the hole is cleaned and the reinforcement is lowered into it. The hole is then concreted and the pile is made.

For stiffer and dense soils the blades of the under-reamer are strengthened suitably to withstand the thrust exerted by neighbouring soil.

The equipments described hitherto are used for construction above the water table. For construction of piles below water table, the equipments listed above may also be used but some more equipments are also used. In case of sandy soils where the borehole is stabilized by pouring drilling mud from top or where bore is full of water but sides do not cave in, the auger boring guide described in Fig. 8.4 may not be used but a boring guide having longer detachable collar with a spout is used. The additional equipments required consist of the other tools, *viz.*

1. A reciprocating type double piston pump.
2. A water swivel with a right angled bend having a U-shaped hook for hoisting.
3. Rubber hose delivery tube about 8 m long with circular clamp.
4. Suction hose pipes about 8 m long provided with foot valve and a clamp for connecting it to the pump.
5. A steel pipe of about 38 mm diameter and length equal to the bore hole with valve at the bottom.
6. Tremie pipe with clamp for hoisting.

The reciprocating pump with suction and delivery hose pipes is used for circulating the bentonite slurry (for properties of bentonite slurry, see Appendix-A) and for cleaning the bore after boring and under-reaming. The delivery hose pipe is connected with steel pipe with valve and lowered into the bore hole. The spout provided in boring guide helps in chanellising the flow of drilling mud to the settling tank. The tremie pipe is 15 to 20 cm diameter pipe having a flap valve fitted at its bottom. The length of pipe is about 50 cm more than the pipe length. It can also be made by concreting pipe pieces of 1–2 m length. The pipe along with concreting funnel is used for concreting. In this case a tripod hoist with winch is necessary for handling the tremie.

8.6 CONSTRUCTION METHODOLOGY OF BORED PILES

The pile construction at site is not difficult. But it requires a lot of patience and alertness with high level of accuracies. It is even easier than construction of an isolated footing, provided the agency is well-equipped with all necessary good quality equipment and also trained manpower. Here are some tips of constructing a good quality pile. Firstly, main equipments required for pile construction should be arranged. These are: Tripod with winch, auger, boring guide, extra rods, top handle, lifting tongue, etc. Beside it, there is inventory of many small equipments.

Generally, we can construct pile from 250 to 1000 mm diameter easily (larger diameter can also be made but with more sophisticated equipments). For making the pile of any diameter, select the auger of appropriate diameter and rotate it (either manually or mechanically) to prepare bore of desired depth. The mix design of concrete should be got done by a reputed lab for the selected quality of cement and coarse aggregates.

If pile has to be made under water table, a lot of precautions have to be made (author has seen at many places, the contractors had made piles below water table without taking extra care for construction of piles. These piles when tested, exhibited one-tenth of the designed load). Below water table, a tremie pipe has to be used during concreting (Fig. 8.6) The advantages of using tremie pipe is that concrete straight goes (under gravity) at desired depth of concreting and no segregation of

Fig. 8.6(a): A view of tremie pipe

Fig. 8.6(b): Use of tremie pipe at site

(Courtesy: M/S SMEC consultants, Roorkee, Ph. 09412071753)

concrete mix takes place due to presence of ground water there. If tremie is not used and concrete mix is simply dropped from top of borehole, it will segregate when it will touch water table and thus segregated concrete mix shall reach the bottom of borehole, and thus we may not get an intact pile made at site. Author's experience is that small contractors are not aware of this technique, neither they adopt this method because of extra efforts.

8.6.1 Construction in Sandy Strata

The sandy strata is collapsible in nature therefore, bentonite slurry has to be used to stabilize the bore hole (refer Appendix 'A' of this book). The bentonite solution density is 1.03 to 1.1 g for each ml of water. The pH value of bentonite is 9 to 11.5 and liquid limit is more than 300% but less than 450% generally.

Bentonite slurry is prepared either in a tub or locally made pond, near to the pile. Every care should be taken that while using the tremie, no bentonite of bore should mix with the concrete, poured for pile, as it may adversely effect the quality of concrete mix. The bentonite slurry should be prepared 24 h before its actual application at site. Many a times, the strata has pebbles, small pieces of gravels, etc. in the borehole. Such material sometimes do not come out through auger. In such cases a circular balls made of paste of bentonite are dropped into borehole, which get stuck to these gravels/ pebbles and then finally come out quite easily through auger only. This method can be adopted to get a clean bore hole. Afterwards, reinforcement cage is inserted and then concreting is done. To achieve a proper cover, the concrete covers are made precast with a hole in centre. These cover blocks (of design thickness) are tied up with Steel cage on their outer sides. Thereafter, the concreting is done. In sandy strata also construction of under-reamed pile are done though with utmost care. The length of under-reamed piles with 2 bulbs may not generally be larger than 10 m in sandy soils, because of difficulty in construction. Instead of bentonite, casing pipe can also be used for bored pile construction. In clayey strata, boring by auger is not generally difficult (unless highly stiff clays of CH category) is there. Such strata is not collapsible in nature, hence pile can easily be made in such strata without much of problem. In such strata, bentonite slurry is also not required. For conventional pile construction, firstly some concrete should be poured into bore hole (may be for 4″–6″ thickness) and after it is set, then only the cage should be lowered, followed by concreting.

When tremie pipe is used for pile construction and bore is filled up with bentonite slurry (in DMC method also), care should be taken that there is always some concrete material into the tremie pipe and pipe is not fully out of concrete during its gradual withdrawl from the borehole. Because if this precaution is not taken, and pipe is pulled out of concrete, then there can be entry of bentonite slurry into pipe and also into the concrete, which may deteriorate the quality of concrete.

Construction of under-reamed pipe is also rather easy, though precautionary, particularly while constructing its bulbs. These piles are highly suitable for high rise structure. In case of clayey soils, construction of 1 to 4 bulbs is easier. However, in case of sands generally two bulbs only are possible to construct. The construction of bulbs is done as per its design. For example, if depth of bore is 8 m from NGL and 3 bulbs have to be constructed and the first bulb is at 2.5 m depth from NGL then for construction of this bulb at 2.5 m, firstly the bore of 2.5 m depth is made and at that depth the bulb is made by under-reamer (Fig. 8.5). In this way the other bulbs of 5 m depth and 7.5 m depth are made respectively. Multibulb piles are highly economical though they require high precision of supervision (*see* Fig. 8.7).

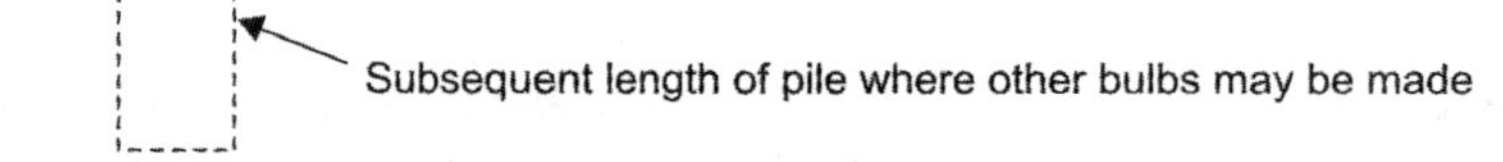

Fig. 8.7: Sequence of under-reamed pile construction

8.6.2 Construction of Compaction Pile (i.e. where Firstly Concrete is Poured, Followed by Insertion of Reinforcement Cage)

The compaction piles are different from normal bored piles. These piles can be "Bored compaction pile" or may also be "Bored Compaction under-reamed pile". The basic difference between compaction pile and ordinary pile is that in compaction pile, firstly concreting is poured into borehole and thereafter the reinforcement cage is inserted into concrete with pressure (Fig. 8.8).

The advantages of "Compaction Pile" is that such piles can take more loads as compared to conventional bored pile of same depth and same diameter. Hence may prove to be economical also.

The only precaution in such pile is that slump of concrete in this pile is kept high. Secondly, immediately after pouring the concrete into bore hole, the cage should be inserted into concrete. Else the concrete in borehole might set and then it may not be possible to drive the cage into concrete, partially or fully both.

Fig. 8.8: Preparations going on for lowering the reinforcement cage into cement slurry, already poured into the borehole

For construction of "bored compaction under-reamed pile", firstly bore of design dia and depth is made with bulbs at specified locations. Thereafter the designed mix concrete is poured. Thereafter the reinforcement cage (as shown in Fig. 8.9) is inserted into freshly filled (green) concrete. This reinforcement cage is made pointed at the bottom (Fig. 8.9), so that it can easily penetrate into concrete. This is however also pushed through in concrete with 200 kg hammer. This hammer strikes a M.S. pipe (4" diameter), fitted with a cast iron shoe (of equal diameter), pointed at its base (quite similar to dynamic cone). The pipe is properly placed inside the cage. The hammer strokes on this pipe easily take down the reinforcement cage into the green concrete.

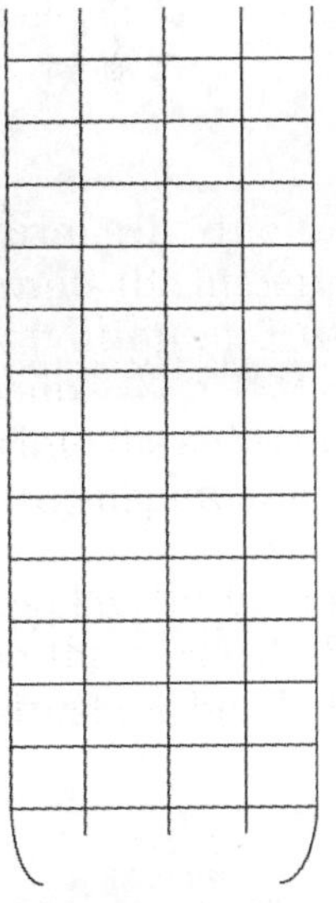

Fig. 8.9: Reinforcement cage with pointed bottom

As stated earlier, such compaction piles cater more loads (as compared to conventional piles), thus may prove to be economical. But its construction requires expertise. Because slump of concrete, used in these piles is kept more, thus, all the preparations of construction of piles are done beforehand. Because, if concrete is set (before driving of cage), the driving of cage may be highly difficult.

8.7 PILE CONSTRUCTION BY DMC (DIRECT MUD CIRCULATION) METHOD

These days DMC* method of pile construction is gaining popularity. In this method pile is made by percussion drilling. In this method, a cutting tool, called **chisel** (Fig. 8.10) is used whose weight is generally double of dia of cutting tool. For example, if dia of borehole is 500 mm then the weight of cutting tool is of the order of 1000 kg. In this cutting tool the pieces of angle iron are welded along its periphery. One socket is welded on its end, which is connected to a hose pipe. The other end of this hose pipe is connected to the pump which is attached with the bentonite tank. Now, the bentonite (mixed with water) slurry is continuously poured into the borehole along with insertion of cutting tool (chisel) into the borehole (cutting tool goes into the pit easily by virtue of its own weight). This bentonite slurry is mixed with bore material and also collected back into bentonite tank. In this way after achieving the required depth of the borehole, the cutting tool is taken out of the borehole. Thereafter the bore depth is once again measured by the engineer-in-charge and the bore is washed again with bentonite mixed with water. Immediately now, the "already prepared" reinforcement cage is lowered into borehole and concreting is done by using tremie pile. The above method is used for sandy strata generally.

Fig. 8.10: Chisel (Cutting tool) used for pile construction

(Courtesy: M/S SMEC Consultants, Mr. PK Jain, Roorkee, Ph: 9411111898)

In some sites where sandy strata is not there but clayey strata is there or cobbles/pebbles are present in the bore, the bore at such sites is made easily by bailor.

---

*_For DMC method, a pit of 20′ × 12′ × 10′ is made at site, which is stabilized by brick masonry all around. This tank is divided into three parts (compartments). In one compartment, bentonite solution mixed with water is prepared. In other two parts recovery material from borehole is collected. When these two compartments are filled up, that material is removed from there and also from the site. The bentonite slurry from the first compartment is poured into bore through a pump._

8.8 PILE CONSTRUCTION BY TMR (*NOT A STANDARDISED METHOD*)

These days some contractor are also using TMR (truck mounted rig) method for pile construction. In this method a rig is mounted on truck which is generally used for making deep boring for tubewell purpose. This equipment basically is not a piling equipment therefore is not approved by CPWD (Central Public Works Department). But still some piling agency, who are not well-equipped or well-acquainted with DMC method, are using it. In this method the auger of required diameter is attached with the truck mounted boring assembly. This auger is taken into the ground by rotating it and along with that the bentonite mixed with water is also poured into the bore through a pump. Thus, the bore is made by using water pressure also.

The drawback of this method is that in this method the recovery material of bore is not taken out from the bore. That material is still leftover in the bottom of the bore. Therefore, because of leftover of loose material into the bore, the pile may not take its desired load on account of end bearing. As it is well, understood that the end bearing pile are supposed to be terminated on firm strata, but in this TMR method the pile is left on loose strata. Therefore such piles, when tested, showed higher settlements as compared to piles made by DMC method.

Chapter

9

Introduction of Well Foundations

(Contributed by Prof. Bhupinder Singh, IIT Roorkee)

9.1 INTRODUCTION

These are foundations constructed on the dry or wet ground in a suitable shape to accommodate the structure and then sunk into the ground to the desired level by excavating earth through dredge holes.

Typical components of a well foundation are shown in Fig. 9.1.

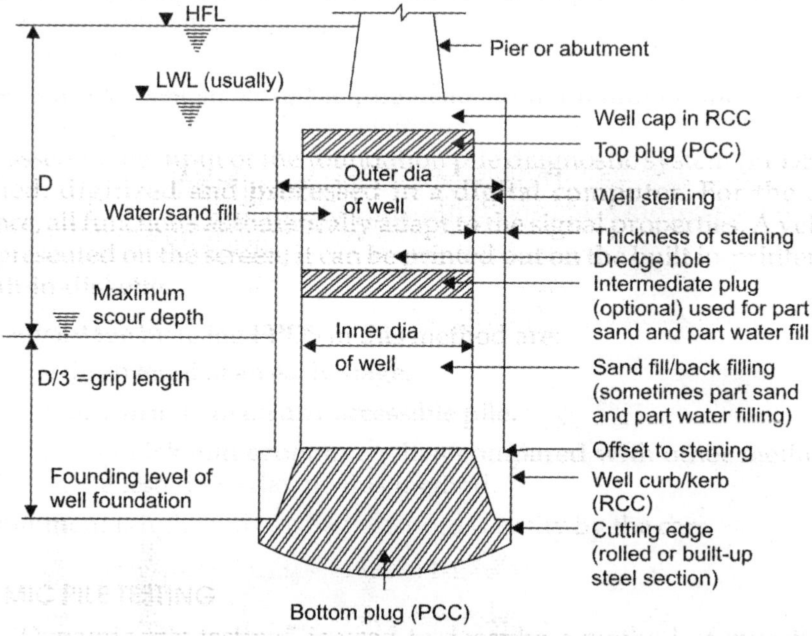

Fig. 9.1: Sectional elevation of a typical well

Relevant Indian Standards codes for design of well foundation are:

1. IRC: 78–1983: Standard specifications and code of practice for road bridges, Section VII: Foundations and substructure.
2. IRC: 45–1972: Recommendations for estimating the resistance of soil below the maximum scour level in the design of well foundations.

The deep foundations are generally:

1. Well foundation
2. Pile foundation

The general difference between well foundation and pile foundation may be understood as follows:

- Wells provide a solid and massive foundation for heavy loads against a cluster of piles which are slender and weak individually.
- Piles are susceptible to damage due to:
  a. Floating trees/debris
  b. boulders rolling in river bed.
- Wells have longer bearing area compared to piles and hence can mobilize higher soil resistance.
- Wells can be provided up to any depth if open—sinking is used and up to 33.5 m if pneumatic sinking is used. Piles are economical for depths up to 27–30 m.
- Size of a well cannot be reduced indefinitely. For relatively smaller loads pile foundations are suitable.
- Due to large section modulus wells can be resist heavy vertical and lateral loads. Section modulus of individual piles in cluster is small.
- Concreting of steining is done under controlled dry conditions.
- There is a flexibility about founding level for well foundations. Difficult to extend short piles or cut long piles in case of difficult strata (driven piles).
- Well foundation is open to inspection during construction whereas pile foundation is not.
- Piles cannot be driven through boulder and difficult strata. Wells can be sunk in such strata.
- Bearing capacity of a pile is uncertain (driven piles). For wells it is possible to examine the soil during sinking of well [open-sinking or pneumatic sinking]. Hence, soil strata actually available at site can be cross-checked with strata assumed for design.
- For cast-*in situ* piles concreting is done below ground or below the water table hence uncertain quality. In driven piles, concrete is subject to severe hammering and stresses during the process of driving the pile.

The general difference between well foundation and open foundation may be understood as follows.

Well Foundations versus Open Foundations

1. Open foundations economical under dry conditions at shallow depths. Even under dry conditions, provide well foundations for depths > 7.5 to 9 m to avoid costly shoring works.
2. For bridge foundations provide well foundation for depths greater than 6 m below spring water level.
3. Open foundations are simple to design and construct and are always open for inspection.
4. In built-up area wells may be preferred to avoid large excavations for open foundations which may endanger nearby structure.

It can more be understood from Fig. 9.2.

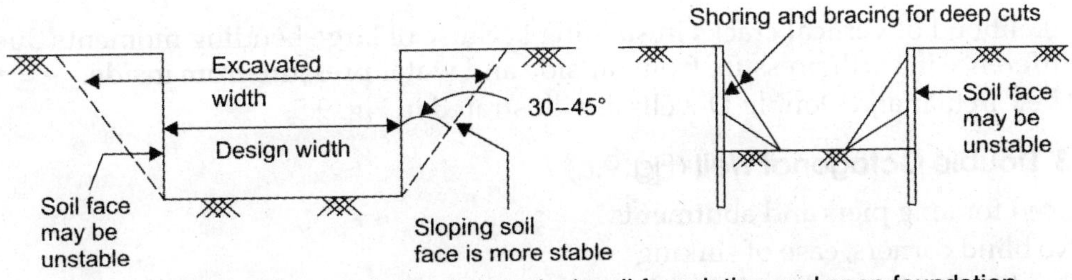

Fig. 9.2: Basic difference between a typical well foundation and open foundation

9.2 TYPES OF WELL FOUNDATIONS: CLASSIFICATION ON THE BASIS OF WELL CROSS-SECTION (Refer Figs 9.3 and 9.4)

i. Circular
ii. Double-D
iii. Double Octagonal
iv. Twin circular
v. Single and double rectangular
vi. Multiple dredge holes.

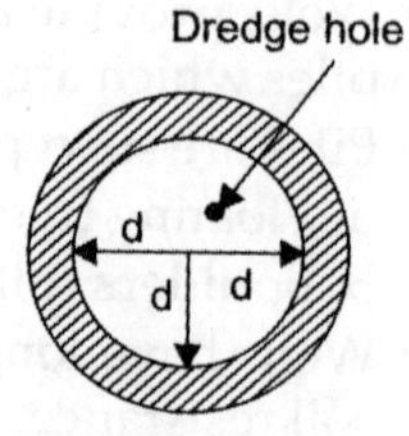

Fig. 9.3: Circular well foundation

9.2.1 Circular

- Most commonly used well.
- Good strength, simple to construct and easy to sink.
- Requires only one dredger for sinking.
- Weight per square metre of surface is highest. Hence sinking effort is high.
- Chances of tilting are minimum.
- Adopted for piers for bridges on narrow roads or single line railway bridges. Not suitable for long piers.
- Maximum usable diameter of circular well is 9.0 m and maximum length of pier resting on this well = 9 + 1 + 1 = 11.0 m.

9.2.2 Double-D Well

- Adopted for bridges with long (wide carriage ways) piers and abutments
- Relatively simple shape and easy to sink.
- Dimensions are so selected that *'l'* ≈ *'b'*.

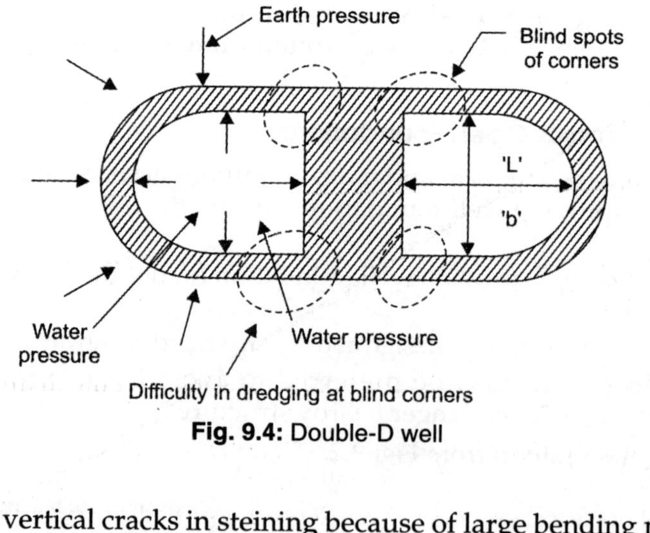

Fig. 9.4: Double-D well

Disadvantages

- Likelihood of vertical cracks in steining because of large bending moments due to difference in earth pressure from outside and water pressure from inside.
- The Circular and Double-D wells are illustrated in Fig. 9.5.

9.2.3 Double Octagonal Well (Fig. 9.6)

- Used for long piers and abutments
- No blind corners, ease of sinking
- Bending stresses reduced in steining

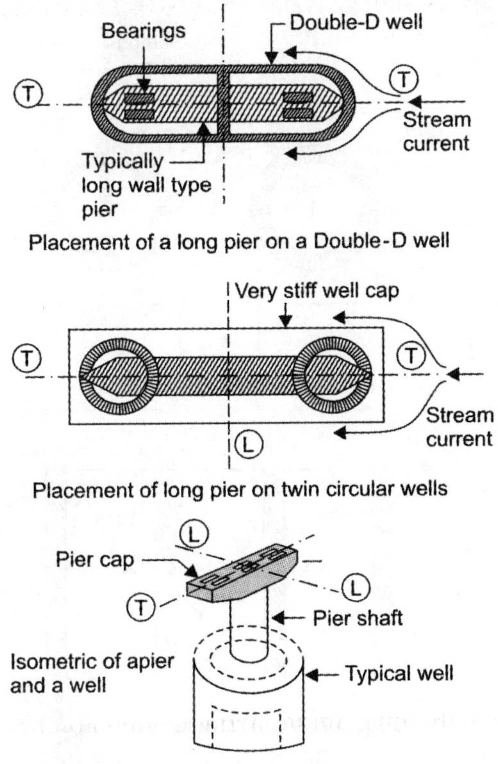

Fig. 9.5: Circular and Double-D well (also seen is isometric view of a well)

Disadvantages

i. Greater resistance to sinking because of larger surface area
ii. Difficulty in form work

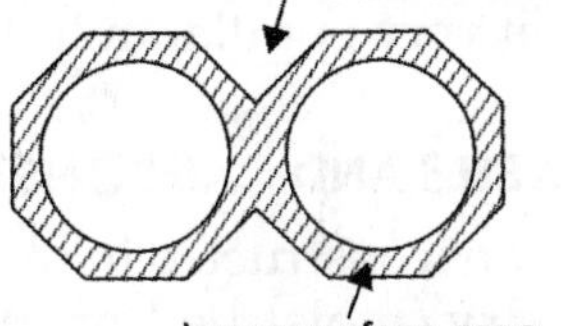

Fig. 9.6: Double octagonal

9.2.4 Twin Circular Well (Figs 9.7 and 9.8)

Advantages for such wells are that they are used for larger pier lengths. Such wells are though very suitable when depth of sinking is small.

Disadvantages

- Wells should be sunk simultaneously to avoid wells tilting towards each other.
- Two parallel sets of dredging equipment (cranes, grabs, etc.) are required.
- Special care for design of common well cap.
- Likelihood of cracks in piers due to differential settlement of individual wells.

9.2.5 Rectangular Well

These wells are adopted for bridge foundations with shallow depths. Can be adopted if foundation has to change from open foundation to well foundation. These wells are generally adopted for depths up to 7.5 m. In these wells, the steining bending stresses are maximum for this shape. But these wells are relatively difficult to sink because of blind corners. However, such well are used for bigger foundations, even used double rectangular well in some specialised situations.

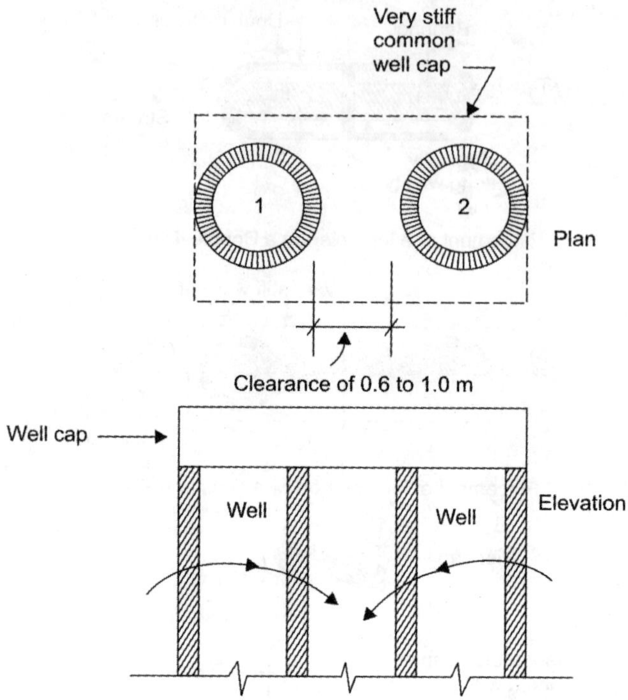

Fig. 9.7: Twin circular well

9.2.6 Wells with Multiple Dredge Holes (Fig. 9.9)

- Used for piers and abutment of very large sizes
- Dredge hole size may be 4 × 4 to 6 × 6 m
- Not very popular in India.

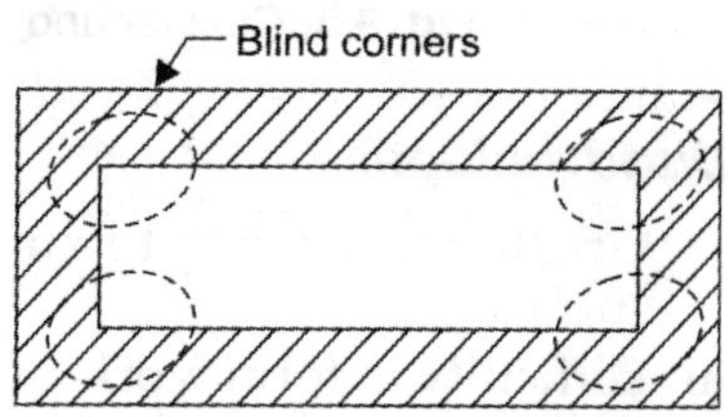

Fig. 9.8: Twin circular well

9.3 WELLS AND CAISSONS (Fig. 9.10)

Wells are constructed in their final position on a dry/moist bed or after making an island in the river at the site of construction. This method is used where maximum water depth does not exceed 4.50 to 6.00 m and velocity of flow is also not high after construction of cofferdam and dewatering. Caisson is a box-type steel shuttering for a well which becomes a part of the well itself. These are shown in Fig. 9.10

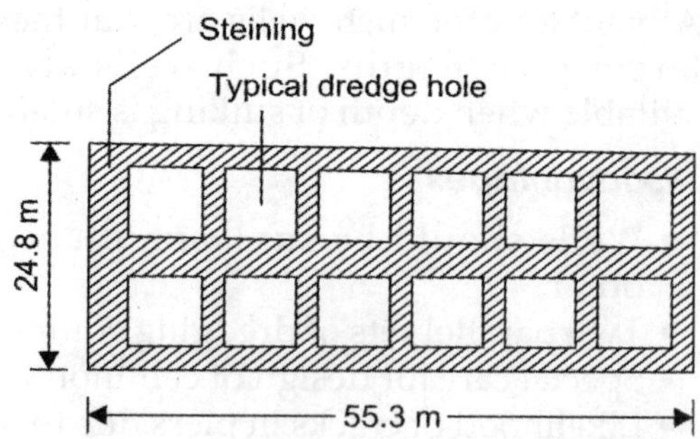

Fig. 9.9: Wells with multiple dredge holes

9.4 COMPONENTS OF A WELL FOUNDATION

9.4.1 Bottom Plug

It transmits the entire load coming on the steining and that of the well infill to the foundation. Refer to the Figs 9.11a and b the different parameters are defined as below:

P: load transmitted by steining

W: weight of sand filling within well

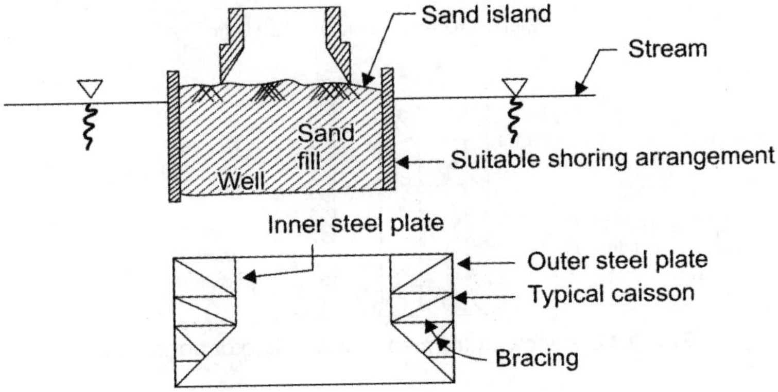

Fig. 9.10: Wells and caissons

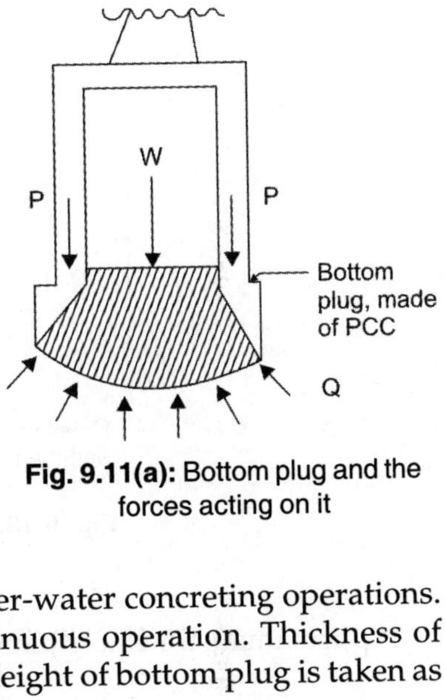

Fig. 9.11(a): Bottom plug and the forces acting on it

Bottom plug acts like an inverted dome supported by the steining on all sides and having the soil reaction as the uniform load acting on it.

If thickness of bottom plug is small it starts acting as a circular slab and bending moments are induced [usually in case of large diameter dredge hole]. It is physically not possible to reinforce the bottom plug.

Minimum concrete grade for bottom plug is M15 though M20 grade is preferred. Maximum water-cement ratio (W/c) = 0.6, with minimum cement content = 350 kg/m^3.

Concreting of bottom plug is done using a "tremie" whose bottom is kept immersed in plastic concrete deposited earlier (Fig. 9.12).

Here, 10% extra cement should be used for all under-water concreting operations. Bottom plug concreting should be done in one continuous operation. Thickness of bottom plug is taken ≈ half diameter of dredge hole. Height of bottom plug is taken as = 0.3 m above top of well curb.

When well is founded on sound rock it should be anchored by taking it 200 to 300 mm within the rock. Positive anchorage of well to rock may be done using dowels (Fig. 9.13).

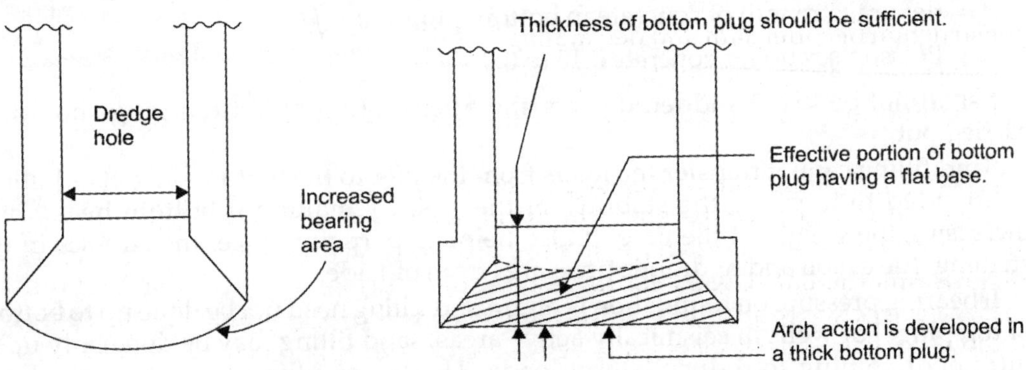

Fig. 9.11(b): Bottom plug and the arch acting on it

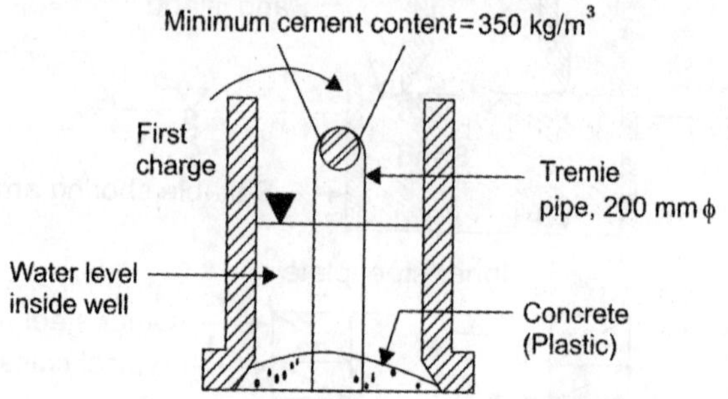

Fig. 9.12: A view of tremie for concreting of bottom plug

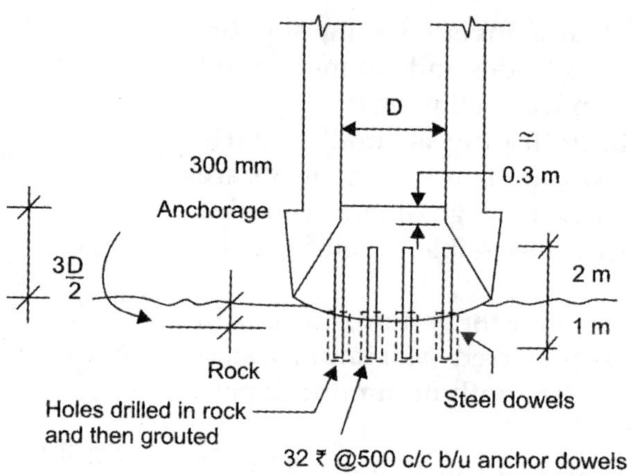

Fig. 9.13: Dowels for anchorage of bottom plug

Approximate thickness of bottom plug may be calculated from the eqn.

$$t^2 = \frac{3W}{8\pi f_c}(3 + v) \qquad \text{... (9.1)}$$

where,

w = total bearing pressure at base of well

f_c = flexural strength of concrete in bottom plug $\approx 0.7\sqrt{fck}$

v = Poisson's ratio for concrete 0.18 to 0.0

t shall not be $<D/2$ sudden dewatering after casting of bottom plug should be carried out.

Sand filling cannot transfer the loads from the pier to the bottom plug. Sand filling is intended to improve the stability of the well by making it bottom heavy and increasing the weight at the base. It also helps to partially cancel the stresses in the steining due to soil and hydraulic pressures from outside.

If bearing pressure on foundation is high, sand filling need not be done up to bottom of top plug. For wells in seismically active areas, sand filling may be done only up to intermediate plug to reduce lateral loads. The water filling may be done above intermediate plug so as to reduce lateral loads.

9.4.2 Top Plug

Top plug is used in small-sized dredge holes to transmit load of pier to steining. The thickness of top plug = D/2 and it is usually unreinforced. (For the case when well cap is not provided).

For larger dredge holes, RCC well cap should be with M15 or M20 grade concrete. When used together with well cap, minimum thickness of top plug = 0.6 m. Intermediate plug, if used may be 0.50 to 0.60 m thick and is made of RCC. This reduces the slenderness of well steining.

9.4.3 Well Cap

Directly supports the pier and transfers load and moments from the pier to the well.

Shape of well cap is same as that of well with a possible overhang of 150 mm all around to accommodate lengthy piers.

Well cap is made up of RCC and designed as a two-way slab with partial fixedity at supports. Thickness is so decided that shear reinforcement should not be required. Minimum grade of concrete: M25.

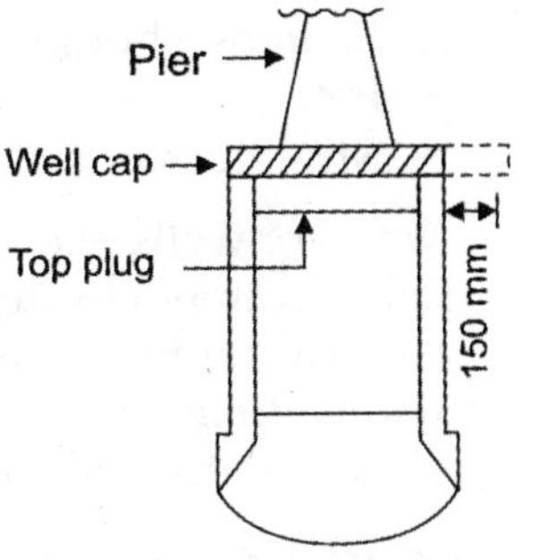

Fig. 9.14: Well cap

For non-perennial rivers top of well cap is kept at stream bed-level whereas for perennial rivers it is kept at stream bed-level whereas for perennial rivers it is kept at LWL (this will require a cofferdam and dewatering during concreting of well cap).

Provide a minimum of 80 kg of steel per m^3 of concrete in the well cap (Fig. 9.14).

Thickness of well cap is usually 1.5 to 2.0 m.

9.4.4 Steining

Steining may be of brick masonry or of RC brick masonry. Minimum grade of concrete in steining shall be M20 with minimum cement content of 350 kg/m^3. To facilitate well sinking, an off-set of 75 to 100 mm is provided in the well steining (Fig. 9.15).

Minimum thickness of steining shall be 500 mm, though the usual thickness is generally 0.75 to 1.50 m.

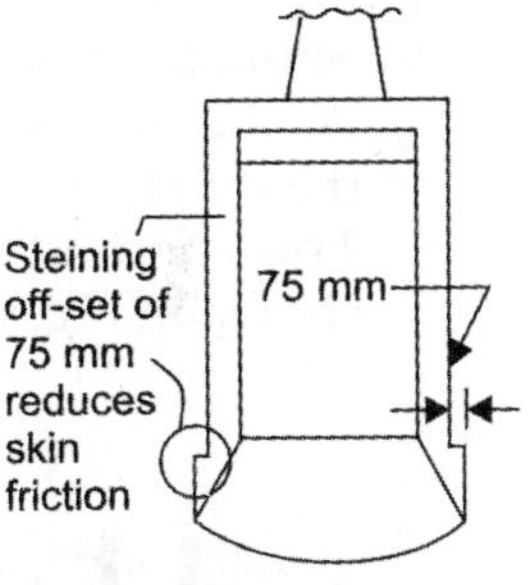

Fig. 9.15: Steining

Thumb rule for steining thickness ≈ 1/10 well diameter

Nominal steining thickness = $t = KD\sqrt{L}$ **(IRC: 78–1973)**

t = Minimum thickness of steining in 'm'

D = External diameter of circular well in 'm'

L = Depth of well in 'm' below LWL or GL whichever is greater.

K = For wells in sandy strata, k = 0.030 for circular well and 0.039 for twin 'D' well. Increase 'K' value by 10% for wells in clayey soil.

Note: For wells with diameter > 8 m minimum steining thickness = 1.0 m

Reinforcement in steining (Uncracked section):

Both faces of steining shall be reinforced (Fig. 9.16).

Minimum vertical steel = 0.125% of gross sectional area distributed equally on both faces.

Minimum hoop steel = 0.004 × volume per unit length of steining.

Minimum cover to longitudinal steel = 75 mm

Thickness of steining is based on following considerations:

i. Enough sinking effort should be generated so as that kentledge should not be required.
ii. Wells should not get damaged during sinking due to external and internal pressures.
iii. It should be possible to rectify tilts and shifts without damaging the well. This will impose additional loads on steining.

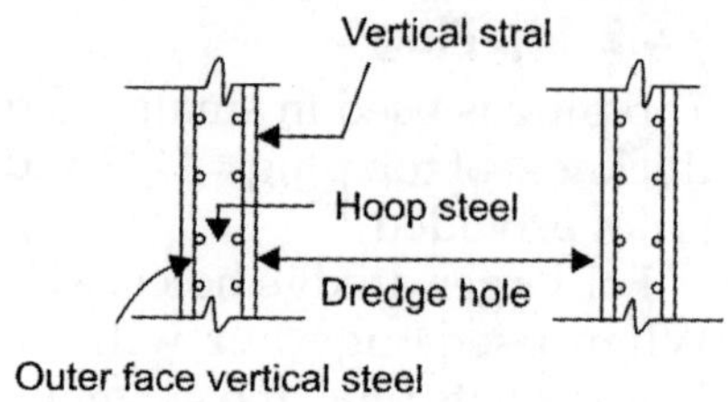

Fig. 9.16: Reinforcement in

iv. Steining should be able to resist all loads transferred to well during its service life.
v. Minimum internal diameter of well as per IRC: 78–2000 = 2.00 m.

Note: For wells of $\phi = 8$ m, minimum steining thickness = 1.00 m.

Reinforcement in steining (RC wells):

Vertical reinforcement = 0.2% of c/s area of steining (on inner face of steining, provide minimum of 0.06%)

Minimum transverse steel = 0.004% of volume of steining per unit length.

9.4.5 Well Curb or Well Kerb

Well curb supports the cutting edge and is a RCC ring beam (Fig. 9.17)

Requirements of a well curb:

i. Its shape should offer minimum resistance to sinking of well.
ii. It should be able to transfer superimposed loads from steining to the bottom plug.
iii. Minimum grade of concrete M25.
iv. Minimum reinforcement shall be 72 kg/m^3. Well curb should be lined with MS plate. [Hoop steel = 60%; stirrups/links = 40%]. The zoomed views of cutting edge are shown in Fig. 9.18.

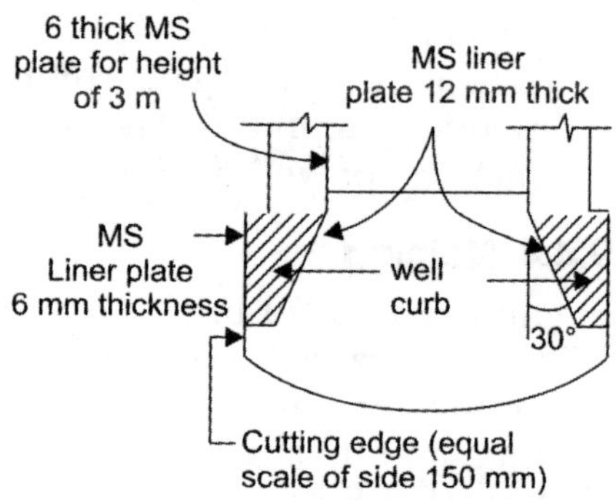

Fig. 9.17: Well curb or well kerb

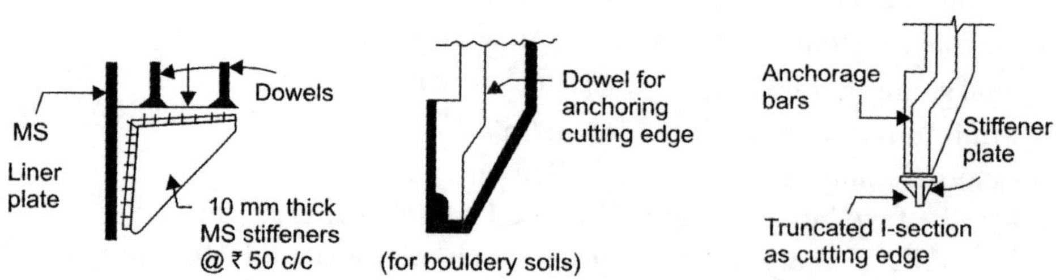

Fig. 9.18: Zoomed views of well curb or well kerb

9.4.6 Cutting Edge

i. Cutting edge is usually a mild steel equal angle of side (Fig. 9.19).
ii. For heavier cutting edges, truncated I-sections are also used.
iii. Cutting edge geometry shall be different for boulder and non-boulder strata
iv. Minimum weight of steel cutting edge = 40 kg/m (combination of angle + steel stiffener plate).

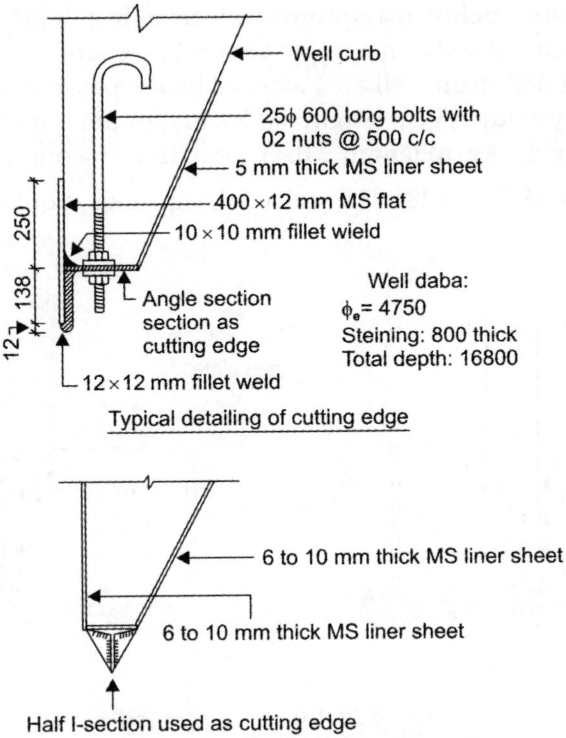

Fig. 9.19: Cutting edge of well

9.5 DEPTH OF WELL FOUNDATION

Depth of foundation is decided with respect to

i. Scour, where applicable

ii. Stability — Overturning; Sliding; Base pressure

Note: Depth of foundation $\nless 1.33 \times$ deepest scour below HFL.

Depth of well foundation in scourable river beds is made up of three parts:

i. *Normal scour*

a. IRC formula

b. Lacey's equation, $d = 0.473\left(\frac{Q}{f}\right)^{1/3}$

Here, 'Q' discharge in cumecs, and 'f' is silt factor.

ii. *Local scour around the pier*

a. at nose of pier, $D = 2d$

b. in straight reach, $D = 1.27d$

iii. *Grip below maximum scour.*

Well should be sunk below maximum scour level to a depth such that:

i. Bearing capacity of soil = maximum bearing pressure.

ii. Lateral forces acting on well are balanced by the passive resistance of soil.

- For railway bridges grip length = 0.5 × maximum scour depth below HFL
- For road bridges grip length = 0.33 × maximum scour depth below HFL

Well on dry ground (Fig. 9.20)**:** The well on dry ground should be laid as illustrated in Fig. 9.20.

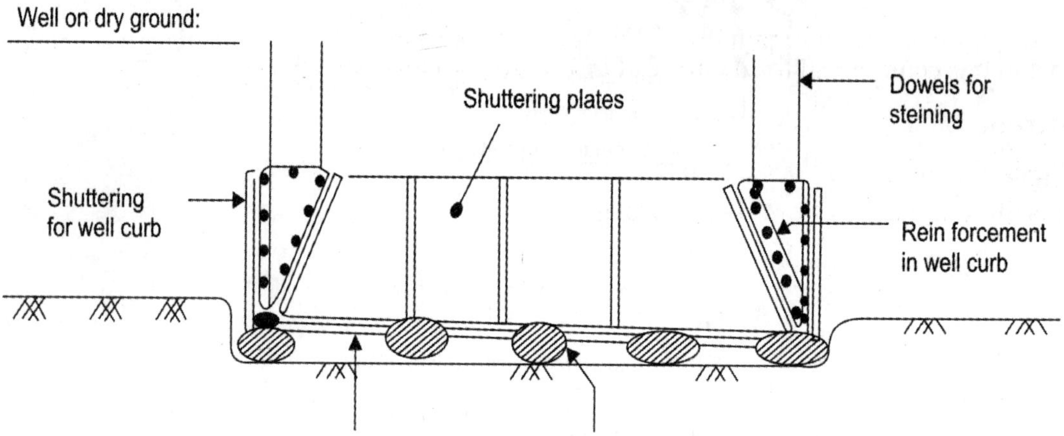

Fig. 9.20: Well on dry ground

Well curb should be allowed to set for one week before starting sinking. First sink the well curb alone (Fig. 9.21).

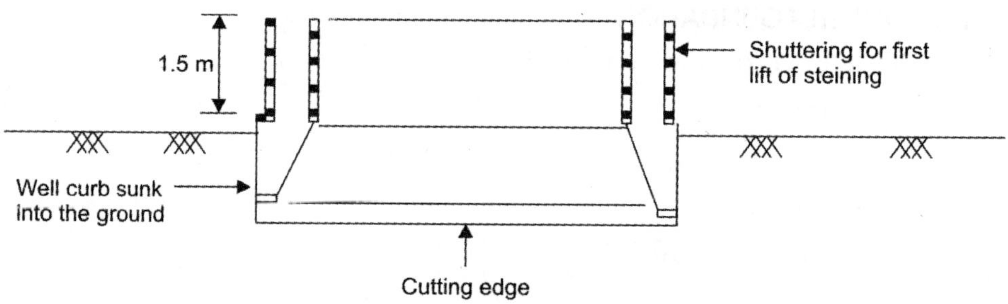

Fig. 9.21: Sinking of well curb

Then raise the steining by 1.5 m at a time. Wait for a setting time of 1 to 3 days (depending on weather) and proceed with sinking.

Once well has sunk for about 6 m, raise steining by 3 m per lift for faster progress. Use smaller lifts near well cap.

Sufficient steining height should be available to take care of quick sand conditions in silt and fine sand.

Plum-bobs should not be used for raising the steining. Instead raise the steining with a straight edge.

Use shear keys while raising steining (Fig. 9.22).

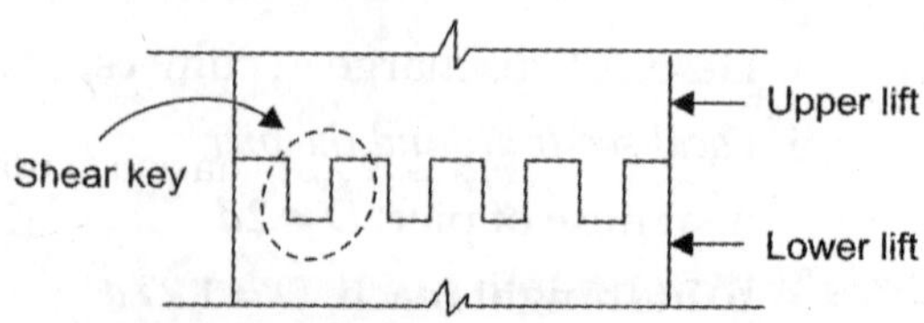

Fig. 9.22: Shear key(s) in steining

9.6 WELL IN WATER

Well construction in water is quite common. Some important aspects of construction are as follows:

- Use of island for casting well curb
- Use of floating caissons which are:
  - i. Fabricated ashore on dry ground in steel
  - ii. Towed to position
  - iii. Grounded into position by pouring concrete.

In streams with maximum flow velocity of 1 to 1.2 m/s and maximum water depth of 4 m it is economical to construct . Otherwise use caissons.

Making an Island

Case 1: Water depth up to 1 m

It is illustrated as follows (Figs 9.23a and b).

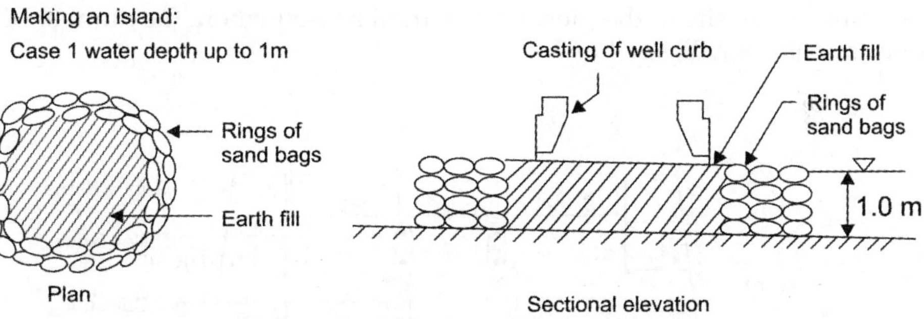

Fig. 9. 23(a): Making of island in water depth up to 1 m

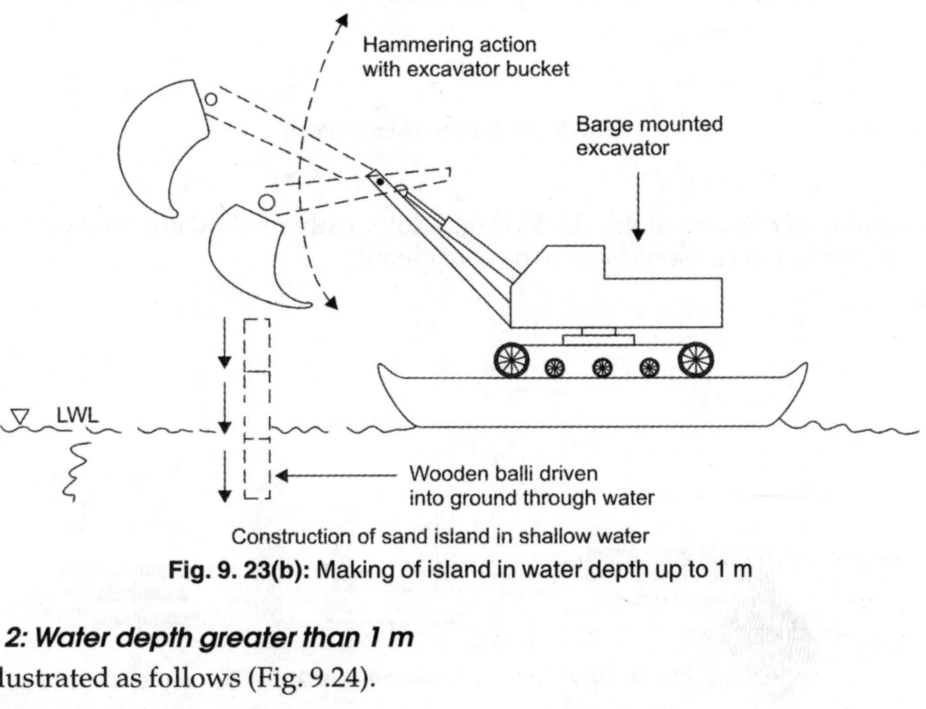

Fig. 9. 23(b): Making of island in water depth up to 1 m

Case 2: Water depth greater than 1 m

It is illustrated as follows (Fig. 9.24).

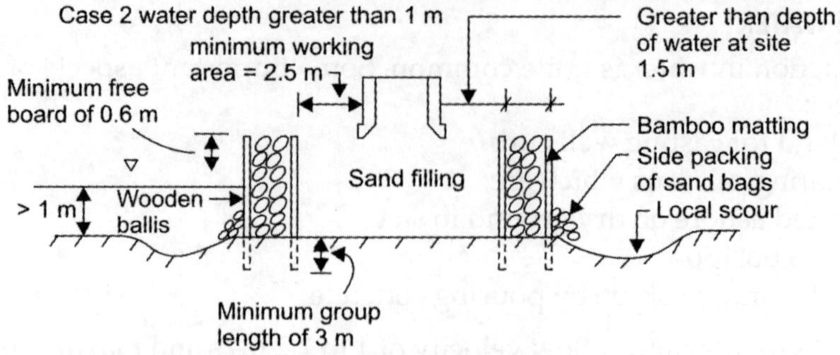

Sectional elevation of a sand island

Fig. 9. 24: Making of island in water depth greater than 1 m

9.7 USE OF CAISSONS

Assembly and launching of caissons (Fig. 9.25):

i. Assembly at the site of the pier: This method is used where the number of caisson foundations is small, say 1 or 2.

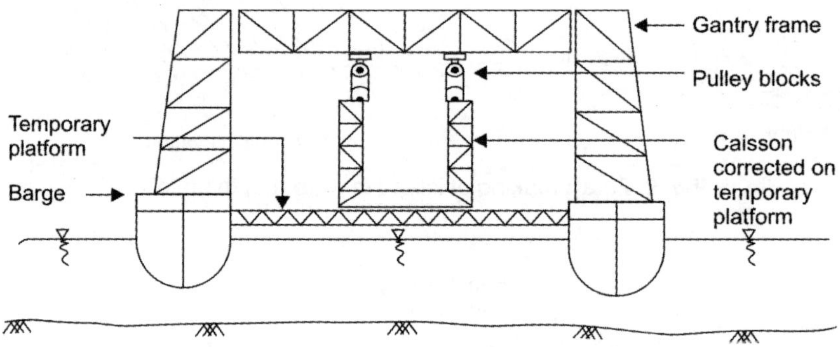

Fig. 9. 25: Caissons Launching

ii. Assembly of caissons at dry docks (Fig. 9.26): Costly method are used only when a large number of caissons have to be considered.

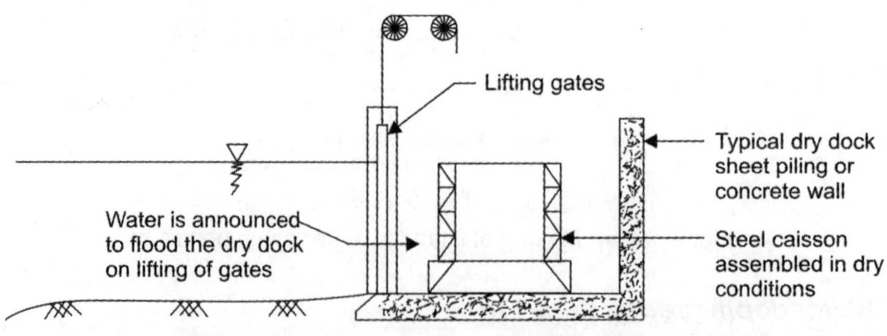

Fig. 9. 26: Assembly of caissons at dry docks

9.7.1 Assembly of Caisson on a Slipway and Launching from a Loose Earth Platform

The assembly of caisson on sloping surface are illustrated in Figs 9.27 and 9.28.

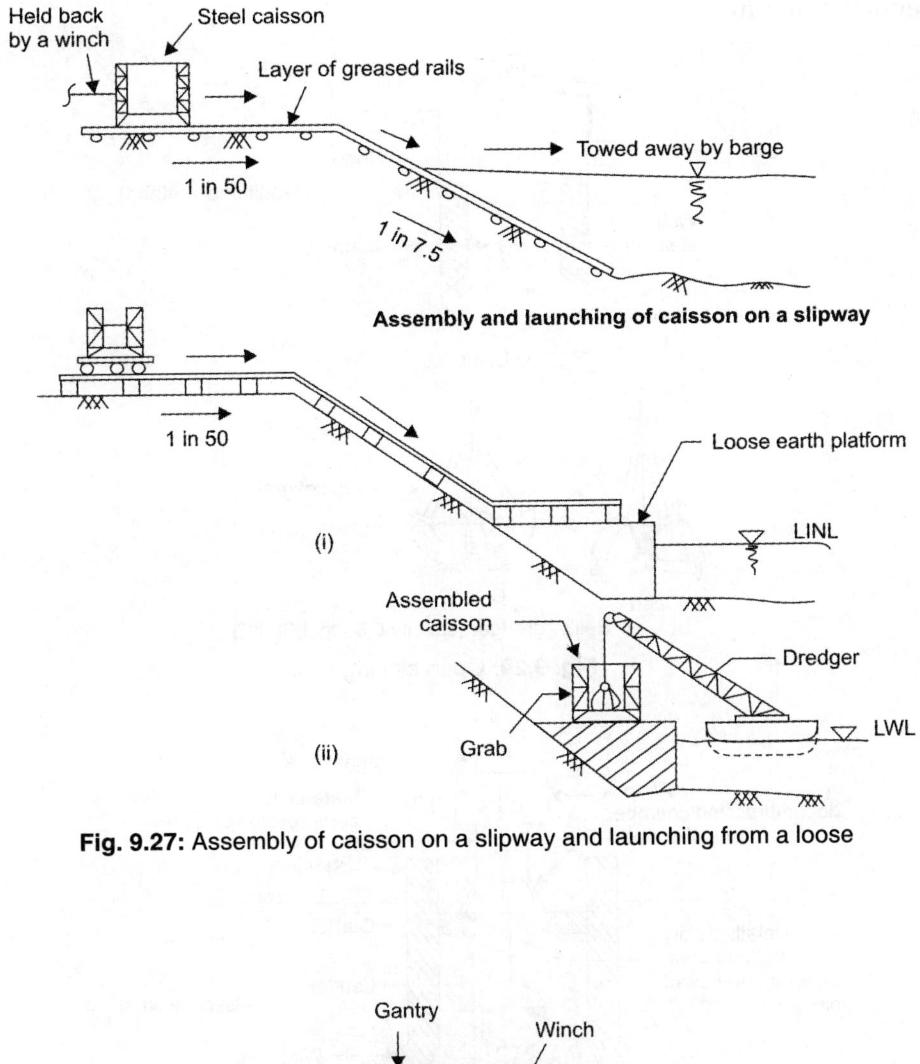

Fig. 9.27: Assembly of caisson on a slipway and launching from a loose

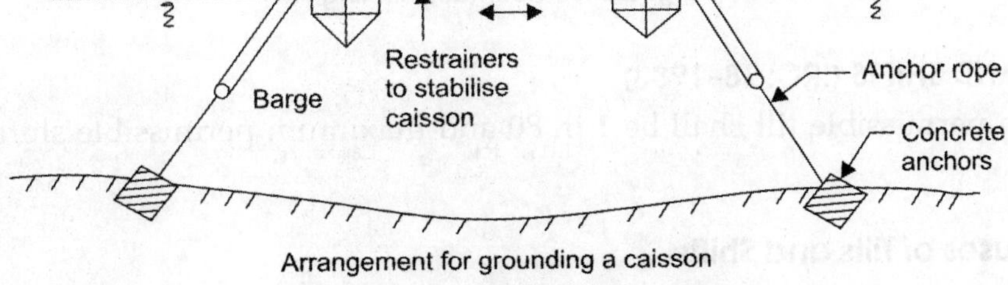

Fig. 9.28: Grounding a Caisson

9.8 SINKING OF WELLS: SOME PROBLEMS AND SOLUTIONS (Figs 9.29 and 9.30)

The sinking is explained in following illustrations:

i. Open sinking:
   - Sinking of small wall using a "Jham" (large sized spade)
   - Sinking using grabs or dredgers.

ii. Pneumatic sinking:

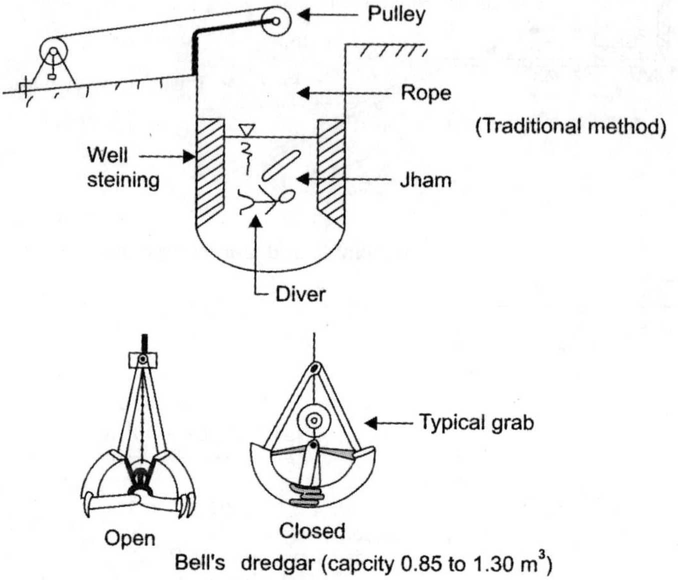

Fig. 9.29: Open sinking

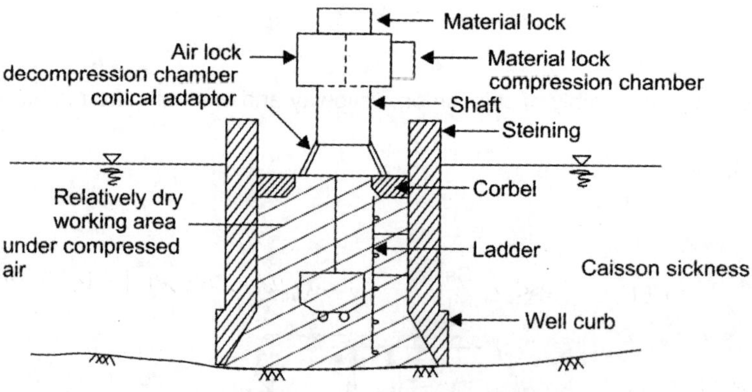

Fig. 9.30: Pneumatic sinking

9.9 TILTS AND SHIFTS (IRC: 78–1983)

Maximum permissible tilt shall be 1 in 80 and maximum permissible shift shall be 150 mm.

9.9.1 Causes of Tilts and Shifts

Causes of tilts and shifts are illustrated in Fig. 9.31.

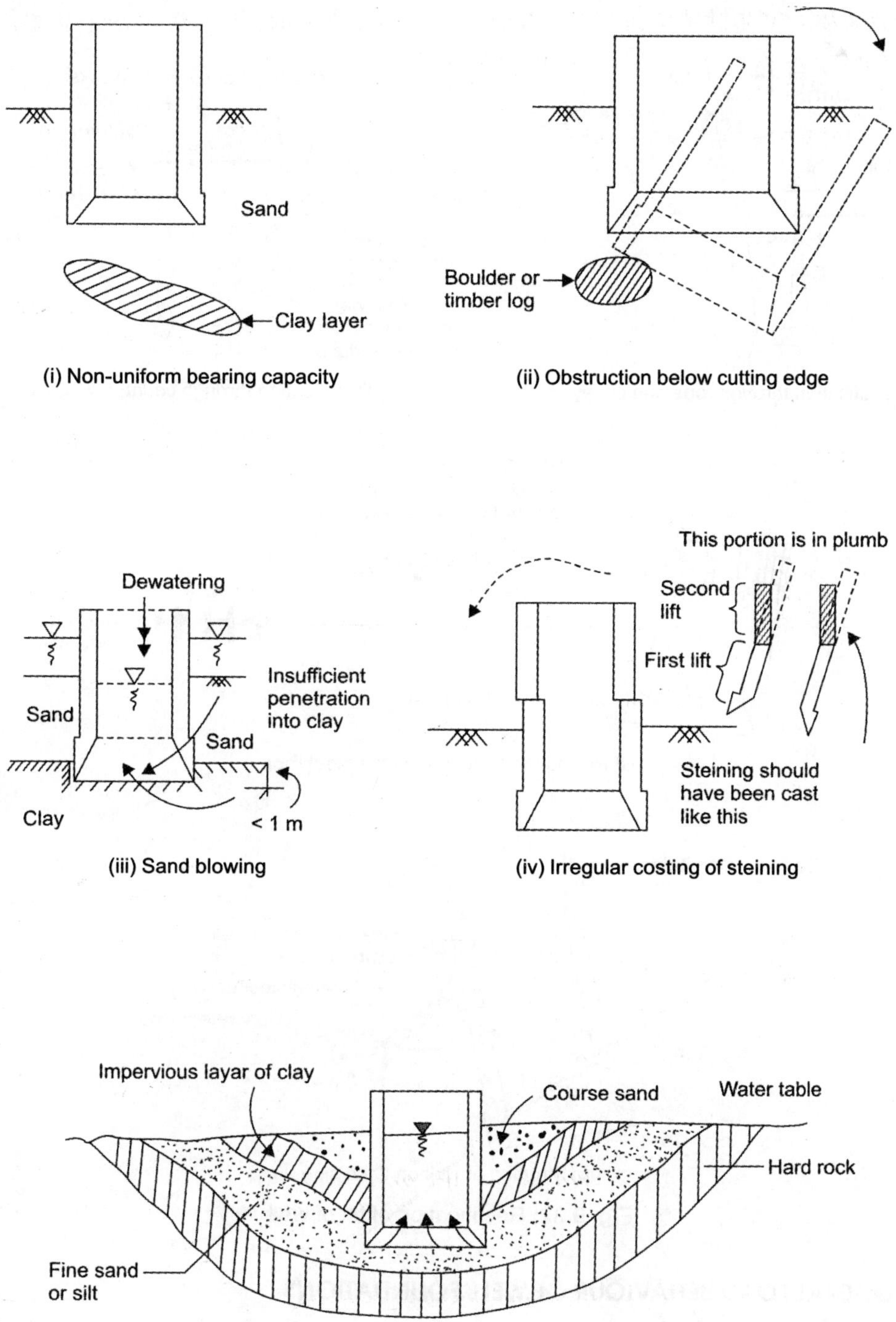

Fig. 9.31: Tilts and shifts in well

9.9.2 Rectification of Tilts and Shifts

The tilts and shifts can be rectified are illustrated in Fig. 9.32.

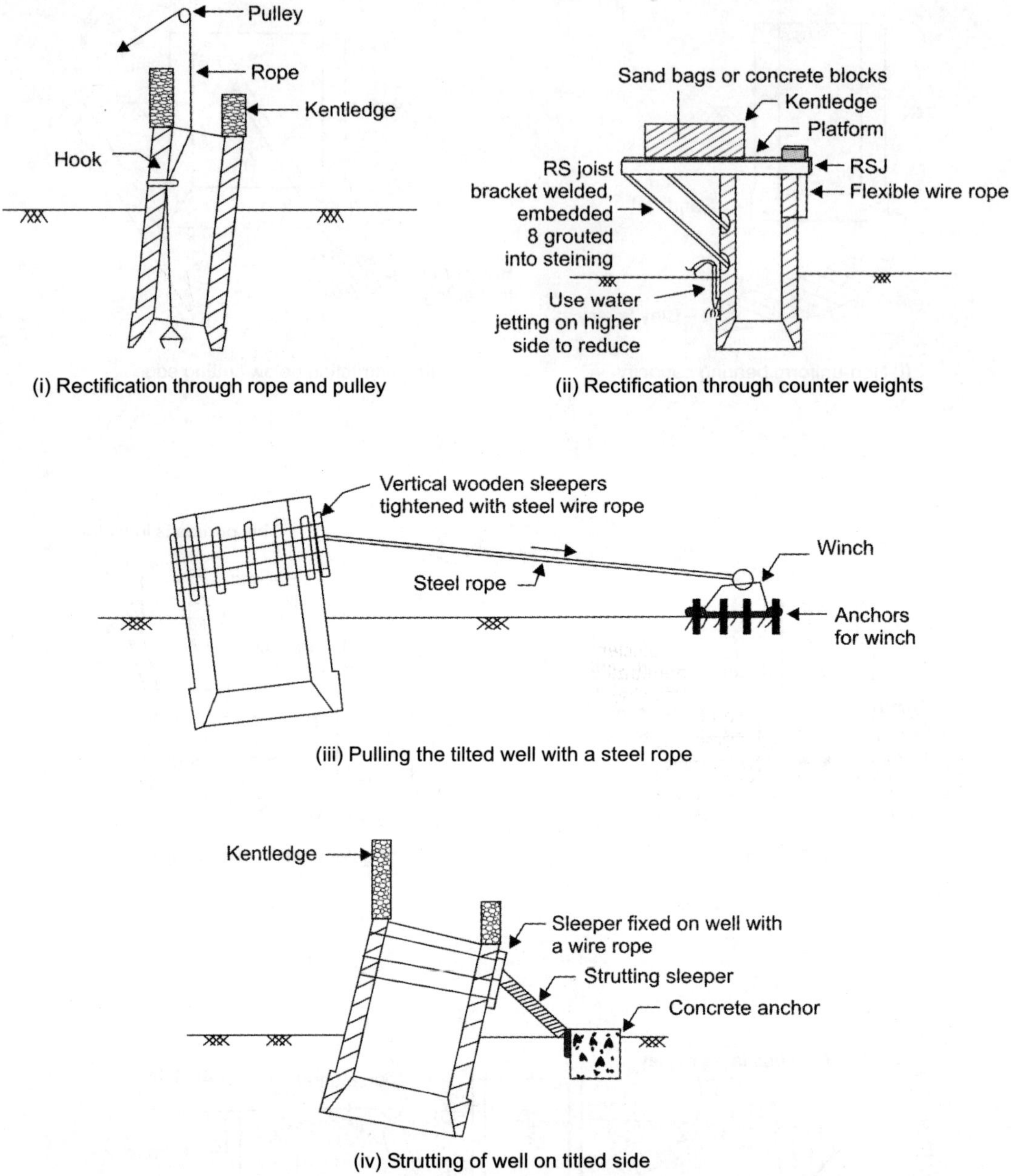

(i) Rectification through rope and pulley

(ii) Rectification through counter weights

(iii) Pulling the tilted well with a steel rope

(iv) Strutting of well on titled side

Fig. 9.32: Rectification of tilt of well

9.10 LATERAL LOAD BEHAVIOUR OF WELL FOUNDATIONS

Analysis of well foundations is carried out using the method of Terzaghi's analysis for free rigid bulkhead. Under lateral loads, as a simplification, the well may be assumed to behave as a rigid wall.

A rigid wall under lateral loads transforms the soil on one side into passive state and on the other side into active state.

Net soil pressure at any depth 'Z' below the surface of soil is given by $\text{Þ} = \gamma Z$ $(K_p - K_a)$, as explained in Fig. 9.33 and 9.34.

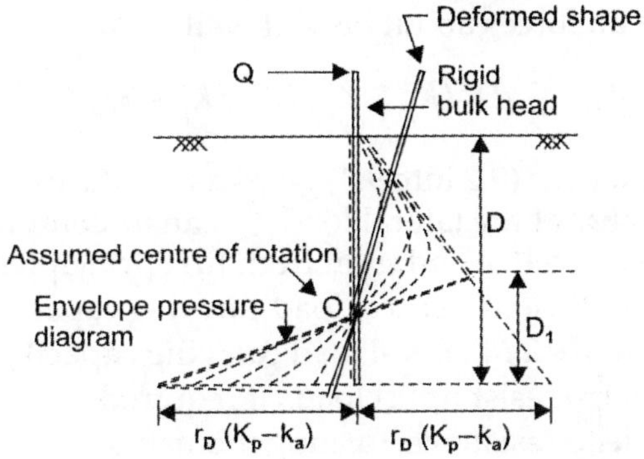

Fig. 9.33: Rigid bulk head under lateral loads

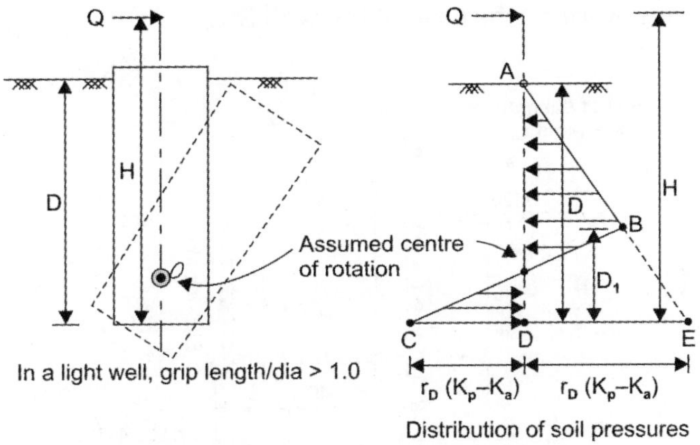

Fig. 9.34: Distribution of soil pressures

Case 1: Heavy Well in Cohesionless Soil

From force equilibrium

$$Q = \frac{1}{2}(K_p - K_A)\,\gamma D^2$$

From moment equilibrium

$$Q = \frac{(K_p - K_B)\,\gamma D^2}{6H}$$

In a heavy–well, grip length/diameter < 1.0.

Case 2: Light Well in Cohesionless Soil

From force equilibrium

Q = Area ABE – Area BCE

$$= \frac{1}{2}\gamma D^2 (K_p - K_a) - \frac{1}{2} \times 2 \times \gamma D (K_p - K_a) D_1$$

$$= \frac{1}{2}\gamma D^2 (K_p - K_a)\,(D - 2D_1) \qquad \text{...(9.2)}$$

Equating moments of all forces about base of well

$$QH = \frac{1}{2}\gamma D^2 (K_p - K_a) D/3(K_p - K_a) D_1 \cdot \frac{D_1}{3} \quad ...(9.3)$$

Substituting for 'Q' from Eqs (9.2 into 9.3) gives quadratic in 'D_1'. Knowing the value of 'D_1' the permissible value of the lateral load 'Q' can be computed.

Elementary analysis of a well foundation from first principles:

i. Well should have sufficient vertical load carrying capacity.
ii. Well should have sufficient lateral load carrying capacity.
iii. Well should be able to resist uplift loads if required.
iv. Well should be able to resist overturning moments.

9.10.1 Vertical Load Capacity

Applied vertical loads on a well are resisted by:

a. Skin friction
b. Bearing resistance of soil at the base of the well (Fig. 9.35).

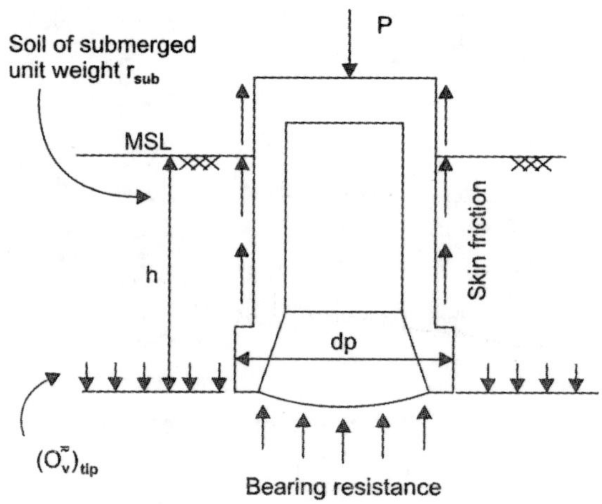

Fig. 9.35: Forces along well

To be on safe side, neglecting skin friction, the bearing resistance of soil at base of well = Q_b.

$$Q_b = \frac{(\overline{\sigma_y})_{tip} \times (N_q - 1) \times A_\Phi}{FOS} \quad ...(9.4)$$

where,

$(\overline{\sigma_y})_{tip}$ = effective vertical overburden pressure at the base of well = $\gamma_{sub} \times h$

N_q = Terzhagi's bearing capacity factor

$A_p = \pi dp^2/4$

FOS = Factor of safety, say 3.0

For safety, $Q_b > P$ = Well is shape for axial compressive loads.

9.10.2 Lateral Load Capacity

If (grip length/diameter of well) > 1.0 the well may be treated as a light well as a light well and assumed to undergo rigid body rotation under lateral loads about a point above its base.

$$\text{Safe lateral load capacity, } H \text{ safe} = \frac{\frac{1}{2}\gamma_{sub}D\,(K_p - K_a)\,(D - 2D1)\,de}{\text{FOS}} \quad ...(9.5)$$

where,

$$D_1 = \frac{3h \pm \sqrt{9h^2} - 6D(h - D/3)}{2}$$

h = Height of line of action of lateral load above base of well

D = Grip length of well

K_p, K_a = Coefficients of passive and active earth pressures

γ_{sub} = Submerged density of soil

De = External diameter of well

FOS = Factor of safety, say 2.0

For safety, H_{safe} > Applied lateral load on well.

9.10.3 Uplift Capacity

The maximum uplift which a well can sustain neglecting skin friction = submerged weight of well.

Design Example 9.1 (contributed by Prof. Bhupinder Singh, IIT Roorkee)

Design a well foundation for Transmission Tower (TT) for 400 KY line (across a river). The site data are as follows. The river cross-section and other details are also illustrated in Fig. 9.36.

Loading details of the transmission tower

Tower type	*Base width at top of pedestal*	*Maximum compression*		*Maximum uplift*		*Leg inclination with vertical*	*Lateral load (kN)*	
		NC	BWC	NC	BWC		NC	BWC
A3 + 25	24000 mm both ways (b/w)	816 kN	1256 kN	724 kN	724 kN	14°15′	76 kN	94 kN

Soil characteristics:

Poorly graded fine to medium sand (SP) up to 30 m from NGL

Average corrected 'N' value = 10

Angle of internal friction, ϕ, = 30°

Stream characteristics:

Maximum flood discharge (50 year return period) = 2010 m^3/s

Maximum stream velocity = 1.94 m/s

RL of HFL = + 149.40 m

RL of surrounding ground = + 150.0 m

RL of LWL = +143.20 m

RL of river bed = +141.20 m

Silt factor = 0.83.

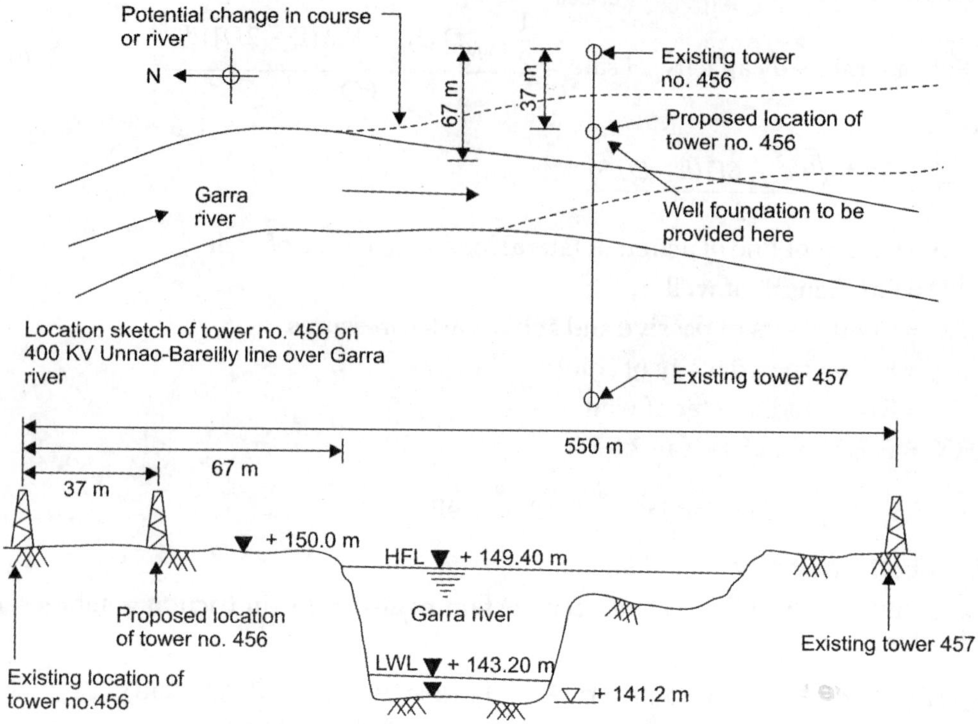

Fig. 9.36: Section across the river

Dimensioning of the Well Foundation

Scour depth: Since the structure is not a bridge, linear water-way and possible construction of stream is not relevant. Hence, scour is calculated using Lacey's formula.

According to Lacey's formula,

$$d = 0.473\left(\frac{Q}{f}\right)^{1/3}$$

where,

d = normal depth of scour, (m)
Q = design discharge, (m^3/s)
f = Lacey's silt factor

Taking value of $f = 0.68$ for the fine sand as a conservative measure based on recommendations of IS: 3955–1967.

In absence of catchment area data, the magnification factor for design discharge is taken as 1.20.

Maximum design discharge = $1.20 \times 2010 = 2412\ m^3/s$

$$\text{Normal depth of scour} = d = 0.473\left(\frac{2412}{0.68}\right)^{1/3} = 7.20 \text{ m}$$

Since tower is to be located in straight reach of river, the maximum scour depth

$$d_{max} = 1.27d = 9.20 \text{ m}$$

Moderate bend = $1.50d$; severe band = $1.75d$; Nose of pier = $2.0d$.

From scour considerations minimum grip length for the well foundation = $0.33 \times$ maximum scour depth = $0.33 \times 9.2 = 3.0$ m. However, provide a grip length of 9.0 m as a conservative measure.

Hence, RL of base of well (i.e. bottom of well curb) with respect to

$$HFL = 149.40 \text{ (HFL)} - 9.2\ (d_{max}) - 9.0 \text{ (grip length)}$$
$$= 131.2 \text{ m}$$

Under existing conditions, well is to be located on dry ground. Keep top of well cap at NGL, i.e. RL of +150.0 m.

Total height of well = 150 – 131.2 = 18.80 m.

Thickness of Well Steining

As per IRC: 78 – 2000 thickness of well steining shall not be less than 500 mm nor less than the following:

$$t = k \cdot de\sqrt{D_e}$$

where,

t = Minimum thickness of well steining, m

de = External diameter of well. No bridge pier is resting on well, however an external diameter of 6.0 m which will give an approximate dredge hole diameter of 4.0 m is proposed. This will ease sinking of the well.

D_e = Depth of well in 'm' below LWL or NGL, whichever is greater = 18.80 m

k = A constant equal to 0.03 for sandy soils

$t = 0.03 \times 6.0\sqrt{18.80} \doteq 0.78$ m

Adopt a steining thickness of 1.00 m for sufficient sinking effort.

Internal diameter of well = 6 – 1 – 1 = 4.0 m > 2.0 m as per IRC: 78:2000. Hence OK.

The thickness of well cap is taken as 1.00 m.

Provide a PCC top plug 0.50 m thick

Take height of well curb = $\frac{1}{2}$ × internal diameter of well = $\frac{1}{2} \times 4.0 = 2.0$ m

Take projection of well curb beyond steining = 75 > 50 mm minimum projection. Adopt sand filling in well. No intermediate plug is used. The proportioned well dimensions are as follows (Fig. 9.37).

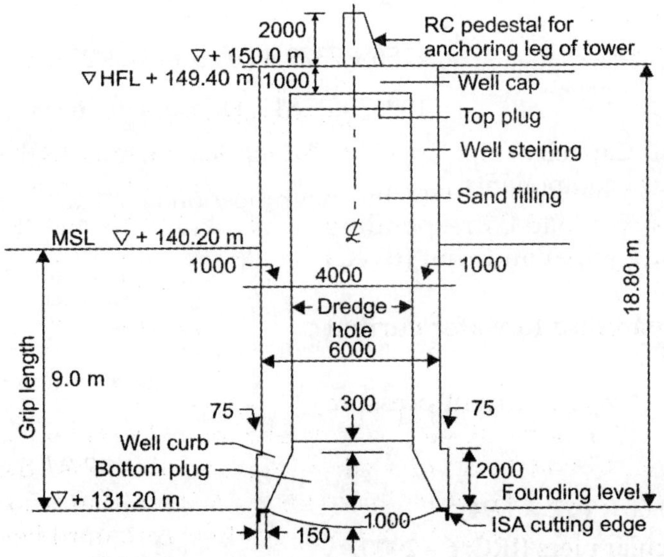

Fig. 9.37: Section elevation of well foundation

Estimation of Well Capacity (Using Simplified Equations)

1. **Uplift capacity** (Fig. 9.38): The uplift loads are assumed to be resisted by the submerged dead weight of the well ignoring skin of the well and ignoring skin friction as a conservative measure.

Maximum uplift load = 724 kN

Submerged dead weight of well = w

$$W = Q_{u1\,safe} = \frac{\pi(6^2 - 4^2)}{4} \times 18.80 \times (25 - 10)$$

$$= 4430 > 724 \text{ kN, Hence OK.}$$

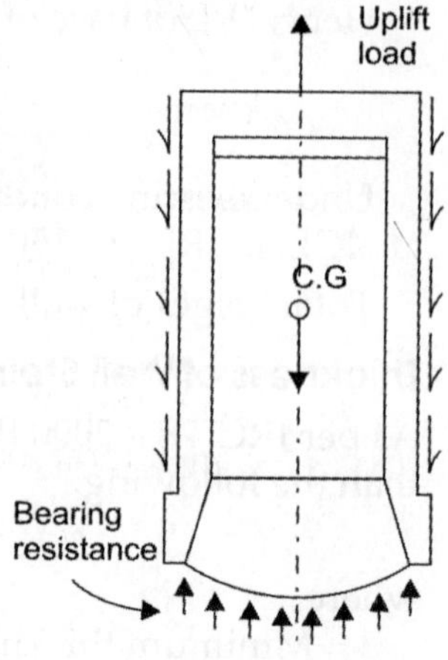

Fig. 9.38: Uplift of well

2. **Safe compressive load:** Neglecting skin friction, the safe compressive load capacity is the bearing resistance at base with a FOS of 3.

Bearing resistance of base

$$Q = Q_{b\,safe} = \frac{\sigma_v \text{ @ base} \times (N_q - 1) \times A_p}{\text{FOS}}$$

For $\phi = 30°, N_q = 22$. Take $\gamma_{sub} = 10 \text{ kN/m}^3$

$$Q_b, \text{safe} = \frac{(140.20 - 131.20) \times 10 \times 21 \times \frac{\pi \times 6^2}{4}}{3(= \text{FOS})} = 17813 \text{ kN}$$

Maximum compression load = 1256 + dead weight of well under dry condition (more conservative case).

$$= 1256 + \frac{\pi(6^2 - 4^2)}{4} \times 18.80 \times 25 + \frac{\pi \times 6^2}{4} \times 1.0 \times 25 +$$

$$\frac{\pi \times 4^2}{4} \times 0.5 \times 24 + \frac{\pi \times 4^2}{4} \times 15 \times 18 + \frac{\pi \times 5^2}{4} \times 2.3 \times 24$$

$$= 1256 + 7379 + 706.5 + 151 + 3391 + 1083$$

$$= 13967 < 17813. \text{ Hence, safe in compression}$$

3. **Lateral Load Capacity** (Fig. 9.39): The lateral load acting on the well foundation consists of two components:

   i. Design lateral load corresponding to breaking wire condition (BWC) = 94 kN

   ii. Lateral load due to water current under HFL.

   Intensity of water current pressure at HFL

   $= 0.52 \text{ KV}^2$

   $= 0.52 \times 0.66 \times (\sqrt{2} \times 1.94)^2$

   'k' for circular piers (IRC: 6 – 2000)

   $= 2.58 \text{ kN/m}^2$

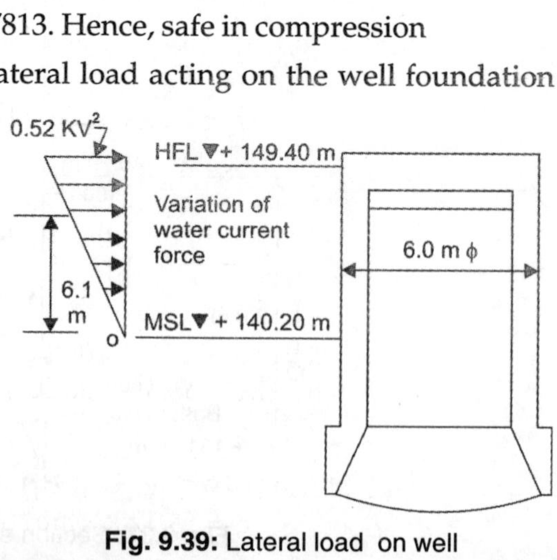

Fig. 9.39: Lateral load on well

Water current pressure at MSL = 0

Water current (lateral) force $= \frac{2.58}{2} \times 6 \times (149.40 - 140.20) = 71.20$ kN

Hence, total lateral load on well = 94 + 71.20 = 166 kN

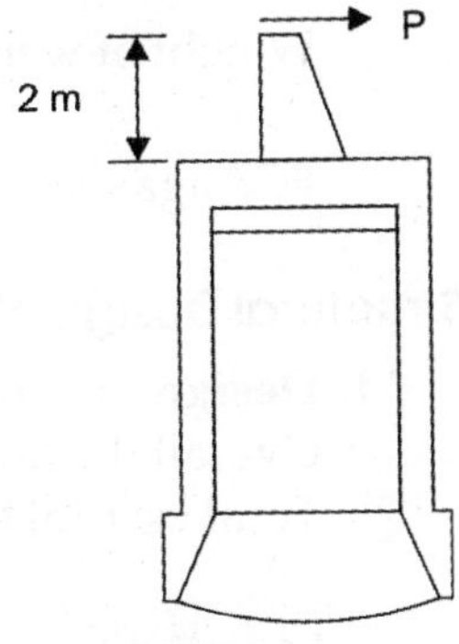

Fig. 9.40: Lateral load on well

Since grip length (9.0 m) is greater than dia. of well (= 6.0 m) the well may be treated as a light well and is assumed to tilt about a point above its base under the action of lateral loads (Fig. 9.40).

Safe lateral load capacity = H_{safe}

$$H_{safe} = \left[\frac{1}{2}\gamma_{sub} D(K_p - K_a)(D - 2D_1) \times de\right]/\text{FOS}$$

$$D_1 = \frac{3h \pm \sqrt{9h^2 - 6D(h - D/3)}}{2}$$

To find 'h' take moments of all lateral loads about base of well.

$$h = \frac{71.20\,(6.1 + 9) + 94(152 - 131.20)}{(71.20 + 94)} = \frac{3030.32}{165.2} = 18.34 \text{ m}$$

$$D_1 = \left[3 \times 18.34 \pm \sqrt{9 \times (18.34)^2 - 6 \times 9\left(18.34 - \frac{9}{3}\right)}\right]/2$$

$$D_1 = \frac{55.02 \pm \sqrt{3027.20 - 828.36}}{2} = 50.95,\ 4.0 \text{ m}$$

$$\therefore \quad H_{safe} = \left[\frac{1}{2} \times 10 \times 9.0\,(3 - 0.3)\,(9 - 2 \times 4) \times 6\right]/2$$

$$= 364.5 \text{ kN} > 166 \text{ kN}$$

Hence, well is safe for lateral loads.

4. **Check on base pressures:** Check is carried out under HFL condition. Overturning moment at base = (71.20 + 94) × 18.34 = 3030 kNm

$$M = 3030 \text{ kNm}$$

Maximum axial load at base = dead load of well under submerged condition + superimposed load = P

$$P = 4430 + 1256 = 5686 \text{ kN}$$

$$A = \frac{\pi d^2}{4} = 28.24 \text{ m}^2$$

$$Z = \frac{I}{y} = \frac{\frac{\pi d^4}{4}}{d/z} = 21.20 \text{ m}^2$$

$$\frac{P}{A} \pm \frac{M}{Z} = \frac{56.86}{28.27} \pm \frac{30.30}{21.20} = 201.13 \pm 142.01$$

$\sigma_1 = 343.14$ kN/m^2 (check with allowable BC at base)

$\sigma_2 = 59.12$ kN/m^2 (No tension is developed at base)

5. **Check against overturning:**
Conservatively, the passive resistance of soil is ignored.
Maximum overturning moment at base = 3030 kNm
Maximum restoring moment due to submerged dead

$$\text{Weight of well} = 4430 \times \frac{6}{2} = 13290 \text{ kNm}$$

$$\text{FOS against overturning} = \frac{13290}{3030} >> 1.5. \text{ Hence OK.}$$

Structural Design of Well Components (Ref. Figs 9.41 and 9.42)

1. **Design of well cap:**
Overall thickness of well cap = 100 mm
Total vertical loading on well cap = superimposed load from tower + weight of RC pedestal + dead load of well cap = P
[Assume RC pedestal as 400 × 700 × 2000 high having a dead load = 0.4 × 0.7 × 2.0 × 25 = 14 kN]
Moment on well cap = Lateral load under BWC × lever arm

$$M = 94 \times 2 \text{ (= height of RC pedestal)} = 188 \text{ kNm}$$

$$\text{Eccentricity } e = \frac{M}{P} = \frac{188}{1977} = 0.09 << \frac{d}{6} = 1.0 \text{ m. Hence OK.}$$

The distribution of bearing pressure on the well steining due to loading from well cap is given by

$$\sigma_{1,2} = \frac{P}{A}\left(1 \pm \frac{6e}{B}\right)$$

$$A = \frac{\pi}{4}(d_1^2 - d_2^2) = \frac{\pi}{4}(6^2 - 4^2) = 15.71 \text{ m}^2$$

$$\sigma_{1,2} = \frac{1977}{15.71}\left(1 \pm \frac{6 \times 0.09}{6}\right) \approx 126 \text{ kN/m}^2$$

[Assume uniform bearing pressure on steining]

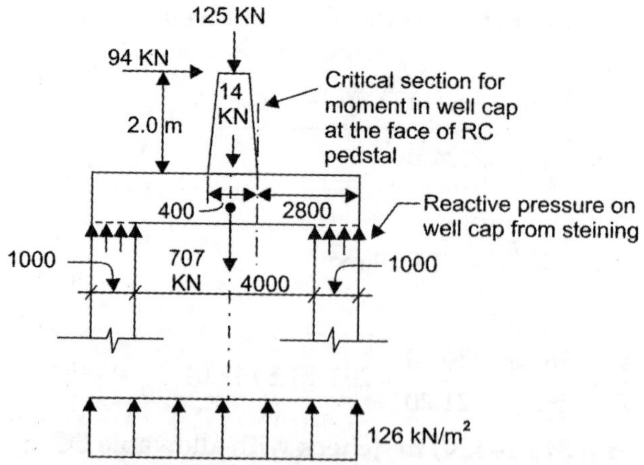

Fig. 9.41: Distribution of bearing pressures on well steining

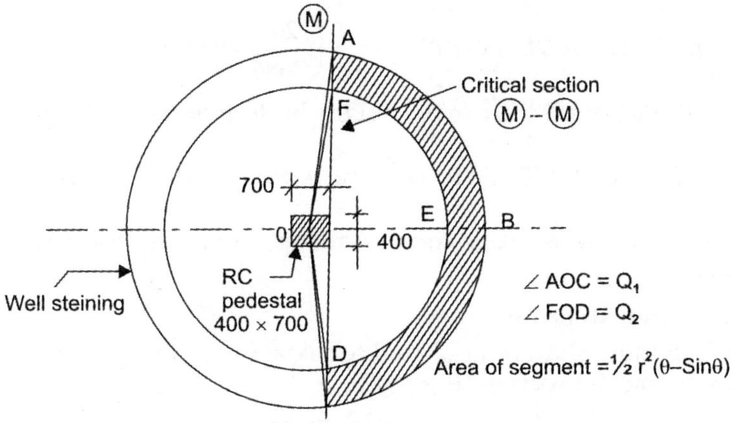

Fig. 9.42: Critical section for moment in well cap (plan view)

$$< \text{AOC} = Q_1 = 2\cos^{-1}\left(\frac{0.35}{3.0}\right) = 166.60° \text{ or } 2.90 \text{ rads}$$

$$< \text{FOD} = Q_2 = 2\cos^{-1}\left(\frac{0.35}{2.0}\right) = 159.84° \text{ or } 2.78 \text{ rads}$$

$$\text{Area of segment ABC} = \frac{1}{2} \times 3^2\ (2.90 - \sin 166.60°) = 12 \text{ m}^2$$

$$\text{Area of segment FED} = \frac{1}{2} \times 2^2\ (2.78 - \sin 159.84°) = 4.87 \text{ m}^2$$

$$\text{Distance of CG of segment ABC from 'O'} = \frac{4R\sin^3 \frac{Q}{2}}{3(Q - \sin Q)}$$

$$= \frac{4 \times 3 \sin^3\left(\frac{166.60}{2}\right)}{3(Q - \sin Q)} = 1.04 \text{ m}$$

Net bending moment at face of RC pedestal

= Reactive pressure × area of segment ABC × CG distance– reactive pressure × area of segment FED × CG distance

= 126 × (12 × (1.47 – 035)) – 126 × (4.87 × (1.04 – 0.35))

= 1693. 44 – 423.40

= 1270 kNm.

(This induces tension on bottom face of well cap)

Assume M 35 grade concrete and Fe-415 grade steel.

$$\text{Minimum depth required} = \sqrt{\frac{1270 \times 10^6}{1.883 \times 6000}} = 7686 \text{ mm}^2$$

$$D_{\text{eff}} \text{provided} = 1000 - 50 - 8 = 942 > 335 \text{ mm. Hence OK.}$$

$$\text{Area of steel required} = \frac{1270 \times 10^6}{200 \times 0.877 \times 942} = 7686 \text{ mm}^2$$

$$\text{Spacing of } 16\phi \text{ bars @ } 150 \text{ c/c} = \left(\frac{201}{7686} \times 6000 = 156 \text{ mm c/c}\right)$$

Spacing of 16ϕ bars @ 150 c/c b/w at the bottom of the well cap as an orthogonal mesh.

Provide 12ϕ bars @ 150 c/c b/w at the bottom of the well cap as an orthogonal mesh.

Because of the relatively light loading, check for two-way shear is not performed.

2. **Design of well steining:**

   The well steining shall be analyzed for axial loads and moments by assuming it to be an uncracked section (Fig. 9.43).

$$\text{Lateral loads acting on well} = \frac{94}{\text{BWC}} + \frac{71.20}{\text{WC}} = 166 \text{ kN}$$

The lateral load of 166 kN is assumed to be balanced by the net lateral resistance of the soil below the maximum scour level.

The resultant lateral earth pressure force at depth 'y' below MSL is given by

$$= \frac{1}{2}\gamma_{sub}\,(k_p - k_a)\,y^2 \times de.$$

Equating the lateral loads at a depth 'y' below MSL gives the location if zero-shear (and maximum moment) section

$$166 = \frac{1}{2} \times 10(3 - 0.3)\,y^2 \times 6 \text{ or } y = 1.431 \text{ m}$$

$$\text{Weight of well steining of 1.431 m height} = \frac{\pi}{4}\,(6^2 - 4^2) \times 1.431 \times 25 = 562 \text{ kN}$$

Moment of lateral forces about section of zero shear $= 166 \times (18.34 - 9 + 1.431)$
$= 1788 \text{ kNm} = \text{M}$

Total axial load at section of zero shear $= 1256 + 14 + 707 + 562$
$= 2539 \text{ kN} = \text{P}$

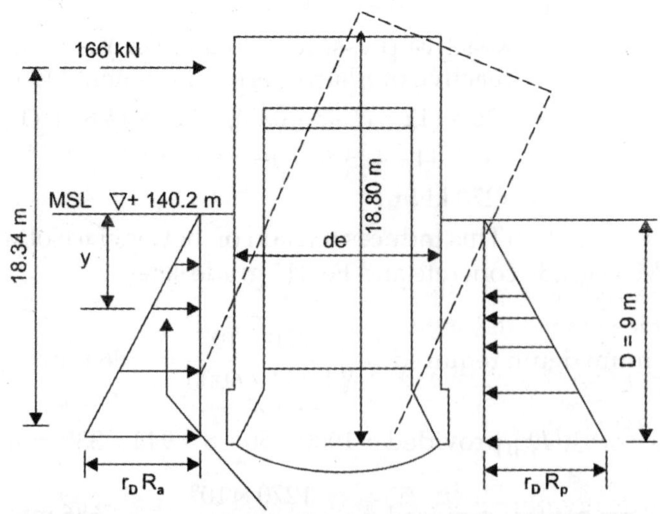

Fig. 9.43: Details of forces on well steining

The stresses in the steining $\sigma_{1,2}$

$$= \frac{2539}{15.71} \pm (6^2 - 4^2) = 15.71 \text{ m}^2$$

$$I_{xx} = I_{yy} = \frac{\pi}{64}(6^4 - 4^4) = 51.05 \text{ m}^4$$

$$y = \frac{de}{2} = 3$$

$$\sigma_{1,2} = \frac{2539}{15.71} \pm \frac{1788}{51.05} \times 9 = 161.61 \pm 105.07$$

If well steining is assumed to be made of M25 grade concrete then

$\sigma_1 = 266.68 = 0.266$ MPa << 8 MPa

$\sigma_2 = 56.54 = 0.056$ MPa > 0. Hence OK.

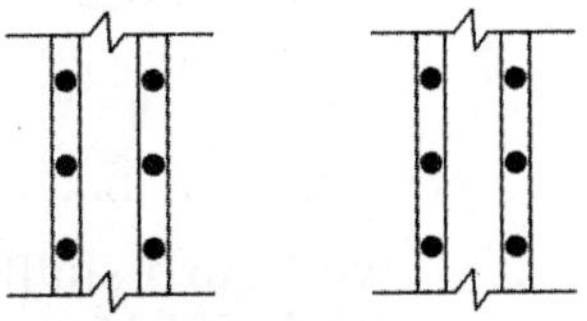

Fig. 9.44: Details of reinforcement

Hence steining section is safe on an uncracked section. Since steining is considered as a plane concrete, uncracked section provide 0.12% minimum steel

Required area of longitudinal steel $= \frac{0.12}{100} \times \frac{\pi}{4}(6^2 - 4^2) = 18850 \text{ mm}^2$

Area of longitudinal/vertical steel on each face $= \frac{18850}{2} = 9425 \text{ mm}^2$ (Fig. 9.44).

No. of 16ϕ bars required on each face $= \frac{9425}{201} = 47$ nos.

Provide 47 nos. of equally spaced 16ϕ bars on both inner and outer faces of steining as longitudinal steel.
Provide hoop steel in steining at 0.04% of volume per unit height of steining.

Volume of hoop steel per 'm' height of steining $= \frac{0.04}{100} \times \frac{\pi}{4}(6^2 - 4^2) \times 1.0$

$$= 6283185 \text{ mm}^3$$

Volume of hoop steel required on each face = 3141593 mm^3
Total c/s area of hoop bars required on each face per metre height of steining

$$= \frac{3141593}{[\pi \times (6000 - 75 - 5)]} = 168.91 \text{ mm}^2$$

Spacing of 10 mm diameter hoops $= \frac{78.5}{168.91} \times 1000 = 464.7$ mm c/c

Provide 10 ϕ hoops in the form of closed rings on both the inner as well as the outer face of the well steining at 300 c/c.

3. **Design of well curb** (Fig. 9.45):

Average width of well curb $= \frac{1.075 + 0.15}{2} = 0.6125$ m

Approximate volume of well curb $= \pi \times [(4 + 2 \times 1 + 2 \times 0.075) - (0.6125)] \times 2 \times 0.6125$

$$= 21.31 \text{ m}^3$$

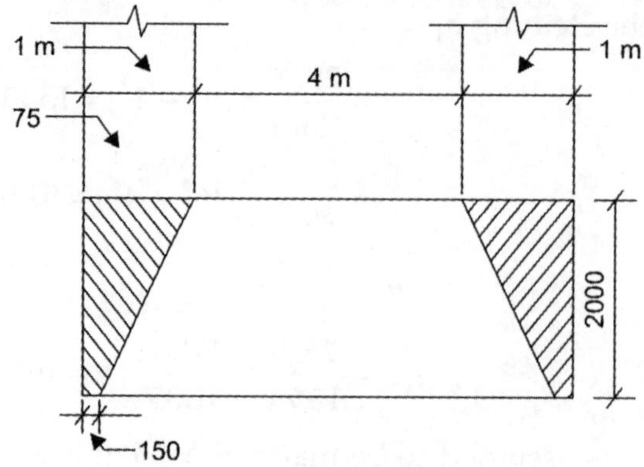

Fig. 9.45: Design of well curb

well curb shall be reinforced with minimum steel of 72 kg per m^3. Total weight of steel in well curb = 72 × 21.31 = 1535 kg

4. Provide 24 nos. of 20 mm diameter closed rings evenly spaced along perimeter of well curb. Assume average diameter of each ring = 5.0 m (Figs 9.46 and 9.47):

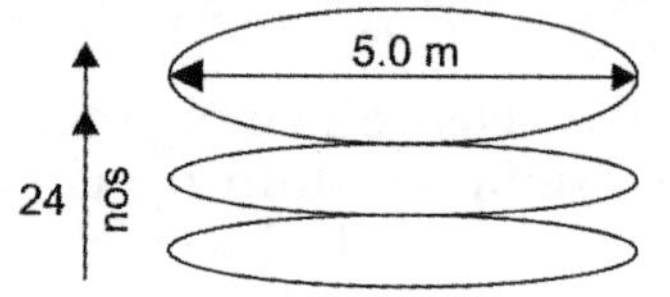

Fig. 9.46: Steel reinforcement rings

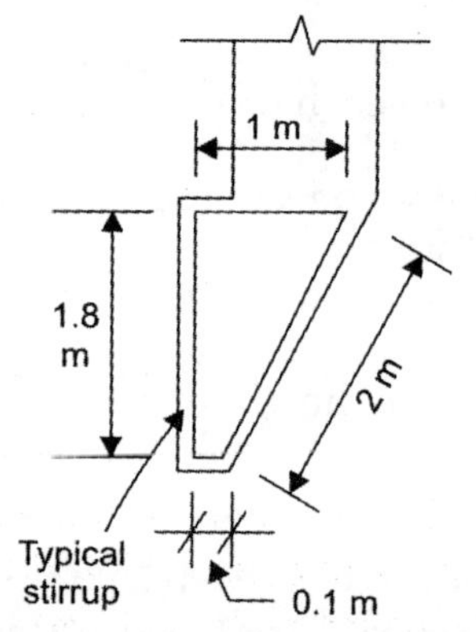

Fig. 9.47: Reinforcement details

$$\text{Weight of 24 nos. of 20 mm closed rings} = \frac{\pi \times 0.002^2}{4} \times 24 \times \pi \times 5 \times 7850$$

$$= 930 \text{ kg}$$

Provide 80 nos. of 16 mm diameter stirrups equally spaced along the circumference of the well curb.

Average approx. length of each 16 mm diameter stirrup = 1 + 1.8 + 0.1 + 2 = 4.9 m

$$\text{Weight of 80 stirrups} = \frac{80 \times \pi \times 0.16^2}{4} \times 4.9 \times 7850 = 618.70 \text{ kg}$$

Total weight of steel in well curb = 930 + 618.7 = 1549 kg > 1535 kg

Hence, design of well curb is OK.

Chapter

10

Well Foundation Construction in Bouldery Bed

Contributed by Er. RK Dhiman, VSM

Well foundation has been used for number of bridge piers in various parts of the world. Construction of well foundation in case of sandy soil strata is easy task, whereas well foundation in case of bouldery bed is really a challenge for bridge engineers. There are number of cases where this type of foundation has been used at bouldery bed sites but it has posed a challenge to bridge engineers due to heterogeneous character of the soil strata underneath. A case study of a well foundation construction for bridges in bouldery bed strata has been discussed here.

10.1 INTRODUCTION

Foundation design and construction in bouldery bed has its own challenges firstly due to nonavailability of any reliable formula for bouldery bed scour and sinking problem there on due to tough strata. Normally, depth is initially finalized as per Indian Road Congress code formula and revision is planned based on tough strata. Foundation depth of well has to be finalized keeping in view the construction technique available *viz*, safety and stability requirement. Infact, in case variation in strata is encountered it should be considered accordingly and revision on either side be considered. Also the tough strata be considered as engineering friendly. Whenever there is change in properties of the soil, the design be amended accordingly for completion of foundation. Based on the experience of the Border Roads Organisation well foundation construction on bouldery bed strata has been discussed here.

10.2 WELL FOUNDATION AN OVERVIEW

Correct assessment of scour is important task for the overall safety of the bridge. Normally, no tension is permitted at the base of well foundation as per codal guidelines. In case the soil strata comprises of rock underneath, the well is sunk into rock and rock level is assumed as scour level and design is finalized as per the codal provisions. However, in case of rock, 20% tension is permitted if need arises and well is suitably anchored. Well foundations are used to ensure design and hydraulic requirement. Shape of well can be square, circular, double-D depending upon the site requirement and construction methodology available, circular well foundation is preferred. Foundations are taken to a sufficient level below the maximum scour level to satisfy design and stability requirement. However during the process of sinking if change in strata is observed, an action is taken according to engineering consideration. There are cases where based on the difficulties faced in sinking the well, pneumatic sinking technique has been used to expedite sinking. Pneumatic sinking is used when the soil strata is tougher and is not possible to excavate with conventional sinking arrangement. Design parameters, to be considered for foundation, are hydraulic and soil parameters. Stratification of soil underneath plays key role in finalizing the design and expediting the construction of foundation. Hydraulic parameters are discharge, velocity, bed slope and soil properties silt factor, bulk density, angle of internal friction, angle of wall friction and cohesion which are assessed based on geological condition of the strata. On finalization of this data, foundation levels are decided. Correct assessment of scour in soil is very important and the same is decided with the help of available formula followed by judicious judgment and if required model study is also carried out to assess the likely scour during the service life of the bridge. This will have vital bearing on completion and serviceability of the bridge at later date.

10.3 CONSTRUCTION OF BRIDGE IN BOULDERY BED

Bouldery bed river strata generally encountered is shown in Figs 10.1 and 10.2. Bouldery beds are generally located in and near foothills. The soil strata is generally heterogeneous is in character. Based on size and shape of the soil strata further action on construction of bridge foundation is taken.

Number of bridges have been constructed and are under construction on bouldery bed by Border Roads Organisation (BRO) in India.

Fig. 10.1: Bouldery bed river

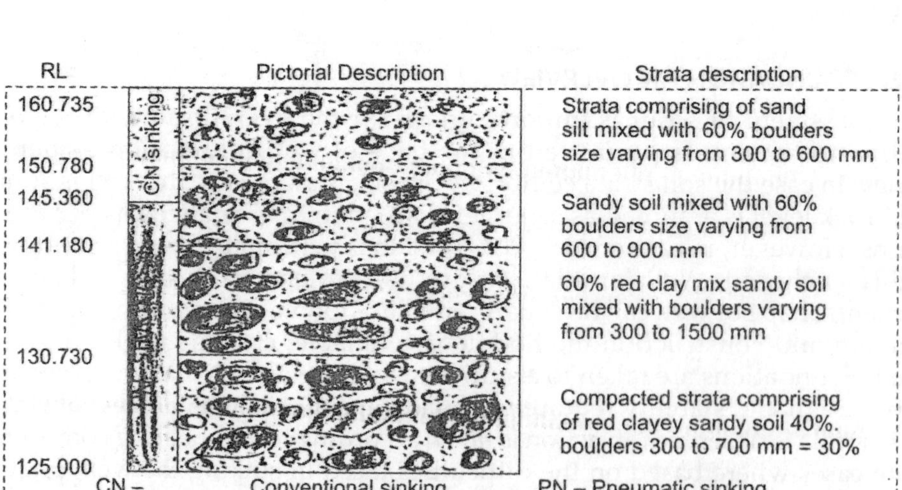

Fig. 10.2: Soil strata generally encountered in bouldery bed

A flow chart of various step being followed for well foundation construction is indicated in Figs 10.3 to 10.7.

Generally, cutting edges are made very strong due to likely blasting anticipated at the site. This is done as per IRC Code and based on past experience of foundation construction in similar strata.

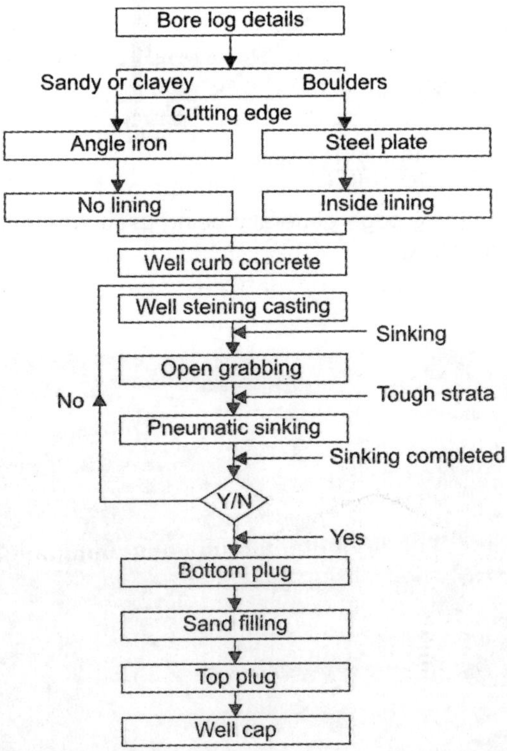

Fig. 10.3: Activities for well foundation construction

Fig. 10.4: Cutting edge being used for well foundation on bouldery bed

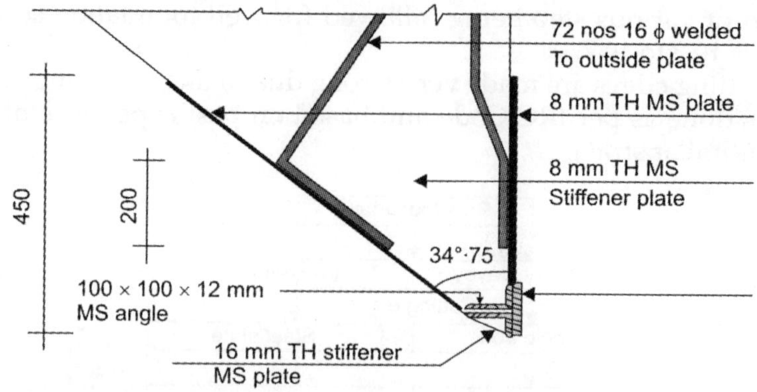

Fig. 10.5: Cross-section of cuting edge generally being used where large size of boulder are anticipated during process of sinking

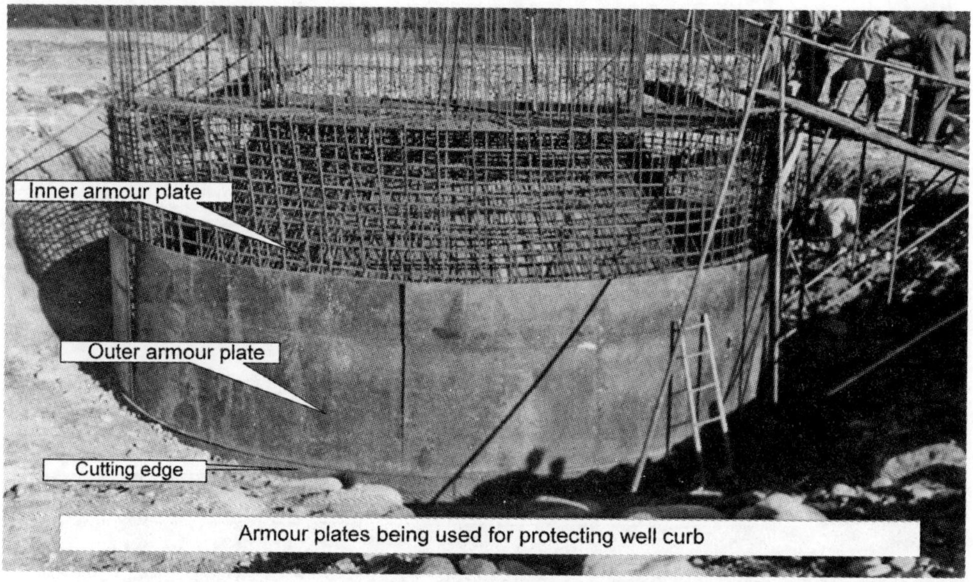

Fig. 10.6: Well curb being used for well foundation in bouldery bed

10.4 WELL FOUNDATION CONSTRUCTION

Based on the construction experience various case studies of well foundation construction have been discussed in this chapter. Case studies includes conventional sinking, pneumatic sinking well launching in water, at Dimwe Dalai and Pasighat bridge. Well foundation at all the locations are single circular well.

10.4.1 Dimwe Bridge

Dimwe bridge is located in Arunachal Pradesh (India) and the span arrangement of this bridge is two spans of length 65.5 m and 22.8 m respectively. Central pier is on well foundation in the main channel of the river and both the abutments are on open foundation. Construction of the bridge took more time than the planned period which included delay due to logistic reasons. Working period was also very less due to heavy

Fig. 10.7: Process of manual sinking during initial sinking of wells

rainfall and unprecedented flood in this area. Salient feature of the bridge are as under:

i. Total length of bridge: 88.30 m (65.50 + 22.80)
ii. Type of structure: Simply supported
iii. Velocity: 5 m/sec
iv. Design HFL: RL 238 m
v. Vertical clearance: 8.5 m
vi. Maximum scour level: RL 222.0 m
vii. River bed level: RL 232.00 m
viii. Foundation level: RL 207.00 m
ix. Silt factor: 1.4
x. Design parameters
   a. Coefficient of lateral friction: 30°
   b. Angle of wall friction: 20°
   c. Bulk density of soil: 1.8 t/cum
   d. Submerged weight of soil: 0.8 t/cum
   e. Buoyancy: 100%

Initial work on this bridge started smoothly. Sinking of well from RL 237.45 to RL 224.00 m was through loose boulders embedded in silty/sandy matrix. Below RL 224.00 to RL 222.59 m rock was moderately hard closely spaced lightly jointed and represented by feebly weathered strained banded quartz, grandiorite, gneisses. Fresh rock occurred below RL 222.59 to RL 219.00 m. Rock was traversed by falling prominent sets of joints (Fig. 10.8).

a. N 30 E S 53 W   Dipping vertical
b. N 70 E S 57 E   Dipping 60 on NE
c. Transverse joints N 20 E, S 20 W subvertical.

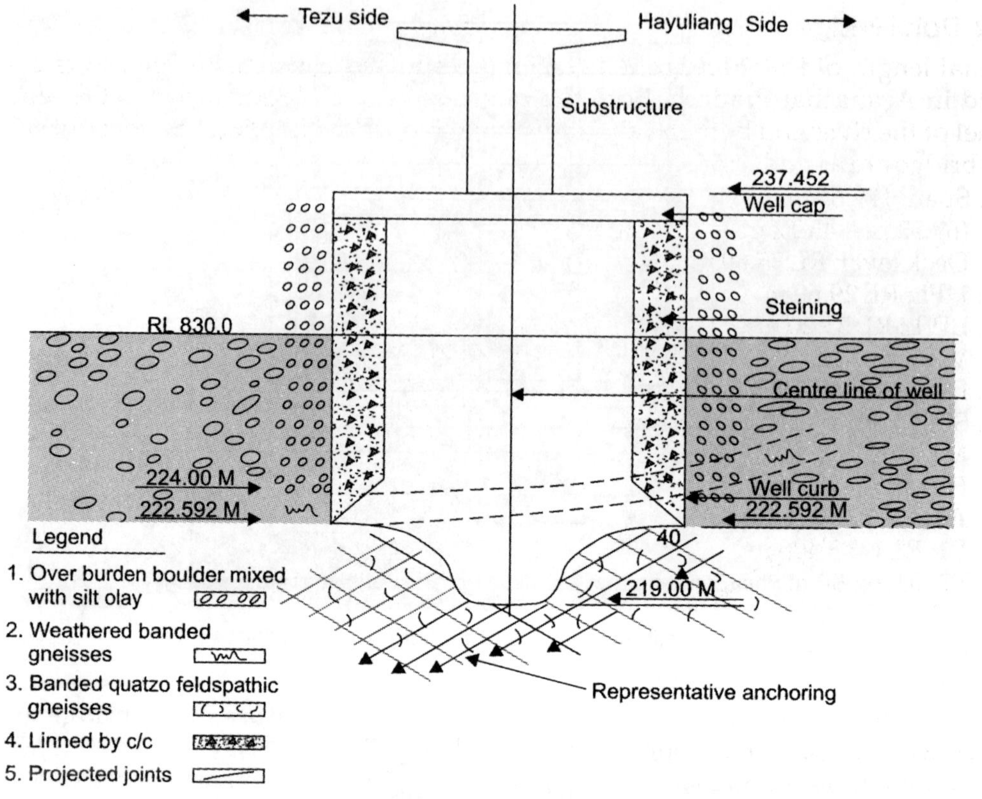

Fig. 10.8: Soil strata encountered during sinking process Dimwe bridge

Besides the two parallel gougy seam of 8 to 9 cm thick striking NE-SW dipping SE were recorded. The intersection of different joints were developing cubical block which were blocked by crushed rock but there were no evidence of crushing. Geologist recommended that anchoring should be provided for the depth of 3 to 5 m in the direction perpendicular to the bending plane (Fig. 10.9). The foundation RL 219.00 m consisting of fresh hard quartz, granodiorite, gneiss lies within the permissible limit. Due to tough strata sinking of foundation became difficult.

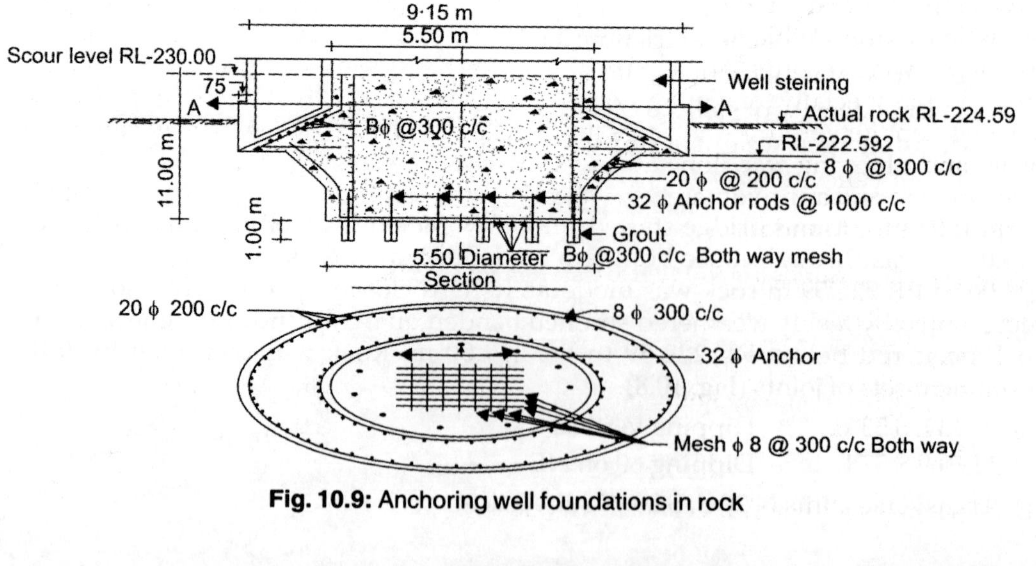

Fig. 10.9: Anchoring well foundations in rock

10.4.2 Dalai Bridge

The total length of Dalai bridge is 129.65 m prestressed concrete bridge and is also located in Arunachal Pradesh. Both the piers are on well foundations in the main channel of the river and both the abutments are on open foundations. Salient features of the bridge are as under:

i. Span: 129.65 m
   (69.52 m + 45.12 m + 15.01 m)
ii. Deck level: RL 35.60 m
iii. HFL: RL 29.60 m
iv. LWL: RL 23.20 m
v. Velocity: RL 8.83 M/Sec
vi. Discharge: 5714 M^3/Sec
vii. Silt Factor (taken for design): 9
viii. Maximum Scour Level: RL 11.90 m
ix. Foundation Level
   A1: RL 29.00 m
   P1: RL (–) 3.400 m
   P2: RL 6.000 m
   A2: RL 29.00 m
x. Diameter of circular well
   P1: 12.00 m
   P2: 9.00 m

Work was started as per approved bridge scheme. Both the abutments A1 and A2 are open foundations and piers P1 and P2 are on well foundation. Initial sinking was started on both the well locations with conventional sinking only. There was no problem in construction of both the abutments and both were completed up to designed RLs. However in case of well foundation, sinking difficulties were faced. Accordingly after reaching RL 18.300 m in case of P1, difficulty in sinking was faced due to tough strata. However, case was examined in greater details in terms of cost aspects and further efforts were made with open grabbing. On further examining, difficulties faced the pneumatic sinking was used and work was restarted accordingly. The scope of work for sinking was also examined in terms of tough strata which is considered to be less scourable and revised RL of foundation were also given to utilize the available strata in a better way (tough strata considered as technically stronger). Pneumatic sinking at this location was started from RL 18.300 m. Soil strata details encountered were recorded. After reaching at RL 12 m, there was problem in sinking the foundation specially for excavating the strata even in pneumatic condition. A close watch was kept for soil strata and SBC got checked at RL 7.490 m which was found out in order and plugging was done at this RL. Since the soil strata was tough, blasting was resorted to and approximately 1.50 tones gelatine was used while sinking. Details of various RLs for foundation and strata report are as under in Tables 10.1 and 10.2.

Table 10.1: Details of foundation level

S1	*Description*	*A1 (RL)*	*P1 (RL)*	*P2 (RL)*	*A2 (RL)*
1	*NIT FL	29.00	(–) 3.40	6.00	29.00
2	Design FL	29.00	(–) 12.00	2.10	–
3	First revision	–	6.00	(+) 8.00	–
4	Final FL	29.00	7.49	11.50	29.00

* NIT: Notice inviting tender, FL: Foundation level

Table 10.2: Strata report of excavated materials: P1 well foundation

S/No	*Date*	*Sample collected at RL*	*Distribution by percentage*	*Remarks*
1	04 Feb 98	19.56	a. Boulder: i. Size 1000 to 2000 mm: 10% ii. Size 500 to 1000 mm: 25% iii. Size up to 500 mm: 25% b. Sand and Gravel: 40%	
2	07 Mar 98	16.90	a. Boulder: i. Size 1000 to 3000 mm: 15% ii. Size 500 to 1000 mm: 30% iii. Size up to 500 mm: 30% b. Sand and gravel: 25%	
3	31 Mar 98	15.47	a. Boulder: i. Size 1 × 2 m: 25% ii. Size 0.5 × 1 m: 25% iii. Size up to 0.5 m: 30% b. Sand and shingle: 20%	Width 20 to 50 cm
4	21 Apr 98	14.50	a. Boulder: i. Size 1 × 2 m: 25% ii. Size 0.5 × 1 m: 23% iii. Size up to 0.5 m: 32% b. Sand ans shingle: 20%	Width varying from 20 to 50 cm
5	19 May 98	13.50	a. Boulder: i. Size 1 × 1.5 m: 20% ii. Size 1 × 0.8 m: 22% iii. Size 0.8 × 0.5 m: 23% iv. Size below 0.5 m: 12% b. Sand and shingle: 23%	

Efforts made during pneumatic sinking viz-a-viz sinking achieved is as under as Table 10.3.

Table 10.3: Details of well sinking

Month	*Manpower*		*Explosive consumed*		*Sinking achieved*
	Sup	*Labour*	*Gel*	*E/D*	*(Mtrs)*
	Man-days		*kg*	*nos*	
May 98	114	2007	44.59	334	1.035
Jun 98	65	1109	13.13	58	0.100
Jul 98	102	1561	58.98	266	0.680
Aug 98	57	1918	46.44	255	0.785
Sep 98	102	1578	29.42	154	0
Oct 98	66	768	11.46	60	0.220
Nov 98	96	1234	23.72	170	0.150
Dec 98	116	1884	43.12	424	0.160
Jan 99	98	1850	21.14	130	0.280
*Feb 99	69	1330	14.17	102	0.400
Total:	**1030**	**17665**	**391.95**	**2540**	**5.120**

After reaching RL 7.50 m plugging of well considered in order.

The soil strata plan at RL 7.50 at foundation level is indicated in Fig. 10.10.

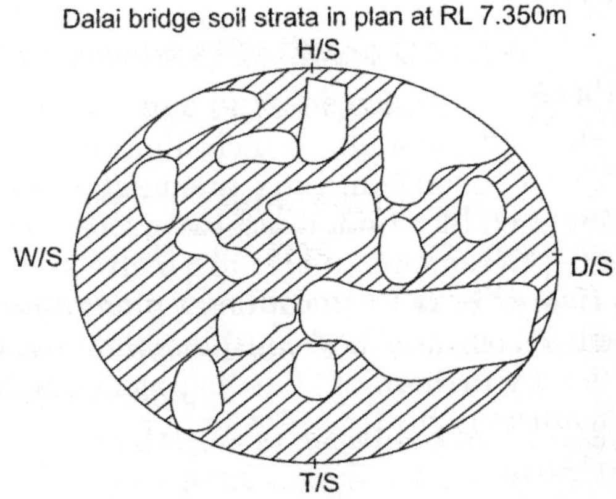

Fig. 10.10: Soil strata at RL 7.50 m

Well curb was anchored with the ground with anchor bar as indicated in Fig. 10.11.

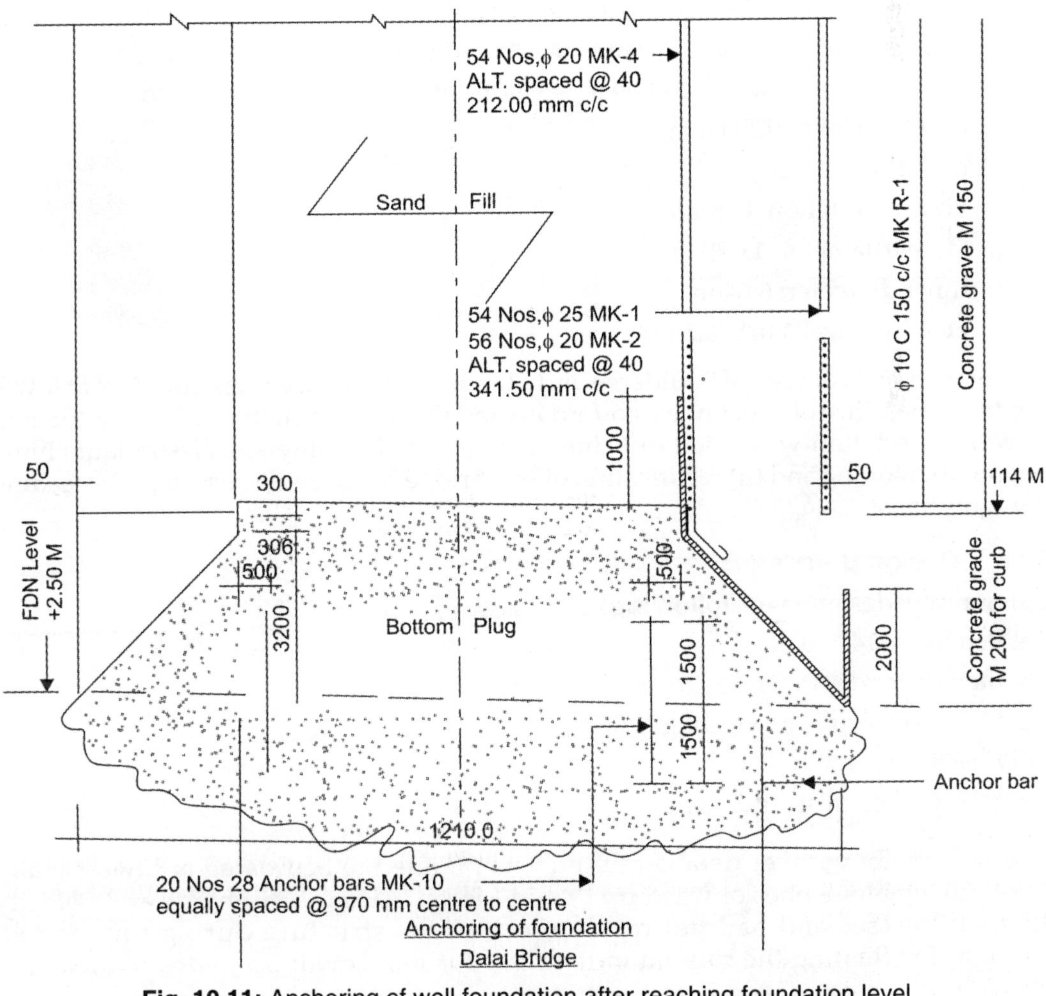

Fig. 10.11: Anchoring of well foundation after reaching foundation level

Well foundation construction at both the location was difficult due to tough strata and it involved lot of efforts to achieve desired level for plugging well foundation.

10.5 CAISSON LAUNCHING

Major bridge having eight well foundations including two abutment wells where all the activities of well sinking, caisson launching, pneumatic sinking and tilt rectification have been discussed. Out of eight foundations, two were constructed in that portion of the river where water depth remains about 7 to 8 metre's even at lowest water level. Maximum flow in this river is near the location of pier P3 for which launching was done. Owing to the river bed condition and considerable water depth at these location, construction of foundation by method of "Floating Steel Caisson" was considered as viable solution. Salient features of bridge are as under:

a. Length of bridge: 703.50 m
b. Maximum discharge: 46000 cum
c. Lowest water level: 153.50 m
d. Highest water level: 164.000 m
e. Velocity of water at HFL: 8 m/sec
f. Top water width at LWL: 400 m
g. Design maximum scour: 133.00 m
h. Foundation level: 125.00 m
i. Foundations
   i. Well foundation: Circular
   ii. Outer diameter: 11.70 m
   iii. Inner diameter: 6.64 m
   iv. Height of well curb: 4.50 m

The river bed consists of boulders-sand-clay-matrix and large size boulders at lower depths. Possibility of sheet piles, cofferdam and sand island method, for construction of well foundation was ruled out due to large sized boulders. Caisson launching, pneumatic sinking and tilt rectification of this bridge has been discussed in succeeding paragraphs.

10.5.1 Design Details

Caisson was designed for following:

a. Bed level: 147.50 m
b. Max water depth while grounding: 7.00 m
c. Max scour while grounding: 1.00 m
d. Free board: 1.50 m
e. Draught for caisson: 2.86 m

Based on above river details caisson was planned to be floated at lowest water level during November onward up to mid March. Also the discharge and velocity 1.0 to 1.2 m/sec and had not much effect on the structure during this period. Draught for floating the caisson in the river for movement up to desired location was 2.86 m.

10.5.2 Steel Caisson Launching

a. Resources used for above event are as follows:

i. Crane: 1

ii. Power generator 55 KVA: 1

iii. Power generator 10 KVA: 1

iv. Concrete mixer: 2

v. Vibrator: 4

vi. Power tug

  i. Power boat with 1 HP engine: 1

  ii. Mallard type 90 HP engine: 2

vii. Barges

  i. For crane mounting: 1 (200 Tonnes capacity)

  ii. For mixer 12 × 14 m: 1 (150 Tonnes capacity)

  iii. For aggregate 12 × 12 m: 1 (130 Tonnes capacity)

  iv. For feeding aggregates: 1 (100 Tonnes cement/HSD at 8 × 4 m capacity)

viii. Buoys

  Main: 5 Nos

  Ordinary: 10 Nos

  Anchor blocks

  RCC blocks: 15 Nos

  PSC blocks: 2 Nos

ix. D shackles 5 to 20 tonnes capacity: 100 Nos

x. Hand Winches (5 tonnes capacity)

  For caisson: 5 Nos

  For Pontoons: 10 Nos

xi. Wire ropes (minimum)

  3/4" wire rope: 3000 RM

  1" wire rope: 600 RM

  1.25" wire rope: 200 RM

xii. Pully blocks

  Single sleave: 5 Nos

  Double sleave: 5 Nos

  Three sleave: 5 Nos

xiii. Welding machine: 3 Nos

10.5.3 Sequence of Activities

Sequence of activities that followed were as under:

a. Area along the U/S and D/S side of the river with reference to proposed bridge axis was surveyed to collect information required for proposed launching. Suitable place for fabrication of caisson and for base line to locate the proposed caisson position was finalized.

b. Cutting edge of the caisson was fabricated in parts at workshop and shifted to proposed location of launching. Caisson was fabricated on the right bank at about 500 m on U/S of proposed bridge location. Fabricated caisson was as under:

   Well curb and Steining up to guage height 9 m (structural steel)

   i. 20 mm PL: 6.50 MT
   ii. 12 mm PL: 8.50 MT
   iii. 10 mm PL: 11.00 MT
   iv. 6 mm PL: 17.60 MT
   v. 3.5 mm PL: 2.20 MT
   vi. 50 × 50 × 6 angle: 2.20 MT
   vii. 90 × 90 × 8 angle: 7.00 MT
   viii. 65 × 65 × 6 angle: 13.50 MT

   Well curb reinforcement

   i. 16 mm diameter MS: 0.60 MT
   ii. 25 mm diameter MS: 17.0 MT
   iii. 28 mm diameter MS: 2.00 MT

c. Height of caisson was 9 m and outer diameter was 11.70 m. Total weight of the caisson was 87.50 MT. Draught required for the caisson was arrived at 2.86 m. After fabrication structure was anchored with the winches placed on right bank. Fabrication of caisson was done at 40 m from river bank, from where the caisson was hauled towards water, with the help of winches and rail track laid on the ground up to 1.5 m from the water edge.

d. After placing caisson near to water edge, jetty was extended which was 20 m wide and 15 m long towards water. Rail track was laid up to jetty for movement of the caisson towards water edge with the help of winches placed on the bank. Caisson was further moved on to the jetty where water depth was about 3.5 m all around. Caisson was kept under control at jetty location.

e. Open grabbing was done on the jetty location so as to render the caisson under floating condition. During the open grabbing cutting edge got punched into the existing bed on one side and the caisson movement got stuck. To overcome this difficulty, further open grabbing was resorted, to bring the caisson, into floating condition. This was achieved with the help of divers and experienced persons. Once the caisson was made to float, it was planned to detach its control from the bank winches and the same was shifted to buoys in floating condition. This was done with the help of steel boat.

f. On floating the caisson, the movement of the caisson towards the proposed location was controlled. Path of travel was kept parabolic towards D/S to avoid the drag of water. Five sets of identical anchor blocks consisting of hexagonal shape were dropped at predetermined locations on U/S and D/S of the proposed location of well P3 attached to the buoys. One set of anchor block was having four blocks of 4 to 5 tonnes each. These blocks were dropped keeping in view the requirement of reaction required so that sufficient resistance is provided during the movement of caisson. Movement was controlled from the winches on the bank till the control was transferred to the buoys in floating conditions. Caisson was moved to D/S in 3 hours and on reaching near the proposed location the movement of caisson was slowed down.

g. At proposed location the caisson was positioned and movement of winches was locked. Theodolites were placed on the left bank on two stations already marked location which is 206 m apart and 103 m on either side of proposed line of bridge. The location of stations were so selected that they should form well-conditioned triangle. The position of well was checked with two theodolites and cross checked with third theodolite placed along centre line of bridge (Fig. 10.12).

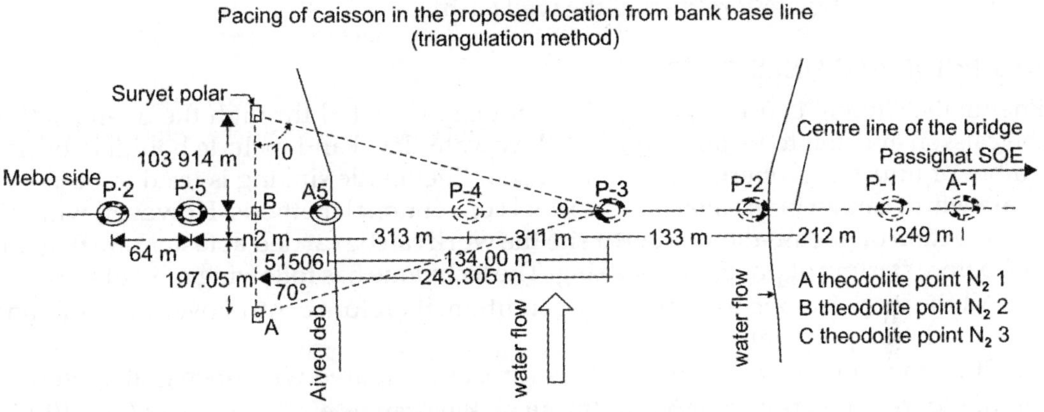

Fig. 10.12: Final placing of caisson

h. After alignment positioning of caisson, concreting of curb/steining was started. Concrete was shifted from the left bank with the help of concrete buckets on a boat.

For this purpose two cranes were used, one was at bank and the other on the floating pontoon. During the process of concreting the self-weight of the structure gradually increased and correspondingly lowering of the caisson was done, till grounding of the caisson. To keep the well in alignment during the course of concreting/sinking the winch anchoring arrangements were retained undisturbed until well was sunk by 3.50 m.

10.5.4 Problems

a. Due to uneven bed profile, it was very difficult to move caisson after fabrication in workshop to the bank from where launching was proposed. Initial fabrication of cutting edge was done in the workshop and remaining portion of caisson was fixed at proposed location from where the launching was monitored. This was done, as it was difficult to move the caisson after erection due to uneven bed surface.

b. Draught calculated for the steel caisson was 2.86 m and accordingly the jetty was prepared where the required draught is available after breaking the jetty. After caisson was placed at jetty location, it got stuck in the sandy bed and was difficult to bring in normal alignment as the soil on which it was placed filled one. Innovative efforts were applied for this, simultaneously horizontal forces and divers were deployed at particular location to make the caisson free for movements.

c. After final placing of caisson at the proposed location, well curb concrete was done from the pantoons only where cement and coarse aggregates were placed.

However, this led to slow down the output as the concrete was poured manually and the space available on the pantoon was less. However, this problem was overcome by shifting the arrangement to bank the remaining concrete for the caisson was done from the bank which was shifted through steel bucket of 1.2 cum capacity by flat pantoon and two cranes were used for this, one at bank location and the other near to the well location on floating pantoons.

d. Due to 9 m height of caisson the meta centric height was more and there were every likelihood of toppling the caisson. However, this problem was overcome by proper control. Better solution could have been part fabrication at bank and remaining at the final location before casting the well curb.

10.6 PNEUMATIC SINKING

Pneumatic sinking is used when soil cannot be excavated through the open shaft in the caisson and soil to be taken out is below water level and include boulders buried timber or unusual properties of the soil itself. Pneumatic sinking is used in bouldery strata which cannot be excavated by normal conventional methods. However, with the introduction of pneumatic sinking at site the work can be carried at faster rate than the rate before starting of pneumatic sinking. The technique has following advantages:

i. All the work is done in near to dry condition, therefore control over the work and foundation preparation are better.
ii. Plumbness of the caisson is easy to control as compared with open grab method.
iii. Boulders and logs encountered during sinking can be easily removed (Fig. 10.13).

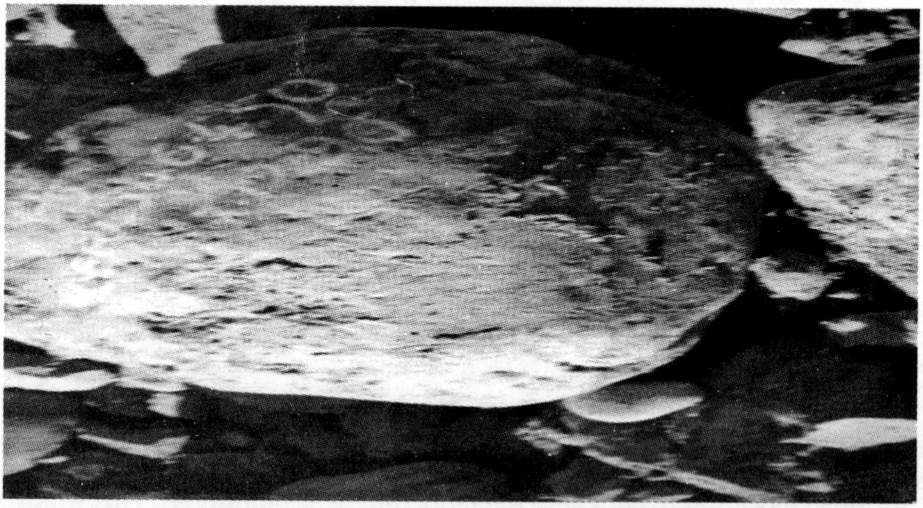

Fig. 10.13: Large size boulder in river bed at 10 to 15 m deep where sinking of well is difficult

10.6.1 Essential Parts and Technical Specification of Pneumatic Caisson

Working chamber: The working chamber is the space at the bottom of pneumatic caisson surrounded by the beveled wall of cutting edge and roof by steel diaphragm or concrete plug (corbel slab). Since it is the space for workmen to excavate the soil, it should be at least 2.5 m high from cutting edge. The side wall and roof are designed to withstand the maximum anticipated air pressure. This air pressure in general is greater than the pressure due to the head of water above the bottom of caisson. (Figs 10.14 and 10.15).

Fig. 10.14: Pneumatic sinking arrangement

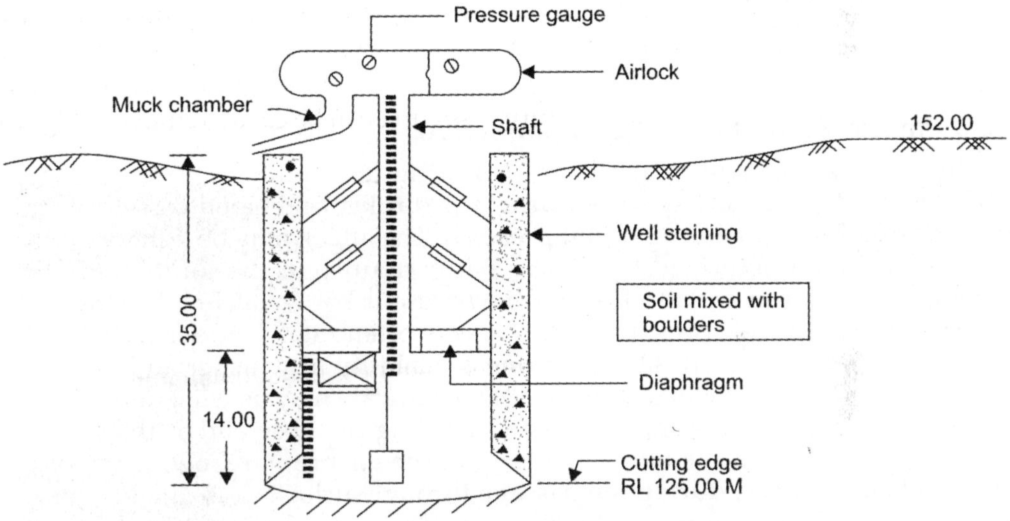

Fig. 10.15: Airlock

Air shaft: Air shaft or shafts are the vertical passage-ways for workmen and materials. In smaller caisson, one shaft may be sufficient with a ladder. One side of shaft is for workmen and the other side for material (removal of soil and placement of concrete). In large caisson, two or three shafts are provided, one for passage of workmen and the other for the materials. Shafts in concrete caisson may be made of steel or may be simply a cylindrical hole. In latter case the concrete must be reinforced to withstand the internal pressure due to the compressed air and the airlock of the shaft must be anchored down to the concrete. Steel shaft are fabricated in short sections of 1.5 to 3.0 m lengths. The points between the section are made airtight with rubber gaskets. As the sinking progresses, additional length of the shaft is added. The shaft must be designed to withstand the internal pressure. Therefore, the common cross-sections are circular, elliptical, or figure of eight. It is ensured that the height of steining is always higher than the LWL.

Airlock: An airlock is required for each shaft. It is mounted on top of the shafts and extends above water level. The function of airlock is to permit the workmen and materials to go in and out of the caisson without releasing pressure in the caisson. Airlock consists of steel chamber with two airtight doors one of which opens to shaft and other opens to atmosphere. When a men enters the airlock through the outside door, the pressure in the airlock is equal to that of atmosphere and then the door is closed and air pressure is allowed to rise slowly. When the pressure in the airlock becomes equal to that of inside of caisson, the door to the shaft is opened and men descend into the working chamber. The procedure is reversed when a men comes out of caisson.

10.6.2 Hours of Work in Compressed Air

Hours of work for workmen who are subject to compression and decompression shall not be more than specified below in any consecutive 24 hours (IS : 4138 : 1977).

Pressure kg/cm²		*Number of hours excluding the period for compression and decompression*
Min	*Max*	
0	1.25	8 (Normal working)
1.25	1.25	6
2.20	3.40	4

10.6.3 Physiological Effects of Compressed Air

Usually, no harmful effects are felt on entering in pneumatic caisson or while remaining inside, but occasionally eardrums are broken and blood vessels ruptured. Problems usually come during decompression. The attack may be light or severe. A light attack is characterized by severe pains mainly in the joints and closely resembles rheumatism and its effects. This disease is known as **bends**, when attack is severe, it usually paralysis, its victim and is often fatal. The sensations felt on entering compressed air are slight giddiness, inability to whisper and a feeling of resistance to movement owing to density of the air. Pain may be felt in the ears which may be relieved by closing the mouth and holding the nose and at the same time trying to expel air from the lungs. On leaving the air pressure, one feels cold, the sensation being keenest during the passage through airlock. It is due to expansion and liberation of gases in the body. To counteract the effects of this cold, airlock should be heated, the men should be given hot coffee to drink on coming out and they should dress warmly. Another sensation often manifested on emerging is itching, pricking feeling under the skin on all parts of the body and this may disappears in few minutes. Even experienced person find it difficult to enter when trouble with a bad cold. In India there had not been any major catastrophes during pneumatic sinking.

10.6.4 Preventions and Cure for Caisson Sickness

If the cause of caisson disease is a mechanical action due to the development of bubbles in the bloods and fluid tissues, which is in turn due to too rapid decompression, then cure is decompression at a slow rate. The length of time will depend upon the amount of the gas in the fluid tissues and upon physical characteristics of person being decompressed. The amount of gas in the fluid tissues will in turn depend upon degree of the pressure in the working chamber and length of time body is under pressure. The length of time taken to saturate the body fluid at any fatness of the subjects depend upon the amount of bodily workdone heat and moisture present. Experiment shows that the fatty tissues absorb about five times as much gas as does the blood and the rate of absorption is much slower, corresponding to the rate of de-saturation will be slow. For this reason the men inclined towards fatness should never be employed for compressed air work. Better the circulation of the blood, the more quickly and easily will the gases be thrown out from the system. For this reason only men in good physical condition should be employed. Old men, those who have abused themselves by excessive drinking should never be allowed in the working chamber. The theory upon which stages decompression based is that, the gas in the blood will not effervesce until a marked decrease of pressure. To the point of effervesce, the gases are discharged at rate varying with some function of change of pressure. The more rapid and lowering of pressure, the more quickly will the blood vessels be freed of the gases contained therein. Apart from the matter of slow decompression, other precautions if taken, will minimize the occurrence of caisson disease. Anything, which tends to lower resistance of the human system tends to promote caisson illness. The only cure for caisson disease is recompression with slow decompression. If the patient can be put into air before the gas bubbles have had a chance to tear the blood vessels and fluid tissues, a cure can usually be effective, but otherwise not. From this reason a medical air lock, large and well-ventilated should always be maintained in readiness and the men should be housed nearby, so that in case of delayed attacks they may be immediately recompressed.

10.6.5 General Precautions to Avoid Caisson Sickness

a. Work under compressed air should be executed under the supervision of responsible engineer.

b. Age of workers under compressed air should be 20 years to 40 years.

c. All employees must be medically examined carefully before they are selected for work under compressed air. They must also be periodically examined (say 15 days) to certify their fitness. When first employed under pressure above 2.10 kg/cm^2 recommendation letter to be taken from authorised doctor after 1st shift.

d. Never go on shift with empty stomach. Also eat moderately.

e. Sleep at least 7 to 8 hours. Keep bowels regular.

f. Liquor should not be given to man suffering from compressed air sickness.

g. Report for medical check-up for any ailment no matter how slight.

h. Take over clothing to avoid becoming chilled during decompression.

i. Move limbs freely during decompression to stimulate circulation.

j. Stay at job site for at least half hour after coming out of working chamber. One man should not work more than one shift.

k. For accurate control of air pressure, a gauge attendant should watch the pressure gauge constantly and the gauge should be accurate and in good working condition. There should be a master pressure gauge at site for periodical check-up of the pressure gauges in the lock.

l. To avoid the air in the working chamber as well as in main chamber becoming stale, fresh air must be circulated in the lock constantly. This can be done by opening a valve in the air lock. In granular soil where certain amount of leakage takes place through the cutting edge and the soil, the air is automatically circulated.

m. Men working under compressed air must be decompressed slowly. If coming out too fast, they are subjected to caisson disease. This disease is due to air bubble formed in the blood and body tissues, which are compressed while working under pressure.

n. The air lock to be protected against temperature rise by means of the sun.

o. The air lock, shaft and the interior of the well are to be provided with sufficient electric lighting. During working, every shift in-charge in the working chamber and every attendant must have sufficient 3 to 4 cell for lighting purpose.

p. The 'IN' and 'OUT' of person should only be monitored by a responsible and experience of lock attendant who is fully conversant with instructions.

q. Unauthorised manipulation of valves of the compressed air pipe is very dangerous for the health of all employees.

r. During locking in, any body felt pain in the ear shall try to remove, the same by continually swallowing as the air pressure is increased or by holding the nose and blowing as this action tends to increase the freedom with which the air can pass through the Eustachain tube into the middle ear there by equalizing the pressure on the inner and outer surface of the ear drum.

s. In case a person feels indisposed during working, he should immediately inform the lock attendant in order to be locked out at once. Before locking out extra cloths are to be put on. Also person who becomes indisposed during locking out must immediately contact the lock attendant who can if necessary again increase the air pressure.

t. Men should not be over crowded to avoid discomfort while entering in the chamber.

u. Persons should not be allowed to carry any matchbox in the air lock nor anybody should be allowed to smoke.

10.7 TILT RECTIFICATION

Caisson as discussed above was launched from November 1990 onwards. Due to flash flood of April 91 the work was severely disturbed and no major event was carried out till monsoon was over after April 1991. The well got over tilted during flood of 1991 monsoon for period of June to August to 1 in 2.34 to u/s on left quadrant (Fig. 10.16). On critical examination it was difficult to make any rectification scheme due to wide

Fig. 10.16: Well in tilted position

spread of water even at LWL condition to a distance of 350 M across the channel. There was no possibility of diversion at this location. However, taking the advantage of bank condition a scheme was made to rectify the tilt.

10.7.1 Design and Layout of Rectification Scheme

The position of the well in tilted condition and soil strata already encountered was examined. Based on present position of well, it was planned to pull the well with horizontal forces to bring the tilt to normal position. For this, pulling arrangement with the aid of prestressing was finalized. Stressing was planned to be done from both ends from D/S. Cable was planned to be taken around a anchor end block to be a part of steining a later date. Important details of the scheme are as under (Fig. 10.17):

a. Tilt of well	: 1 in 2.34
b. Weight of well (after adding block)	: 2407 MT
c. Weight of well (submerged in water)	: 1407 MT
d. CG of well above base	: 7.76 M
e. Force required to over come tilt	: 330 MT
f. Design pull of six cables.	: 520 MT

Based on the analysis of the site condition and force required to pull the well, two anchor blocks were made on both the banks at downstream of the bridge as per following details:

a. Left bank at 260 Mtrs D/S
b. Right bank at 315 Mtrs D/S

Left bank anchor block location was located on a bed having deposits of wooden logs subsequent to flood and also the area was low lying compared to same on right bank which was full of boulders at higher level. Concrete block was made to resist design force of 501 tonnes against requirement of 330 tonnes only. Space in front of the

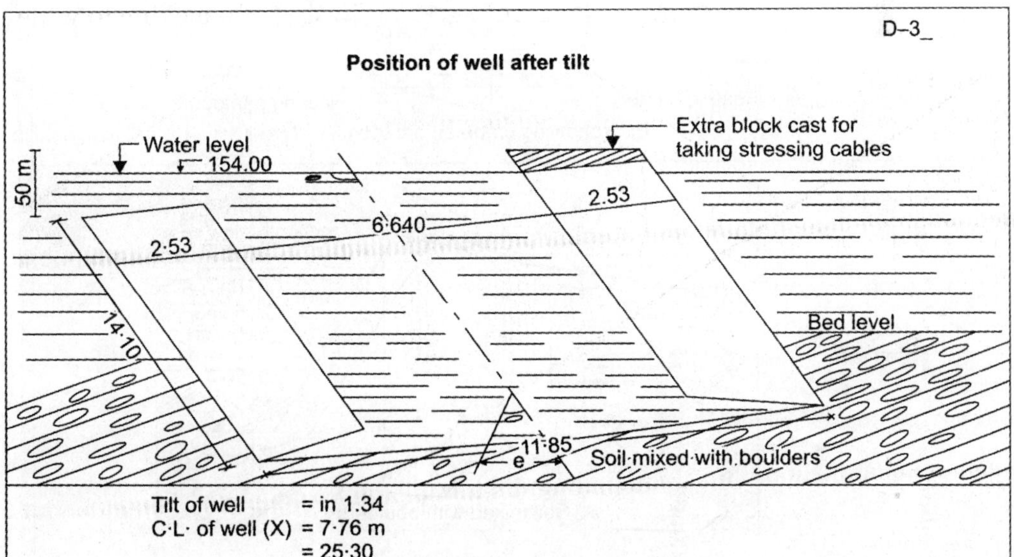

Fig. 10.17: Well in tilted position

blocks were filled with boulders to avoid resistance in horizontal direction. Arrangement for till rectification was implemented systematically on ground. (Fig. 10.18).

On completion of ground work for rectifications scheme, action was taken to improve the tilt.

a. Sag of cable was removed and grabbing was on downstream of the well.
b. Stressing of cable was done simultaneous in stages to have the loads up to 301. This was done in following stages:

Stage	*Force applied*	*Improvement in tilt*
1st stage	50 T	1 in 3.0
2nd stage	100 T	1 in 7.0
3rd stage	150 T	1 in 10.0
4th stage	200 T	1 in 12.0
5th stage	250 T	1 in 14.0

c. After three days, efforts the tilt which was 1 in 2.34 was reduced to 1 in 15. Subsequently, one lift of concrete was added to increase the height of steining. After this tilt rectification was done to the desired limit.

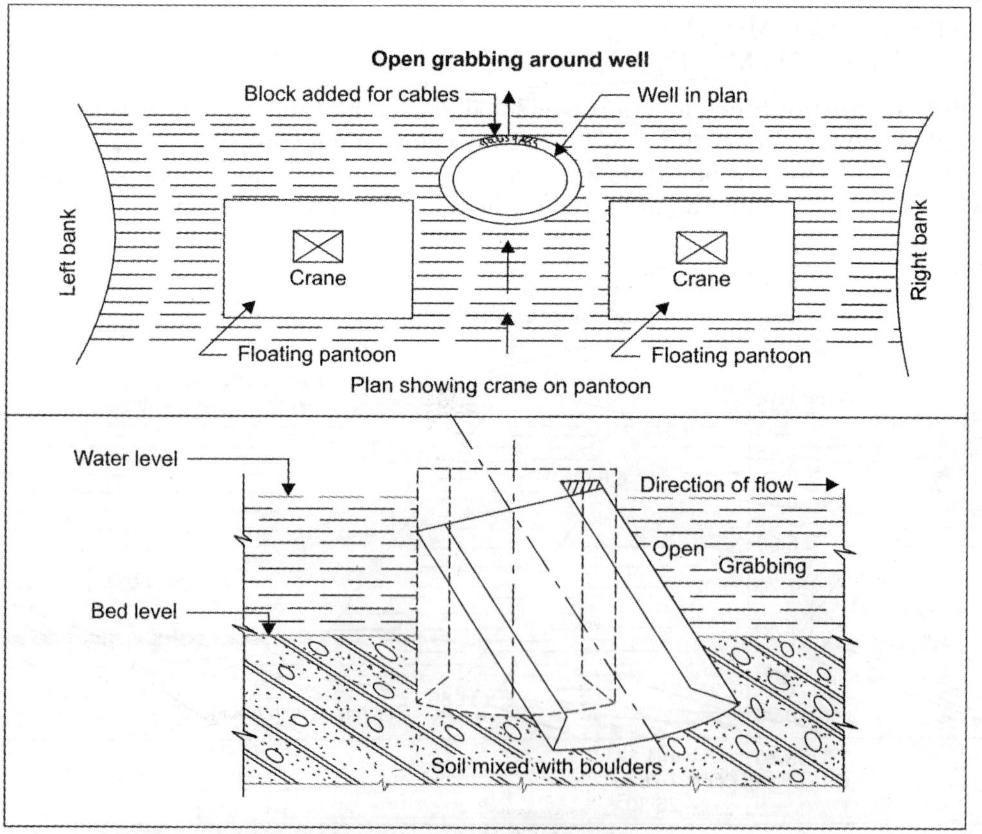

Fig. 10.18: Addition of block on the top of the steining for catling around

10.8 MAJOR PROBLEMS ENCOUNTERED

a. Due to high rainfall in the area the time period available was very less. There was urgent need to add additional steining after rectification of the tilt.
b. There was a problem of logistic due to remote location of site.
c. Since distance between two banks was large, coordination was not possible without Walkie-talkie. However, two boats were kept busy to ensure better communication across two banks. Also signaling system between banks for indication was ensured with the help of flags (Fig. 10.19).

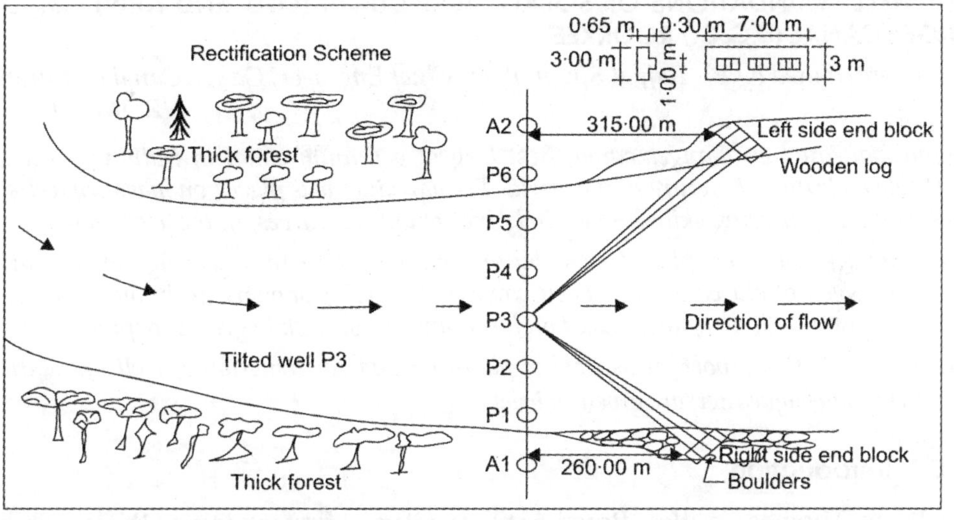

Fig. 10.19: Cable stressing from end for rectification of over tilled well

10.9 RECOMMENDATIONS AND CONCLUSIONS

All the bridge schemes be finalised based on the construction method planned or equipment/plant to be deployed. However, the following points be given due consideration while planning well foundation in bouldery bed.

a. Bridge scheme be finalized based on examination of the pros and con of the data and guidelines exists. Numbers of foundation be so decided that there is less sinking efforts involved considering all factors. Construction system/ methodology available should also be kept in view while planning a major bridge.
b. Scour depth in bouldery bed, should be finalized based on scan pattern data of existing bridge and same should be analyzed. **The existing bridges on bouldery beds should be considred as live model for studying scour pattern.**
c. Change in soil strata properties be considered or examined with due consideration. All the soil strata actually encountered should be recorded per metre wise and same should be compared with soil strata anticipated during investigation stage. Tough strata should be considered as engineering friendly.
d. Construction difficulties be given greater attention from execution point of view.
e. Sinking should be progressed in a systematic way to avoid the over tilting of any well foundation during the sinking process. Tilt and shift should be checked regularly.
f. Pneumatic sinking technique be used judiciously wherever applicable as this is very costly technique and require lot of precautions.

g. Soil strata likely to encountered in bouldery beds are large size boulder which will delay the progress of well foundation construction. Special attention should be given to the initial studies being carried out for all projects on such strata.
h. Overall depth of well foundation should be so designed that it is safe, economical and be feasible on ground as per available construction methodology. Experienced gained at one site be dove tailed for other location to tackle the similar problems in a better way. However, the certain construction difficulties are likely to occur in a way which can only be solved with systematic planning and skilful execution coupled innovative efforts.

10.10 WELL FOUNDATIONS OF SOLANI AQUEDUCTS (OLD AND NEW) FOR UPPER GANGES CANAL (U.G.C.) ROORKEE

(Contributed by Er. Rajpal Singh, Retd. Chief Engineer, Ganga Canal and Professor, WRDM, IIT Roorkee)

Solani aqueduct on Ganges canal, first largest perennial canal in India was rated as a remarkable hydraulic structure in the world. The aqueduct was placed on masonry wells, built through local engineering skills and entirely indigenous resources, in the year 1854.

New Solani aqueduct, placed close by on new parallel canal, during rehabilitation of vulnerable reach of old canal, has been constructed using modern technology, with large prestressed beams of long spans, placed on large size wells, sunk to greater depths.

Here, some of the important aspects of planning and construction of well foundations of above prestigious aqueduct, are given in brief.

10.10.1 Introduction

Capt. Proby Thomas Cautley, Royal Artillery Officer, first examined the possibility of projecting a perennial canal in 1838, and submitted a feasibility report in 1841 which was recommended by the committee of three Military Engineers, with Cautley as one of the members and it got the approval of the then Government for initiating related activities in 1842. The detailed project report was prepared solely by Sir Cautley, which was ultimately approved in March 1847 at Roorkee, by Lord Hardinge, the Governor General of East India Company in India. The Ganges canal, the largest artificial stream on the earth (discharge 6750 cusecs) in its magnitude and the use over large agriculturable land, took off from the Ganga at Haridwar and terminated through one branch tailing into the same river at Kanpur (U.P. India), while the other branch with its out fall into Yamuna river near Hamirpur (upstream of Allahabad).

Total length of conveyance network of main canal and branch canals was more than 650 miles (nearly 1000 km) cost amounting to nearly Rs 15.0 million in the year 1847. The project construction works were completed in scheduled time and within the sanctioned cost and the canal was opened for use in April 1854. The upper (head) reach of the canal from Haridwar to Roorkee (20 miles) was quite vulnerable, as it passes through difficult terrain, crossing four wild and mighty torrents (streams), the Solani stream was the largest of all, passing by the side of Roorkee. Two upper streams were crossed through major cross-drainage works of masonry-arch superpassages, the third one by major level crossing involving escape dam of nearly 80,000 cusecs capacity and the fourth one on Solani river, by Solani aqueduct. The total cost of canal from Hardwar to Roorkee was nearly 30% of cost of the project and nearly half of the above amount was spent on Solani crossing, through earthen embankments protected by masonry revetments (length nearly 5 km, height 8 to 10 m) and the Solani aqueduct.

10.11 SOLANI AQUEDUCT (OLD)

The area of Solani basin lying above the canal works is equal to 216 square miles, its length may be estimated at 27 miles, and its average breadth at 8 miles; of this area, 50 square miles lie in the Sivalik hills, the remainder on rapid but gradually decreasing slopes, until it reaches the Ganges canal works, where the fall in the bed of the river is 5.08 ft per mile. The Solani river bed consists of fine sand with average diameter 0.20 mm and carries a flood flow (highest estimated) of 80,000 cusecs, but at other periods its stream is quite insignificant varying from 10 to 20 cusecs.

The river itself is crossed by a brick masonry aqueduct, which was not merely the largest work of its kind in India, but one of the most remarkable for its dimensions in the world. The original design of the aqueduct (prior to floods of 1844) was to give a clear water way of 500 ft broken into 10 spans of 50 ft each. However, after very high flood of year 1844 (observed), the water way was increased from 10 to 15 bays of 50 ft having a total width of 750 ft. By adding counter-arch and otherwise strengthening the floorings and by remodelling the transverse section of the superstructure, to meet the increase of head-water, the capacities of the opening for the torrent were sufficient, for floods of even greater magnitude than those of 1844.

The design of foundations was adopted to a water way of 15 bays of 50 ft each, with piers of 10 ft and abutments of 21 ft width, the latter being flanked by the wing buildings, which were connected to embankments of earthen aqueduct. The extreme width of channel for carrying the canal water over the Solani including their walls or revetments was 192 ft. The total superficial area that the body of aqueduct covered was 178, 944 square feet (932 ft in length and 192 ft in breadth). The total height of the structure above the valley of the river is 38 ft. It may not be an imposing work when viewed from below, in consequence of this deficiency of elevation, but when viewed from above, and when its immense width is observed with its line of masonry channel (nearly 3 miles in length), the effect must be most striking.

The piers rest upon blocks (wells) of masonry, sunk 20 ft deep in the bed of the river and normally being cubes of 20 ft wide, piered with four well each, and under sunk in the manner practiced by natives of India in constructing their wells. These foundations throughout the whole structure, were secured by every device that knowledge or experience could suggest, and the quantity of the masonry sunk beneath the surface was scarcely less than that visible above it (Figs 10.20 and 10.21).

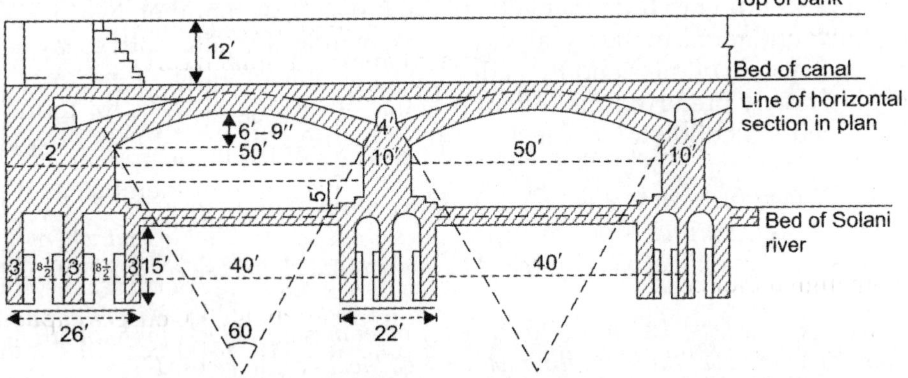

Fig. 10.20: REF: Cautley Atlas (all dimensions in feet)

Solani aqueduct (1848–54)

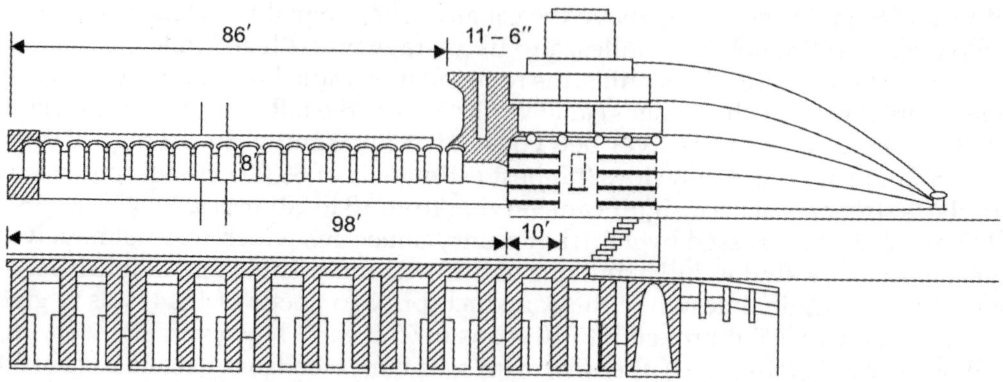

Fig. 10.21: Half front section through spandril showing arrangement for ventilation and drainage

Detached from the series of wells of foundation, which were distinctly for the support of the superstructure, there were advanced lines of blocks in substitution of sheet piling which had been originally proposed, as well as blocks at the terminal point of piers for the support of cut waters. These with the blocks for support of wings and lines of revetment forming the flank approaches to the masonry from earthen aqueduct, constituted the work to be done in block sinking.

The observations of record plan of similar (Cautley's Atlas plate XXVII) are interesting and explanatory of very small change that has taken place in sinking large masses of masonry to a depth of 20 ft, whilst in the smaller masses especially in those where there is an aberration to any extent from the perfect square, the movement is more considerable. The method of sinking was practiced by the natives of India, (in N–W Provinces). It was universally adopted in wells in sinking purposes and was largely used by their Engineers in obtaining support for their masonry works, when the soil was either sand or a treacherous nature. In such cases, cylinders of masonry were sunk either at distances apart or close together and upon these tops which were arched over and connected by masonry or other bonds, the superstructure was raised.

In dealing with the progress of well-sinking, it could be observed that in a general way, the difficulties of sinking at all, and the danger of blocks going down unevenly increases, as form wanders from that a square or the circle.

From the character of bed and slopes and the existing springs on the surface in rivers and stream descending to Indo-Gangetic plain, it was clear that there was no method of laying the foundations in the usual way by excavation. The alternatives lay between the European system of piles and that adopted by people of India, of masonry under-sinking. Sir Cautely preferred the Indian practice of sinking of wells for foundations of Solani aqueduct. Following excerpts from Sir Cautley's report further explain the issue.

"No body, I imagine, can argue that a building supported on a series of sharp points, standing on silts as it were, as is the case where piles are used, be equal to that where it stands on a broad base, as the result of block system. On this ground alone, I select blocks in preference to pile.

I have never felt the least distrust as to the selection of masonry blocks for foundations of the aqueduct, nor would I ever think of piling where such blocks could be used".

As regards the practical method of sinking the blocks, which form the support of the aqueduct, use of wooden curb (Neemchuk) or frame was placed in the position in which the block was to be sunk, the masonry was then either built up to its full height or only partly built. Normally, each block was perforated by four cylindrical holes, an arrangement which could be changed into square or octagon ones, giving more space for under sinking. A common trestle windlass being placed over each hole. The soil being simultaneously removed by means of Jham; lot of field experiments were done on the perfection of practice of sinking and also making it more economical for such large work. Great attention was required to see that four holes are equally worked, and that more soil is not taken from one hole than from another as a guide to the well sinkers, strings and plummets were appended to each side of the block so that a departure from perpendicularity was easily detected.

It was found that on a number of blocks the average rate of sinking in one day was 0.6 to 0.8 ft above 10 ft in depth, and 0.25 to 0.50 ft below 10 ft in depth. The use of Strapiron (or wine plate) was largely made in early part of block building but as the masonry, both of brick and mortar, was of a very superior quality, the use of strap was subsequently discontinued to cut down the expense. The distance between the large class of blocks did not exceed 2 ft 3 inches; than between the smaller ones, 2 ft 11 inches. The first thing that was done before a commencement was made to the floorings, was to drive pile 12 feet in length, so as to fill up the spaces that existed between the blocks.

As the depth of foundations of Solani aqueduct was limited to 20 ft, adequate arrangements of protection of foundation against serious scour during very high flood were made, by providing an independent row of masonry block on the upstream and downstream end and also by row of sal balla piles and cribwork with boulders in apron, both on upstream and downstream of foundations. Investigations at site during mid 1970's indicated that the above protection works have been standing in good form, even after a use of more than 125 years and after passing the designed discharge (80,000 cusecs) in 1924 in the river at this work. The details of weighting frame used to facilitate the foundation, block sinking are shown in Fig. 10.22.

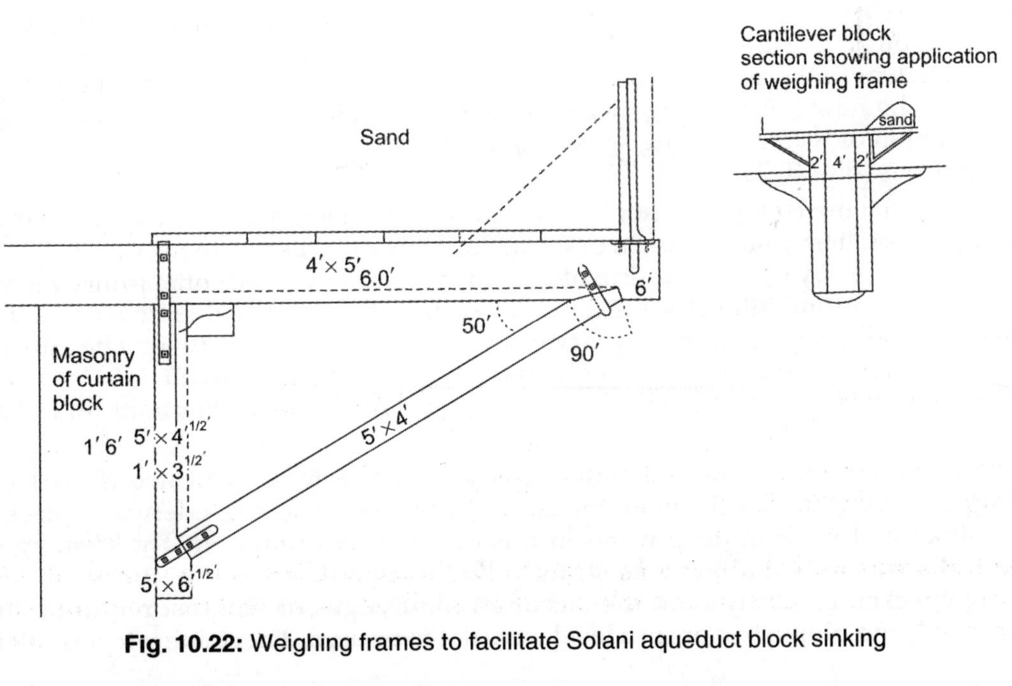

Fig. 10.22: Weighing frames to facilitate Solani aqueduct block sinking

10.12 NEW SOLANI AQUEDUCT

The new aqueduct is located at km 28.7 of parallel Upper Ganga Canal, against the project of rehabilitation of upper vulnerable reach of Ganges canal along with four major cross-drainage works. Its construction work was taken up in year 1984. As the old aqueduct was placed on masonry wells sunk to 20 ft depth, it was a serious concern to ensure the safety of this old structure, during construction of new aqueduct, proposed to be placed on large size wells to be sunk to 25 to 30 m depth. After careful considerations, the new aqueduct was located at a distance of 150 m from old aqueduct, on downstream side. Also, in view of limited headway in old aqueduct during high floods, it was also considered necessary to control the interference in the flow of high floods through the old aqueduct to the lowest possible limit. This was achieved by placing the piers of new aqueduct just opposite the alternative pier of old aqueduct; without going into the in-depth analysis of most economical span for the new aqueduct. The layout is shown in Fig. 10.23.

The new aqueduct is designed with a water carrying capacity of 310 cumecs (11,000 cusecs) with in-built additional capacity of 20% over load, for long future use, to serve the prestigious canal, commanding extensive fertile tract of the state. The aqueduct has 8 bays of 36.6 m centre to centre and is founded on 9 nos. R.C.C. drum-bell shaped wells of size 29.5 m × 9.5 m × 25 m to 29 m deep. The abutments, transitions and piers are of R.C.C. construction. The length of the aqueduct from abutment to abutment is 320.8 m, exactly matching with the total water way of old aqueduct (Fig. 10.24).

The main superstructure of the aqueduct is of prestressed concrete having two independent trough, 10 m wide and 5.12 m deep having a maximum water depth of 4.93 m, the prestressed concrete super structure consists of 35.2 m long and 7.0 m deep main-girders, 1.0 m deep cross girders and 0.35 × 0.60 m top lies at a spacing of 2.0 m c/c. The centering has been done with the help of 4 nos 91.2 m long centering girders and other supporting arrangements in two spans of a trough. The load is carried by prestressed concrete framed structure and water is carried in reinforced concrete twin troughs, placed inside it. The structure has been designed on no tension, no crack basis, taking into account dynamic analysis of the structure. One end prestressing of the cables has been adopted as per the Freyssinet system. The new Solani aqueduct is the unique symbol of prestressed concrete construction of water carrying structure, placed on very large size and deep well foundations (Fig. 10.25).

The arrangement of foundation wells in piers, abutments and transitions is shown in Fig. 10.24. The foundation work consist of wells, taking them down to the indicated levels by open dredging of the soil and other ingredients, plugging the bottom and top and filling the inside not by earth and sand available from dredging, but by the sand arranged from quarries in river Ganga at Haridwar. The cutting edge was fabricated from M.S. Angle, flat and plate as per construction drawings and specification and fixed to the RCC curb and staining by anchor bolts and bond rods.

When the curb was to be laid in dry beds, the site was to be excavated up to 0.3 m above sub-soil water level, before the cutting edge was placed. Underwater placing and sinking of wells in deep water in this case was not required. The steining of the wells was to be built in one straight line from bottom to top, the work was being checked carefully with the aid of straight edges of lengths, required and approved, and plumb bob or sprit level were not to be used. The steining to be built in

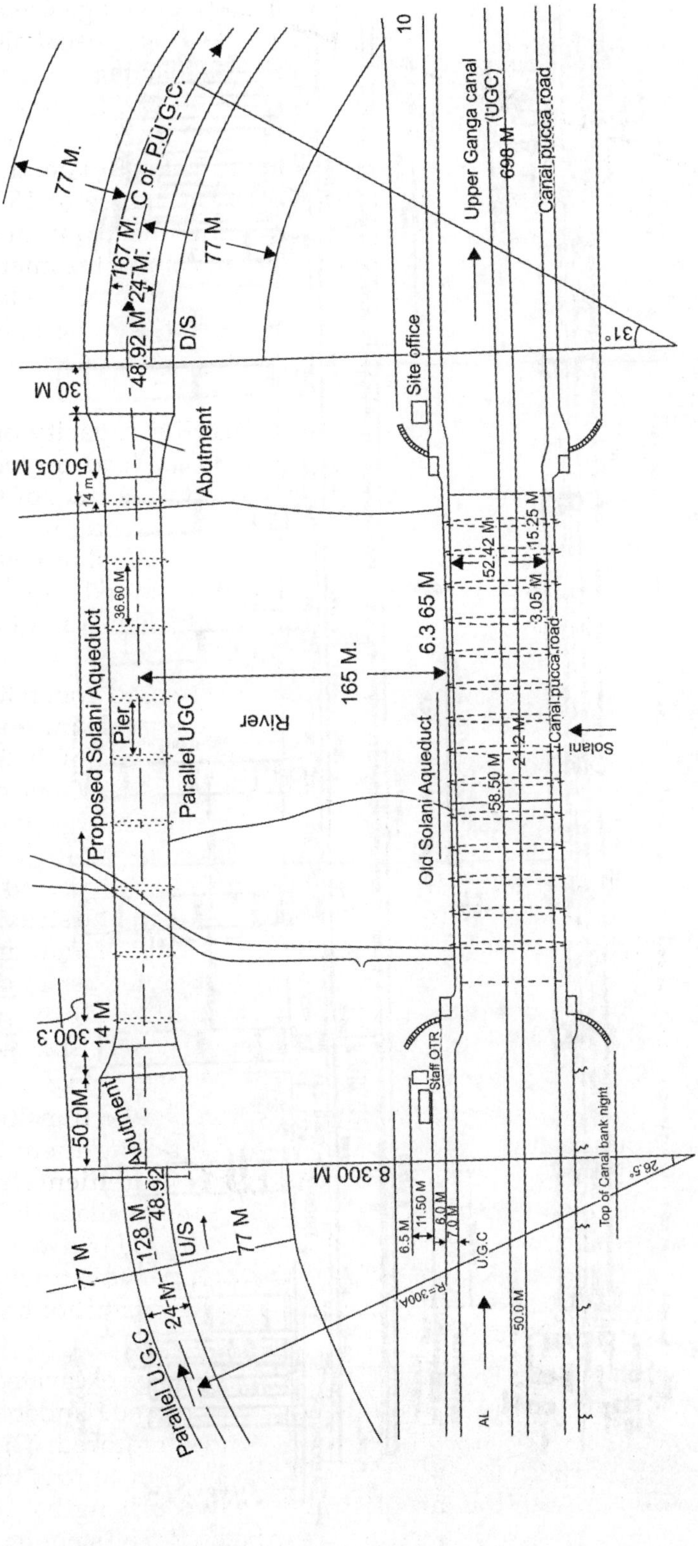

Fig. 10.23: Lay-out plan of Solani aqueduct (scale –1 CM = 10.0 M)

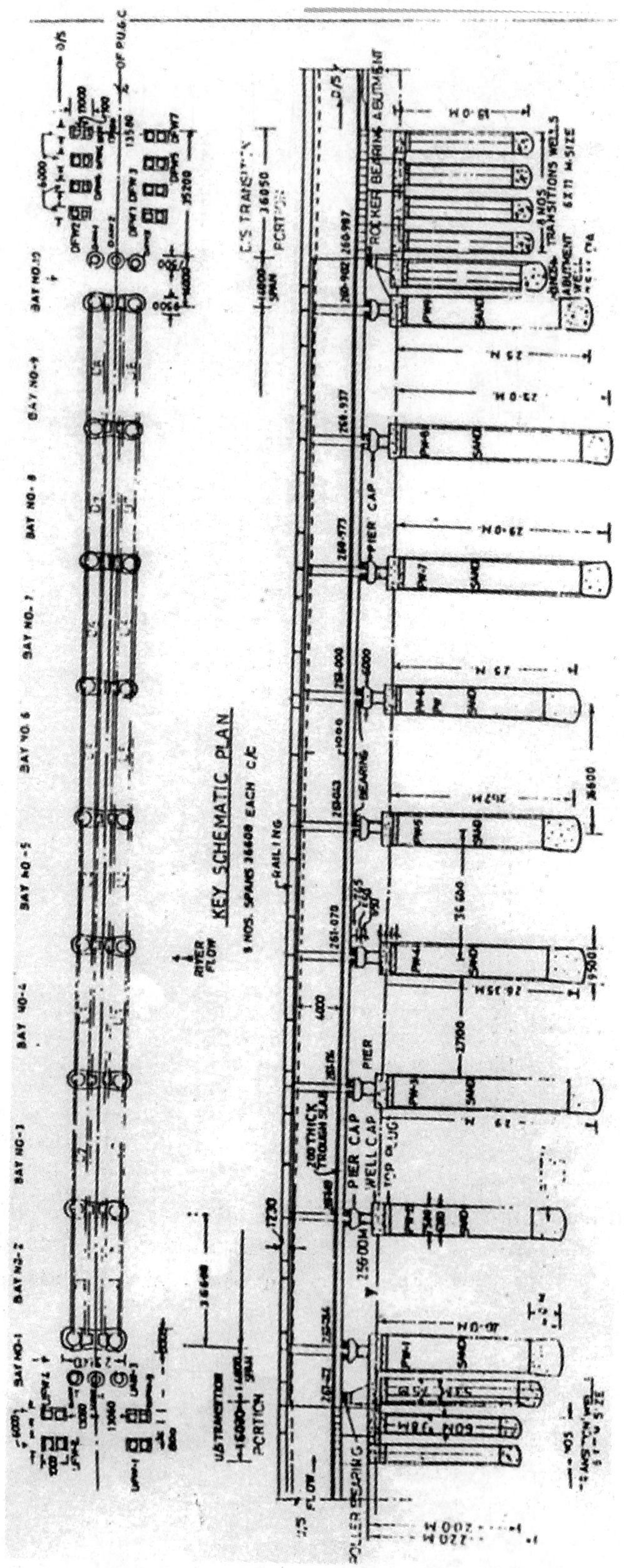

Fig. 10.24: Schematic configuration of new Solani aqueduct

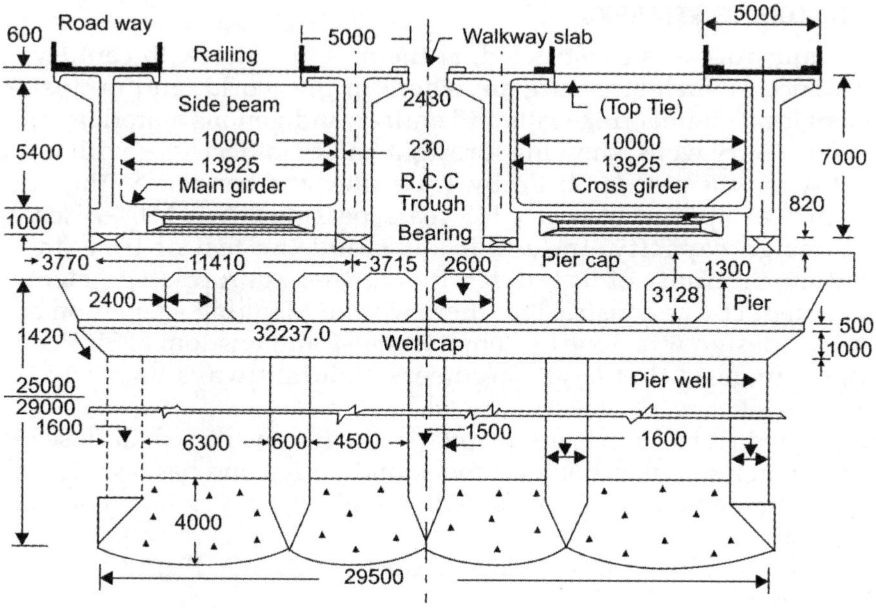

Fig. 10.25: Section through pier of new Solani aqueduct

first stage was not to exceed 2.0 m in height and in subsequent stages it was not to exceed the diameter of the well or depth of well sunk below the adjoining bed level at a time.

During sinking of wells, simultaneous and even dredging were carried out in all dredge holes. Sinking and loading of the well with kentledge were allowed to commence only after steining had been cured for at least 48 hours. The process of sinking included pumping out of water and disposal of dredged materials in depressions of river bed. Dewatering was to be avoided in sinking and in no case differential water head was allowed in two dredge holes of the well.

When the wells had to be sunk close to each other and the distance between them was not greater than the diameter of the wells, sinking was taken up on all the wells simultaneously and they were sunk alternatively so that sinking of well could proceed uniformlzy and together. Whenever pier wells were located very close to the abutment wells with different levels, the pier wells were sunk first. The tilt and shift of each well were measured regularly during entire sinking and necessary corrective measures were taken to contain the tilts and shift within permissible limits. The tilt of any well was not to be exceed 1 in 80 and the shift was not to exceed 1% of outside diameter of circular well or maximum dimension of the well for any other shape, to a maximum of 150 mm.

No major event of any accident or objectionable dislocation of any well could be noticed in the huge work of foundation well. The work of completion of foundation was completed in nearly 2 years as planned. In order to ensure the dependability as well as durability of such a major structure and to be fully satisfied with the soundness of design and quality level of construction of wells, one well of a pier was tested for maximum expected load (9000 MT) coming on the well during full use of the work. Everything was found to the entire satisfaction of the construction engineers. The total cost of new Solani aqueduct has been assessed to the order of Rs. 460 millions.

10.13 CONCLUDING REMARKS

Old Solani aqueduct was constructed, at the mid of nineteenth century, without any precedence of such structure anywhere in the world, and exclusively as a ingredient of local engineering skills and entirely indigenous materials, and placed on shallow masonry wells, showing foresight and bold decisions with utmost care in the quality, on the part of Sir Proby T. Cautley and has served the prestigious canal system over a period of nearly 165 years, passing nearly 150% discharge than its initial design capacity. In fact, it has stood the test of time and clearly established the ingenuity of its builder. It is further appropriate to mention here, when old aqueduct was constructed, there was no Manning's equation or Lacey's theory. All the design was done by common sense and wisdom of Sir P.T. Cautley. It definitely suggests that Civil Engineers should always be ready to accept challenges.

The new aqueduct has been carefully planned and constructed, in close vicinity of old one, based on modern technology and sound design and has been tested during 2002, by running the new canal almost with its full capacity, without any noticeable problems, so far.

The photographs of magnificent old aqueduct are given in Fig. 10.26 and a view of new aqueduct is given in Fig. 10.27.

(a) (b)

(c)

Figs 10.26 a, b and c: Panoramic views of Solani aqueduct in Roorkee, constructed during British era (1845–47)

Fig. 10.27: A view of Solani aqueduct (New)

REFERENCES

1. Dhiman RK. "Bouldery Bed Scour-Proposed Formula" Indian Road Congress (IRC) Journal 65 Vol. 3 Paper No. 508. Nov 2004.
2. Dhiman RK. "Pneumatic Sinking; A Case Study" Indian Road Congress (IRC)–Indian Highway, Feb 1996.
3. Dhiman RK. "Caisson Launching A Case Study" Civil Engineering And Construction Review (CE&CR) 1996.
4. Dhiman RK. "Foundation Level for Bridges; A Programmatic Approach" 1996 New Building Material And Construction World (NBM & CW).
5. Dhiman RK. Essence of Silt Factor for Scour Around Bridge Pier"; International Conference on Scour of foundation (ICSF-I) held at Texas USA (17–20 Nov 2002).
6. Dhiman RK. "Construction Problem of Bridges in Hilly Region; A Review", 1997 International Association of Bridges and Structural Engineers (IABSE)
7. Dhiman RK. "Dimwe Bridge Foundation; A Case Study" 4th International Seminar on Bridge and Aqueduct 1998
8. Dhiman RK. "Well foundation Construction in Bouldery Bed; A Case study" International Association of Bridges and Structural Engineers (IABSE) Colloquium Feb 1999.
9. Dhiman RK. "Affect of Flash Flood; A Case Study", Disaster Mangement NERIST. Itanagar-1999
10. Dhiman RK. "Caisson Sickness During Pneumatic Sinking" International Symposium at University of Dundee Scotland (UK); Sep 2003.
11. Dhiman RK. "Bridge Construction Problems and Solutions; A Review" 17th National Convention of Civil Engineering at Bhubneshwar 2001.
12. Dhiman RK. "Extension of Span Ranga II Bridge; A Case Study", 17th National Convention of Civil Engineering and Seminar on Modern Trend in Construction and maintenance of Roads, Flyovers and bridges, Bhubaneshwar, Nov 2001.
13. Dhiman RK. "Tilt Rectification of Well Foundation; A Case Study", Indian Road Congress (IRC) Indian Highways May 2002.

14. Dhiman RK. " Essence of Training of Manpower for Concrete Technology", Sixth International Confrence on Concrete Technology (61CCT) at Aman (JORDAN) Oct. 2003.
15. Dhiman RK. "Construction Challenges for Bridges in Hilly Area; An Over View"; Indian Road Congress (IRC) Indian Highways Jan 2004.
16. Dhiman RK. "Damages to Bridges due to Flash Flood-A Case Study"; Indian Road Congress (IRC) Indian Highways Oct 2004.
17. Dhiman RK. "Effective Construction Management fo Bridges" International Association of Bridges and Structural Engineers (IABSE) Dec 1996.
18. Special issue on "Round Table Conference on Scour Around Bridge Pier"; 1993 Indian Institute of Bridge Engineering (IIBE), New Delhi.
19. Model Study Report of Pasighat Bridge by UPIRI, Roorkee, 1984.
20. Indian Road Congress (IRC) 78:2000.

Appendix

A

The Drilling MUD (Bentonite)

A1 INTRODUCTION

Lining tubes or casings to support the sides of pile boreholes are a requirement for most of the bored pile installation methods using the equipments described in chapter 7. Even in stiff cohesive soils, it is desirable to use casing for support since these soils are frequently fissured or may contain pockets of sand which can collapse into the boreholes, resulting in accumulations of loose soil at the pile toe, or discontinuities in the shaft.

Casing can be avoided completely (except for a short length that is used as a guide at the top of the hole) by providing support to the pile borehole in the form of a slurry of bentonite clay.

A2 PROPERTIES

The bentonite suspension used in bore holes is basically a clay of Montmorillonite group having exchangeable sodium cations. Because of the presence of medium cations, bentonite on dispersion will breakdown into small plate like particles having a negative charge on the surfaces and positive charge on the edges . When the dispersion is left to stand undisturbed, the particles become oriented building up a mechanical structure of its own, the mechanical structure held by electrical bonds is observable as a jelly like mass or jell material. When the jell is agitated, the weak electrical bonds are broken and the dispersion becomes fluid.

A3 FUNCTIONS OF BENTONITE

The action of bentonite in stabilizing the sides of bore holes is primarily due to the thyrotrophic property of bentonite suspension. This property permits the material to have the consistency of a fluid when introduced into the excavation and when undisturbed forms a jelly which when agitated becomes a fluid again.

In the case of granular soil, the bentonite suspension penetrates into the sides under positive pressure and after a while forms a jelly. The bentonite suspension gets deposited on the sides of the hole and makes the

surface impervious and imparts a plastering effect. In impervious clay, the bentonite does not penetrate into the soil but deposits only a thin film on the surface of the hole. 'Under such circumstances, stability is derived from the hydrostatic head of the suspension.

A4 SPECIFICATIONS OF DRILLING MUD

The bentonite suspension used for piling work should satisfy the following requirements:

a. The liquid limit of bentonite shall be more than 300% and less than 450%.

b. The sand content of the bentonite powder shall not be greater than per cent.

c. Bentonite solution should be made by mixing it with fresh water using pump for circulation. The relative density of the bentonite solution should be between 1.034 and 1.10.
d. The differential free swell shall be more than 540%.
e. The pH value of fresh slurry should be 9 to 11.5.
f. Marsh cone viscosity of slurry should be 30 to 90 seconds.

A5 PREPARATION OF SLURRY

For preparation of slurry, the bentonite powder should be thoroughly mixed with water. It should be left overnight to stabilize. Before applying to the borehole, it should be thoroughly mixed and churned. While using the same slurry several times (like in case of wash borings, etc.), before reusing, it should be led through settling tanks and loss of bentonite is made usable by adding fresh mix.

In general five per cent mix of good quality commercial bentonite is found adequate. Sometimes local clay and other additive can also be used with the bentonite. In India, bentonite of various grades is available. Bentonites from Rajasthan and Gujarat are very often with good swelling properties.

Generally, a bag of 50 kg of bentonite is sufficient for two under-reamed piles of 30 cm stem diameter and of about 3.5 m length.

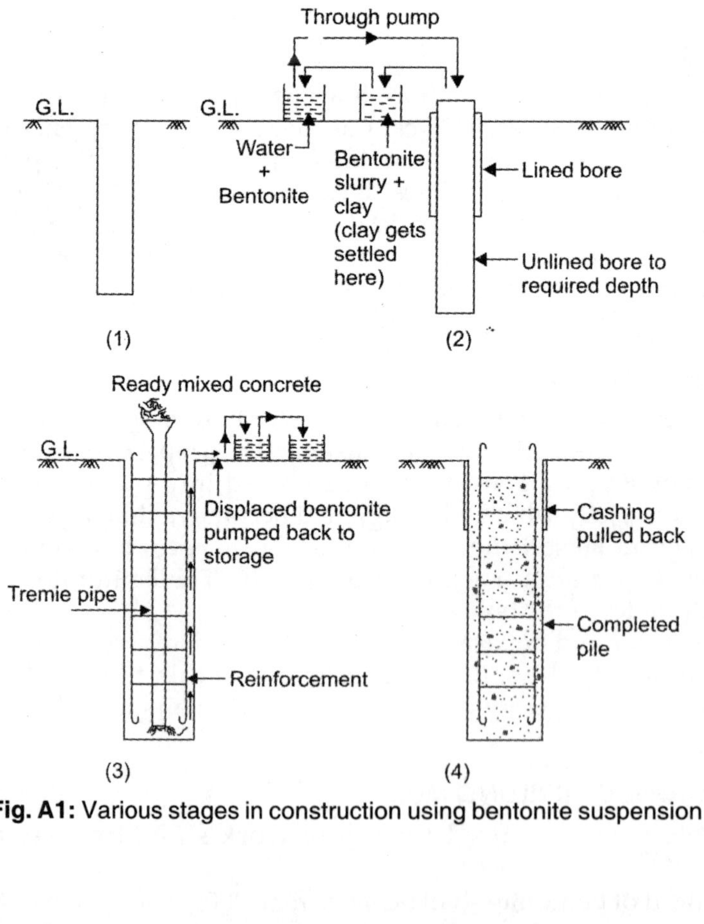

Fig. A1: Various stages in construction using bentonite suspension

Appendix B

Concrete Grade

For driven—Pre-cast concrete piles, following grade of concrete shall be applicable.

Situation

1. For hand driving where driving stress exceeds 100 kg f/cm^2.
2. For easy driving, where driving stress is less then 100 kg f/cm^2.

Grade

Not less than

M-20

Not less than

M-15

Clean water, free from acids and other impurities shall be used in the preparation of the concrete.

Appendix

C

Foundation Pile Diagnostic System

C1 INTRODUCTION

A sophisticated instrument has been developed by M/s TNO institute for building materials and structures, Delft, Netherlands, for pile testing, pile driving analysis, bearing capacity analysis and driveability studies. The instrument is completely self contained and is very quick in installation and operation. The quality of the piles can be assessed very quickly by this instrument. The testing of the pile called "**Integrity Testing**" provides information on pile length, cross-sectional variations, cracks and other discontinuities.

The method involves striking the pile head with a hammer and picking up the response of the pile by means of an accelerometer pressed onto the top of the pile. The signal is passed to the input of the foundation pile diagnostic system (FPDS), where it is amplified, digitized and processed in a digital computer. For the operator's convenience, all functions automatically adapt to the signal properties. A velocity-time record is presented on the screen; it can be printed out on the built in-printer; or stored on the built-in-diskette.

The main advantage by using FPDS in this method are:

1. Defects are discovered at an early stage.
2. Testing can be carried out on any accessible pile.
3. The method is quick and economic when compared with other methods. Up to 100 piles can be tested in a day by this method.
4. The equipment is light and can be transported easily by the car.

C2 DYNAMIC PILE TESTING

The term "Dynamic pile testing" is used to describe a method of investigating the properties of an installed foundation pile, by a heavy impact.

To carryout such a test a heavy weight is dropped onto the specially prepared pile head and the applied force and the displacement or acceleration in the region of the pile are recorded.

The following data can be directly obtained on site from the measurements taken by FPDS.

1. The force applied as a function of time.
2. The displacement, velocity and acceleration as a function of time.
3. The dynamic resistance due to shaft friction.
4. The dynamic resistance due to to resistance.
5. The permanent settlement of the pile following the impact.

This method may be used for the following purposes:

a. To predict the behaviour of a pile in terms of
   - Static shaft friction
   - Static toe resistance
   - Static load/settlement behaviour
b. To compare the dynamic behaviour of a series of piles of similar dimensions.
c. To check a pile for:
   - Cross-sectional variations
   - Length
   - Lack of homogeneity
   - The presence of transverse cracks and their significance
   - The load transmission across a crack.

Appendix

D Various Soil Properties and Relationships

D1 PROPERTIES OF SOILS

Type of soil	*Angle of repose (approx.) degrees*	*Weight in kg/ litre*
1. Fine dry sand loose	31–37	1.6
2. Sand-dry and consoli dated	35	1.92
3. Sand-wet	32	1.84
4. Sand-very wet	26	1.93
5. Vegetable earth-dry	29	1.44
6. Vegetable earth-moist	45–49	1.6
7. Vegetable earth very wet	17	1.68
8. Vegetable earth-consolidated and dry	491.6	1.6
9. Loamy earth-consolid a ted and dry	401.3–1.6	1.3–1.6
10. Clay dry	29	1. 76
11. Clay damp (well-drained)	45	1.84
12. Clay with gravel	38	2.08
13. Clay-saturated	15	1.92
14. Gravel-clean	48	1.79
15. Gravel with sand	26	1. 92
16. Loose shingle	29	1.84

D2 SAFE BEARING PRESSURE OF SOILS (AVERAGE VALUES)

Type of soil	*Pressure in tonnes/m^2*
1. Soft clay	10 to 20
2. Ordinary clay (dry), dry and mixed with clay	20 to 30
3. Dry sand and clay	30 to 40
4. Hard and firm clay or coarse sand	40 to 60
5. Firm coarse sand and gravel	60 to 80
6. Rock	150 and above

D3 IMPORTANT RELATIONSHIPS OF VARIOUS SOIL PARAMETERS

i. Relationship between void ratio (e), specific gravity (G), water content (w) and degree of saturation (S_r)

$$e = \frac{w.G}{S_r}$$

ii. Relationship between dry density (γ_d), G and e

$$\gamma_d = \frac{G.w}{1+e}$$

or $$e = \frac{G.\gamma_\omega}{\gamma_\delta} - 1$$

also $$\gamma_d = \frac{G\gamma_\omega(1-n)}{1}$$

where $$n\text{-porosity} = \frac{e}{1+e}$$

iii. Relationship between γ_d, Bulk density (γ) and w

$$\gamma_d = \frac{1+w}{\gamma}$$

Relationship between γ_d, G, w and S_r

$$\gamma_\delta = \frac{G.\gamma_\omega}{1+\dfrac{\omega.G}{S_r}}$$

Appendix

E

Useful Conversions

To convert	*Into*	*Multiply by*
	A	
acres	sq. ft.	4.35×10^4
acres	sq. metres	4.047×10^4
acres	sq. miles	1.562×10^{-3}
acres	sq. yards	4.840×10^3
acre-feet	cu. feet	4.356×10^4
acre-feet	gallons	3.259×10^5
atmospheres	tonnes/sq.in.	7.348×10^{-3}
atmospheres	tonnes/sq.foot	1.058
atmospheres	cms. of mercury	7.6×10
atmospheres	ft. of water (at 4°C)	3.39×10
atmospheres	in. of mercury	2.992×10
atmospheres	ponds/sq. in.	1.47×10
atmospheres	kg/sq. metre	1.0333×10^4
	B	
bars	atmospheres	9.869×10^{-1}
bars	dynes/sq.com.	1×10^4
bars	kg/sq.metre	1.020×10^4
bars	pounds/sq.ft.	2.089×10^3
bars	pounds/sq.in.	1.45×10
	C	
centigrade (degree)	Fahrenheit (degrees)	$(°C \times 9/5) + 32$
centilitres	litres	1.0×10^{-2}
centimetres	feet	3.281×10^{-2}
centimetres	kilometres	1×10^{-5}
centimetres	miles	6.214×10^{-6}
centimetres	yards	1.094×10^{-2}
centimetres	microns	1.0×10^4
centimetres of mercury	atmospheres	1.316×10^{-2}
centimetres-grams	metre-kg	1.0×10^{-5}
centimetres of mercury	ft. of water	4.461×10^{-1}

centimetres of mercury	kg/sq. metre	1.36×10^{-2}
centimetres of mercury	pounds/sq.ft.	2.785×10^{1}
centimetres/sec.	kilometres/h	3.6×10^{-2}
centimetres/sec.	miles/min.	3.728×10^{-1}
cubic centimetres	cubic in.	6.102×10^{-2}
cubic centimetres	cubic ft.	3.53×10^{-5}
cubic centimetres	cubic yards	1.308×10^{-6}
cubic centimetres	litres	1.0×10^{-3}
cubic feet	cu. cm	2.832×10^{4}
cubic feet	cu. metres	2.832×10^{-2}
cubic feet	litres	2.832×10
cubic feet	cu. yards	3.704×10^{-2}
cubic feet/min.	gallons/sec	1.247×10^{-1}
cubic inches	cu. ft.	5.787×10^{-4}
cubic inches	cu. yards	2.143×10^{-5}
cubic inches	gallon	4.329×10^{-3}
cubic inches	litres	1.639×10^{-2}
cubic metres	cu. ft.	3.531×10^{1}
cubic metres	cu. inches	6.1023×10^{4}
cubic metres	cu. yards	1.308
	D	
days	seconds	8.64×10^{4}
days	minutes	1.44×10^{3}
days	hours	$2. 4 \times 10$
dynes	grams	1.02×10^{-3}
dynes	kilograms	1.02×10^{-6}
dynes	pounds	$2. 248 \times 10^{-6}$
	E	
ergs	gram–cm	1.020×10^{-3}
ergs	joules	1.0×10^{-7}
	F	
feet	centimetres	3.048×10
feet	kilometres	3.048×10^{-4}
feet	metres	3.048×10^{-1}
feet of water	atmospheres	2.95×10^{-2}
feet of water	in. of mercury	8.826×10^{-1}
feet of water	kg./sq. cm.	3.048×10^{-2}
feet/min.	cm./sec.	5.08×10^{-1}
feet/min.	km./hr.	1.829×10^{-2}
feet/min.	metres/min.	3.048×10^{-1}
feet/min.	miles/hr.	1.136×10^{-2}
feet/sec.	km./hr.	1.097

feet/sec.	miles/min.	1.136×10^{-2}
foot-pounds/sec.	kilowatts	1.356×10^{-3}
	G	
gallons	cu. cm	3.785×10^{3}
gallons	cu. feet	1.337×10^{-1}
gallons	cu. inches	2.31×10^{2}
gallons	cu. metres	3.785×10^{-3}
gallons	cu. yards	4.951×10^{-3}
gallons	litres	3.785
grams	dynes	9.807×10^{2}
grams	milligrams	1.0×10^{3}
grams	pounds	2.205×10^{-3}
grams/litre	pounds/cu. ft.	6.2427×10^{-2}
grams/sq. cm.	pounds/sq. ft.	2.0481
	H	
hectares	acres	2.471
hectares	sq. feet	1.076×10^{5}
horse power	kilowatts	7.457×10^{-1}
horse power	watts	7.457×10^{2}
hours	days	4.167×10^{-2}
hours	weeks	5.952×10^{-3}
	I	
inches	miles	1.578×10^{-5}
inches	yards	2.778×10^{-2}
inches of mercury	atmospheres	3.342×10^{-2}
inches of mercury	feet of water	1.133
inches of mercury	kg./sq. cm.	3.453×10^{-2}
	J	
Joules	ergs	1.0×10^{7}
Joules	watt-h	2.778×10^{-4}
	K	
kilograms	dynes	9.80665×10^{5}
kilograms	pounds	2.2046
kilograms/sq. cm.	inches of mercury	2.896×10
kilograms/sq. cm.	pounds/sq. ft.	2.048×10^{3}
kilograms/sq. metre	atmospheres	9.678×10^{-5}
kilograms/sq. metre	bars	9.807×10^{-5}
kilograms/sq. metre	feet of water	3.281×10^{-3}
kilograms/sq. metre	inches of mercury	2.896×10^{-3}
kilograms/sq. metre	pounds/sq. ft.	2.048×10^{-1}
kilograms/calories	kilowatt-h	1.163×10^{-3}

kilolitres	cubic yards	1.308
kilolitres	cubic feet	3.5316×10
kilometres	inches	3.937×10^4
kilowatts	horse power	1.341
kilowatts	watts	1.0×10^3
kg/sq.cm	Mega Pascal	1.0×10^{-1}
	L	
light year	miles	5.9×10^{12}
litres	cu. cm	1.0×10^3
litres	cu. ft.	3.531×10^{-2}
litres	cu. inches	6.102×10
litres	cu. metres	1.0×10^{-3}
litres	cu. yards	1.308×10^{-3}
lux	foot-candles	9.29×10^{-2}
	M	
Mega Pascal	t/m^2	100
metres	inches	3.937×10
metres	yards	1.094
miles/h	ft./min.	8.8×10
miles/h	ft./sec.	1.467
miles/h	km/min.	2.682×10^{-2}
millimetres	miles	6.215×10^{-7}
millimetres	yards	1.094×10^{-3}
minutes (angles)	degrees	1.667×10^{-2}
minutes (angles)	radians	2.909×10^{-4}
	N	
newton	dynes	1.0×10^5
	O	
ounces	grams	2.8349×10
ounces	pounds	6.25×10^{-2}
	P	
pounds	dynes	4.448×10^5
pounds	grams	4.5359×10^2
pounds	kilograms	4.536×10^{-1}
pounds	tonnes	5.0×10^{-4}
pounds	feet of water	2.307
pounds/sq. in.	inches of mercury	2.036
pounds/sq. in.	kg/sq. cm.	7.03×10^{-2}
	Q	
quadrants (angle)	minutes	5.4×10^3
quadrants (angle)	radians	1×571

	R	
radians	degrees	5.7296×10
radians	minutes	3.438×10^3
radians/sec.	revolutions/min.	9.549
reams	sheets	5.0×10^2
revolutions	degrees	3.60×10^2
revolutions/min.	degrees/sec.	6.0
	S	
seconds (angle)	degrees	2.778×10^{-4}
seconds (angle)	minutes	1.667×10^{-2}
seconds (angle)	radians	4.848×10^{-6}
square centimetres	sq. feet	1.076×10^{-3}
square centimetres	sq. miles	3.861×10^{-11}
square feet	acres	2.296×10^{-5}
square feet	sq. inches	1.44×10^3
square inches	sq. cm.	6.452
square inches	sq. ft.	6.944×10^{-3}
square kilometres	sq. cm.	1.0×10^{10}
square metres	acres	2.471×10^{-4}
square metres	sq. miles	3.861×10^{-7}
square metres	sq. millimetres	1.0×10^6
square metres	sq. yards	1.196
square miles	sq. km.	2.590
square yards	acres	2.066×10^{-4}
square yards	sq. cm.	8.361×10^3
	T	
tonnes (metric)	kilograms	1.0×10^3
tonnes (metric)	pounds	2.205×10^3
tonnes/m^2	Mega Pascal	10^{-2}
	V	
volt/inch	volt/cms.	3.937×10^{-1}
volt/inch	abvolts	1.0×10^4
	W	
watt-hours	horse power-hours	1.341×10^{-3}
watt-hours	kilowatt-hours	1.0×10^{-3}
weeks	seconds	6.048×10^5
	Y	
yards	centimetres	9.144×10
years	days (mean solar)	3.65256×10^2
years	hours (mean solar)	8.7661×10^3

Appendix

F

Pile Instrumentation

By Vijay Kumar, FIE, CEO, Record Tech Electronics, India, vijay.kumar@recordtek.com

F1 INTRODUCTION

With a multifold increase in infrastructure development throughout the world, many of innovative structures are being constructed globally. But the assurance of performance of these structures, after their construction still remains a question mark. With increasing knowledge of instrumentation these days, it can be possible to periodically validate the behaviour of the structure even after several years. At the time of construction of piles at site the initial load test and routine load test are conducted on representative piles. Their results are compared with the predicted load capacities, computed from analytic methods. After that the superstructure is constructed. It has been reported many a times that the actual superstructure sometimes may demand changes in architectural patterns of the structure. It has also been reported that the promoters of the project increase the no. of storeys of the super structure or increase the height of structure. All these things might cause an extra load on the pile, which though is overlooked by the developers. At a later stage when some distresses are noted in the structure, the experts are consulted who may like to probe into the reasons for the distress. At that time it is almost impossible to determine how much actual load is being catered to by the pile.

However, this task might be possible if the pile had been instrumented at the time of its construction. The advantages of instrumentation in pile at its pre construction stage is that its behaviour could be watched even after several years of its construction. For this purpose, a range of geotechnical instruments are manufactured by few companies such as **Geokon, USA** based on **VIBRATING WIRE Strain Gauges**.

Through instrumentation, it is possible even to monitor the performance of pile in its different segments, i.e. across its different sections. The instrumented pile may also give the details of end bearing component of load and also the friction load component of pile. The aim of the present chapter is to impart the basic knowledge of instrumentation and create an awareness towards importance of instrumentation in modern day construction industry. The discussion is however focussed to pile foundation only. Effort has been made to explain it further through some successful case studies.

The purpose of the instrumentation is two-fold—first, to instrument a test pile, and then do the loading/unloading as per ASTM standards to the desired extent in order to evaluate a pile and soil as a system based on the data so collected from the instruments.

On a site, this will give clear picture as to condition of soil and the pile, taking into account all factors including pore pressures.

Then, having tested the sample "Test Piles", actual working piles too are instrumented to keep watch on the structure in future. **It greatly helps evolving mitigation plans in case of any alternation is done to the structure in future.**

In engineering designs, we design on the basis of design inputs. In case of Piles, these are designed to carry the load of the structure for the life of this structure. A design engineer has to take into account all the ground conditions and the typical requirement of loads on the structure and the future possibilities.

A structure is as good as designed, and as good as your design inputs/assumptions used in the designing.

Another golden principle of designing is the verification of design—of scaled down models or life-sized models. The verifications are done by measurement of relevant parameters.

It is important to know what is relevant in specified application? And how to measure effectively by remaining alert to the fact that data so measured is correct and dependable?

So, as applied to the piles, we need to ensure that these are designed sound enough to carry the load and these stand on the sound enough ground that these will not sink or sway or crumble.

That is, these piles should remain as they were cast and carry the load for the life.

The above discussion gives us—one requirement very promptly:

- **"That we need to measure, then continue to measure (monitor) for a very long time."**
- Long time, in this case means, several years. And there should be a stake holder to do the analysis of these data, or else these are just junk/meaningless numbers.
- The use of piles for support of buildings, infrastructure, bridges, tall structures is continuously increasing. Having designed a pile for a particular location, it is imperative to know how good is the pile at a given location. Normally, it is a cluster of piles for a structure.

We normally do one or more of the following tests on a pile:

- Vertical load test (static/compression)
- Vertical load test (static/tension)
- Lateral load test (static)
- Dynamic load test

A general requirement is to find the settlement with respect to test load (2 times the design load)

In order to do that, we measure the following:

1. Strains at various cross-sections of a pile, along the length of the pile.
2. Deformations/Contraction/sinking of pile due to loading from the top.
3. Load applied, load at various cross section and load at the tip of the pile.
4. Rebound after loading is removed.

To determine stresses, all we need to know is the young's modulus of the material and the strain values at a point corresponding to the load.

Following test procedure is implemented.

F2 TEST PROCEDURE FOR VERTICAL LOAD TEST FOR INDIVIDUAL TEST PILE OR A PILE GROUP

We should properly select the following:

F2.1 Loading apparatus
F2.2 Equipment/instruments to monitor movements
F2.3 Instruments to monitor load and strain
  F2.3.1 Load monitoring
  F2.3.2 Strain monitoring
F2.4 Procedure to apply load
  F2.4.1 Compression pile load test
  F2.4.2 Compression testing load test for a group of pile
  F2.4.3 Tension pile load test
  F2.4.4 Lateral pile load test

A typical instrumentation scheme of a vertical compression test of pile is shown below:

Figure F1 below, Geokon USA suggests:

1. Displacement transducer (Fig. F6)
2. Load cells (Fig. F3)

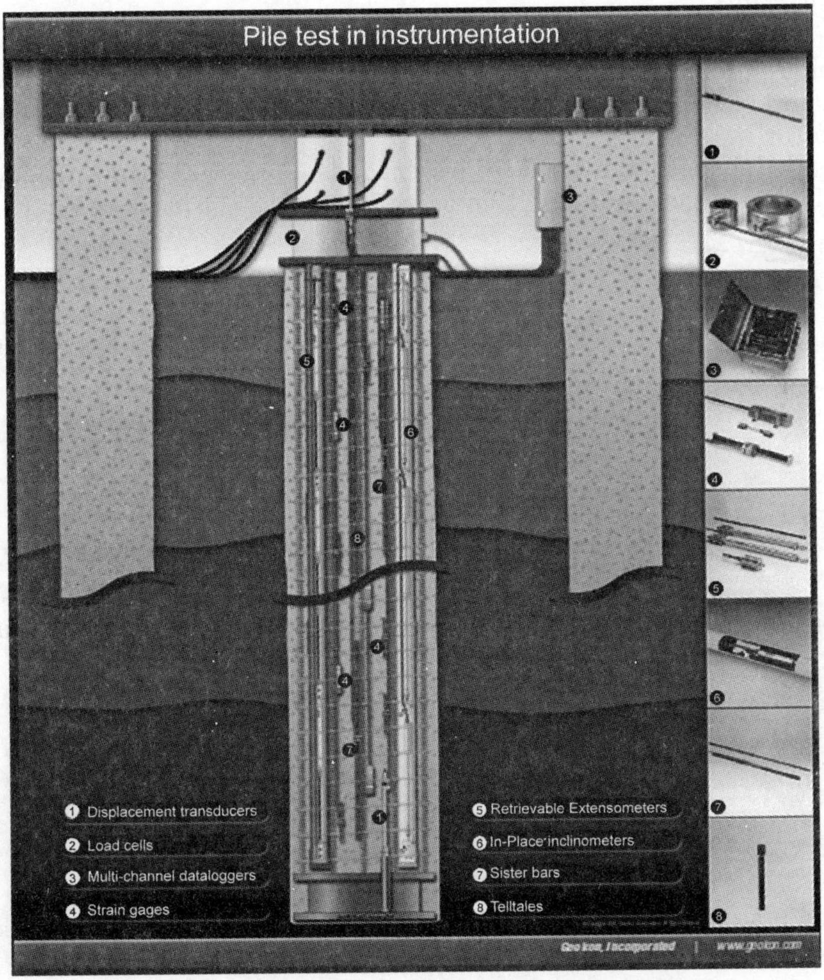

Fig. F1: A typical instrumentation of a pile test plan

3. Multichannel data loggers
4. Strain gauges Figs F20 and F21
5. Extensometers/retrievable Fig. F8
6. In place inclinometres Fig. F31
7. Sister bars Fig. F17
8. Tell tales Figs F9 and F10

F2.1 Loading Apparatus

Load is applied using a **One or many Hydraulic Jacks** of appropriate capacities.

If two or more jacks are to be used to apply the test load, they shall be of the same RAM diameter, connected to a common manifold and pressure gauge, and operated by a single hydraulic pump.

RAM of the hydraulic cylinder is restrained so that load is applied on the pile(s) by reaction.

There could be 2 ways of **restraining** the RAM

- Kentledge method/Dead weight on a loading platform, and
- Reaction beams

In Kentledge method, a proper steel platform is made over the pile under test by leaving enough room for placing the hydraulic jack(s) and a **Load cell** between the pile head and steel plateform (Fig. F24). Distributed dead weight is placed on the steel platform.

The Kentledge method is used for lesser loads. For larger loades reaction beams are used. But for larger loads also, Kentledge Method can be used *see* Fig. F2.

Fig. F2: A typical "Type 1" Kentledge loading platform

Fig. F3: Model: 4900 load cell, Geokon USA *www.geokon.com*

Should the ground conditions or site constraints preclude the use of reaction piles. A frame is assembled over the pile to be tested on top of which an amount of weight (a minimum 110 to 120% of maximum test load) is safely stacked. This generally takes the form of concrete blocks of regular dimensions and weight although steel ingots can be used provided that their weight can be assessed with reasonable accuracy.

A care must be taken to ensure "Stability" to avoid any failure of loading arrangement.

The size of the testing apparatus is generally a function of the pile size and loading to be applied. In broad terms an area of at least 15 × 15 m is required for the test (*see* Fig. F4).

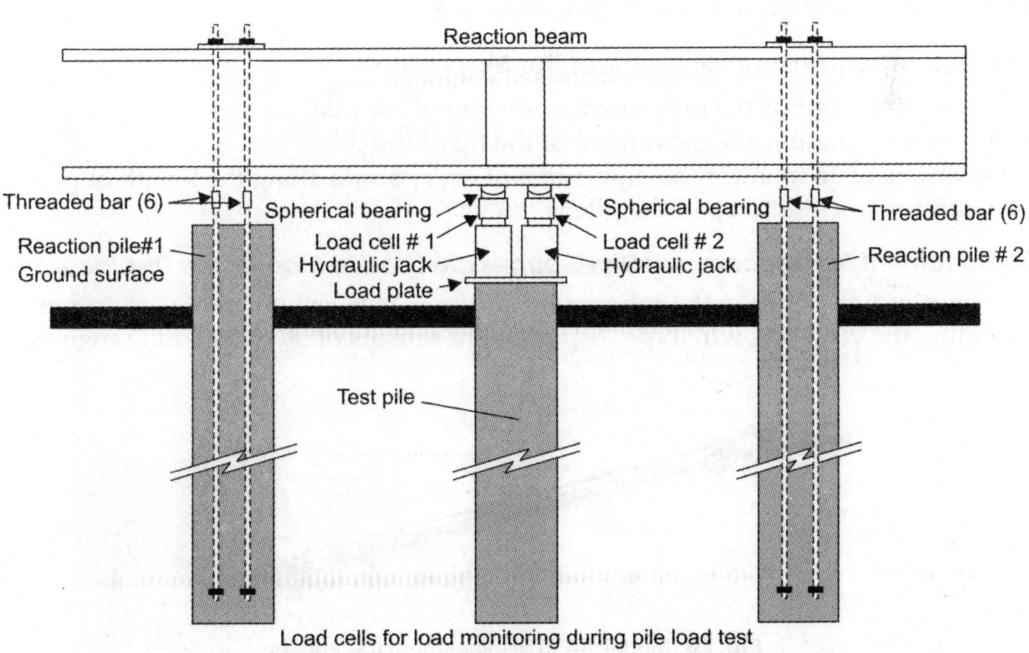

Fig. F4: Type 2: Reaction beam loading

Fig. F5: Courtesy: Strainology Pte LTD, Singapore

The steel girders will be laid across the test pile up to the anchor piles with system set-up similar to that shown in Fig. F7

Hydraulic Jacks

Axial compressive test: One or many (generally limited to 4, lesser the better) hydraulic jacks of appropriate capacity are provided on top of the pile head as loading apparatus, with a plate below the jack(s) (Fig. F24).

Ball Bearing

To provide non-eccentric load to the pile head, a ball bearing shall be inserted in between the reaction beam and the hydraulic jack.

F2.2 Equipment/Instruments to Monitor Movements

- We need to monitor the movement of the top of the pile.
- We need to monitor the movement of the tip of the pile.
- We also need to monitor the movement of every strata along the length of pile, i.e. at various cross-sections of the pile.

Displacement Transducers and Dial Gauges are Used to Monitor the Displacements

It is required to monitor the movement of the pile head at 4 points 90° apart by providing the vibrating wire type displacement sensors of 50 or 100 mm range , such

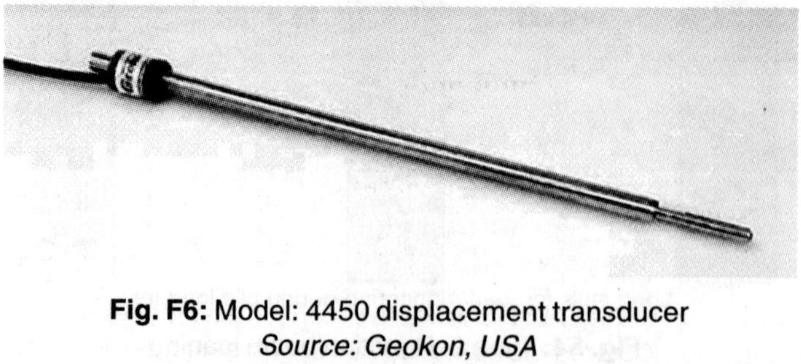

Fig. F6: Model: 4450 displacement transducer
Source: Geokon, USA

Geokon Displacement Sensor
Project location: Bangladesh

Fig. F7: Geokon 4450 measuring with reference to reference beam

as, model 4450–50 of Geokon, USA with an accuracy of 0.01 mm by mounting these between the pile head and reference beams (Figs F6 and F7).

During the process of setting up, Mitutoyo dial gauges can be used for quick check to monitor the pile movements the micrometer too should have a range of 0–50 mm and an accuracy of 0.01 mm.

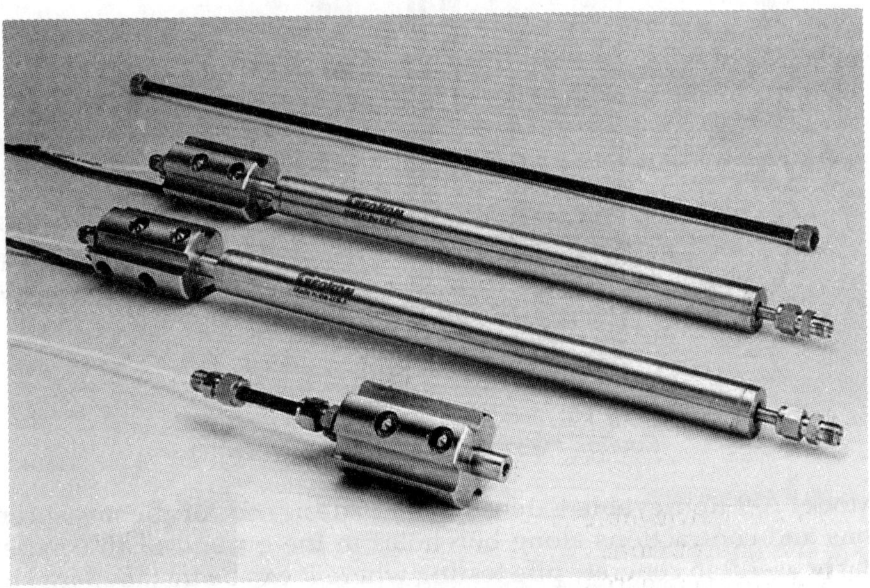

Fig. F8: Retrievable Extensometer Model A-9
Courtesy: Geokon USA, For Repeated use in piles

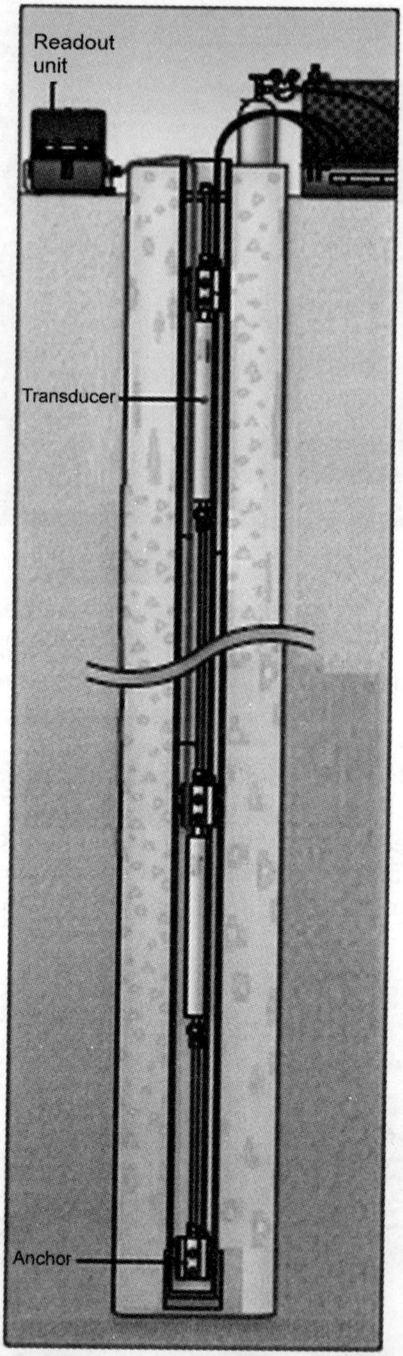

Fig. F9: Model A9 Installation in a Pile,
Courtesy: Geokon USA www.geokon.com

The Model A-9 Retrievable Extensometer is designed for the measurement of extensions and contractions along boreholes in the ground or in concrete. It is particularly useful in concrete pile testing where it can be installed inside a PVC or steel pipe cast in the pile: in this application it is a substitute for either tell-tales (Fig. F9).

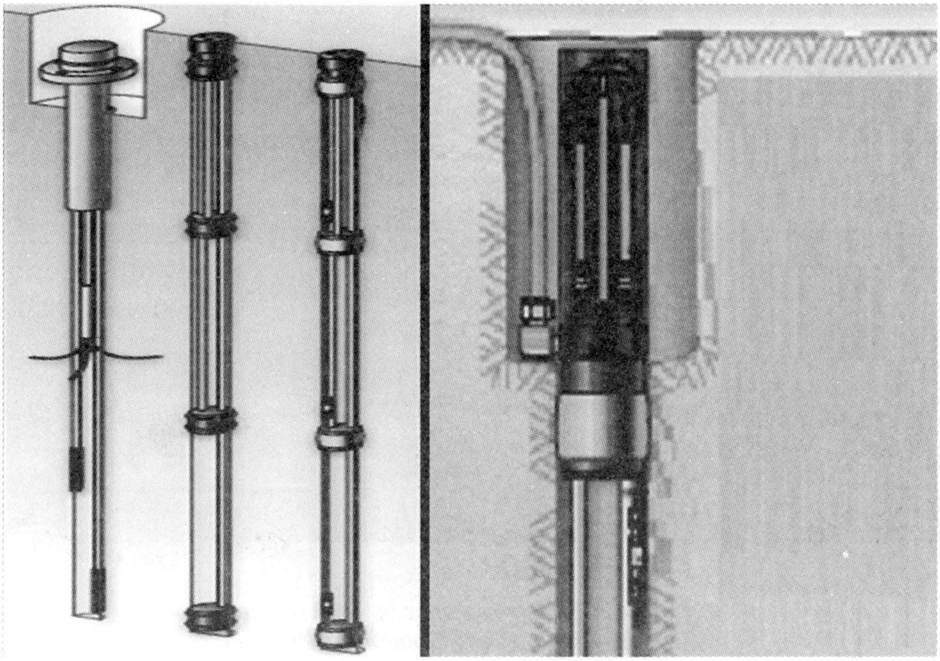

Anchors head assembly

Fig. F10: Model A-3 Multiple point extensometers
Courtesy: Geokon USA www.geokon.com

Fig. F11: Extensometer pipe
Courtesy: Record Tech Electronics, Roorkee

Extensometer anchor installed at pile tip

Anchor

Extensometer rod

Fig. F12: Extensometer installed at pile tip
Courtesy: Record Tech Electronics, Roorkee

Head assebly for 3 point extensometer at test pile

Project location: Iskcon Temple Vrindavan

Fig. F13: Geokon head assembly for 3 point extensometer
Courtesy: Record Tech Electronics Roorkee Mail: sales@recordtek.com

Reference Beam

Two reference beams (channels) will be cross-connected and laid on support, firmly embedded in ground with one end fixed and the other end free to take care of any thermal expansion. These should be so placed that the loading does not affect the position of these reference beam.

Fig. F14: Displacement sensor mounted on pile top
Courtesy: Record Tech Electronics, Roorkee

F2.3 Instruments to Monitor Load and Strain

F2.3.1 Load Monitoring

Load at the pile butt should invariably measured using a loadcell, because load determination based on hydraulic jack pressure are seriously prone to large errors, in excess of 20%, or even higher. The jack pressure is simply unreliable.

Evaluation of results of Load tests in axial compression are seriously hampered if tip load is not known, because total applied load cannot be separated in anyway into point resistance and skin friction.

It is, therefore, strongly advised to use pile tip pressure cells to measure directly tip load, i.e. point resistance (Figs F15 and F16).

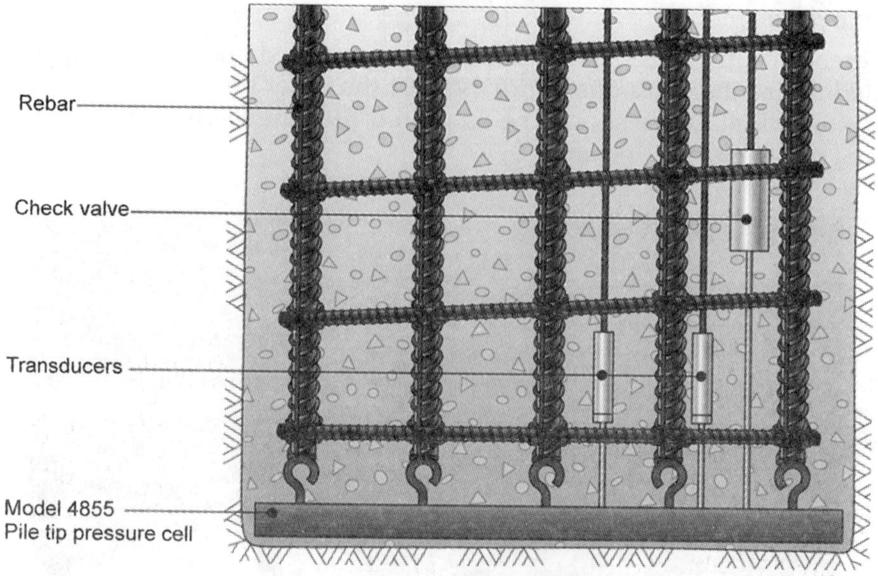

Fig. F15: Pile tip pressure cell
Courtesy: Geokon USA

Fig. F16: Pile tip pressure cell at a US project Location
Courtesy: Geokon USA and Record Tech—India

F2.3.2 Strain Monitoring

Vibrating wire type sister bars type 4911 of Geokon should be used, 4 at each cross-section along the length of pile. A soil investigation report can be used to decide at how many cross-sections these sister bars (Fig. F17) will be installed.

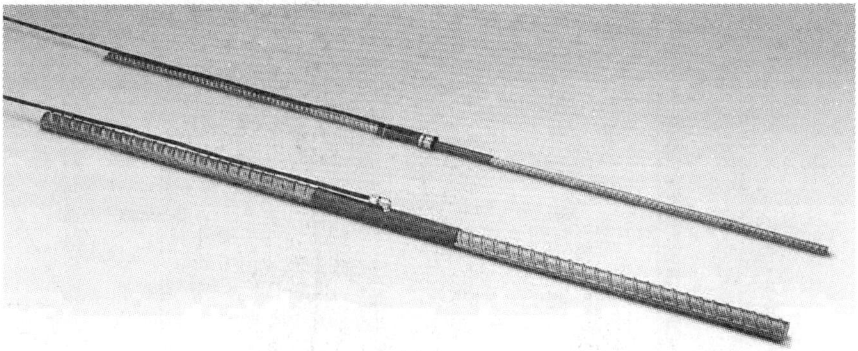

Fig. F17: Model 4911 sister bars
Geokon USA

Due to cost constraints, 2 sister bars opposite to each other can also be used with a little sacrifice on accuracy.

Average of these 4 or 2 sister bars will provide strains at designated cross-section. Young modulus times the cross-section area, when multiplied with strain will directly give load at that cross-section.

The Figs F18 and F19 next show a typical arrangement of tying up the sister bars 4911 to opposite bars of pile cage at multiple levels.

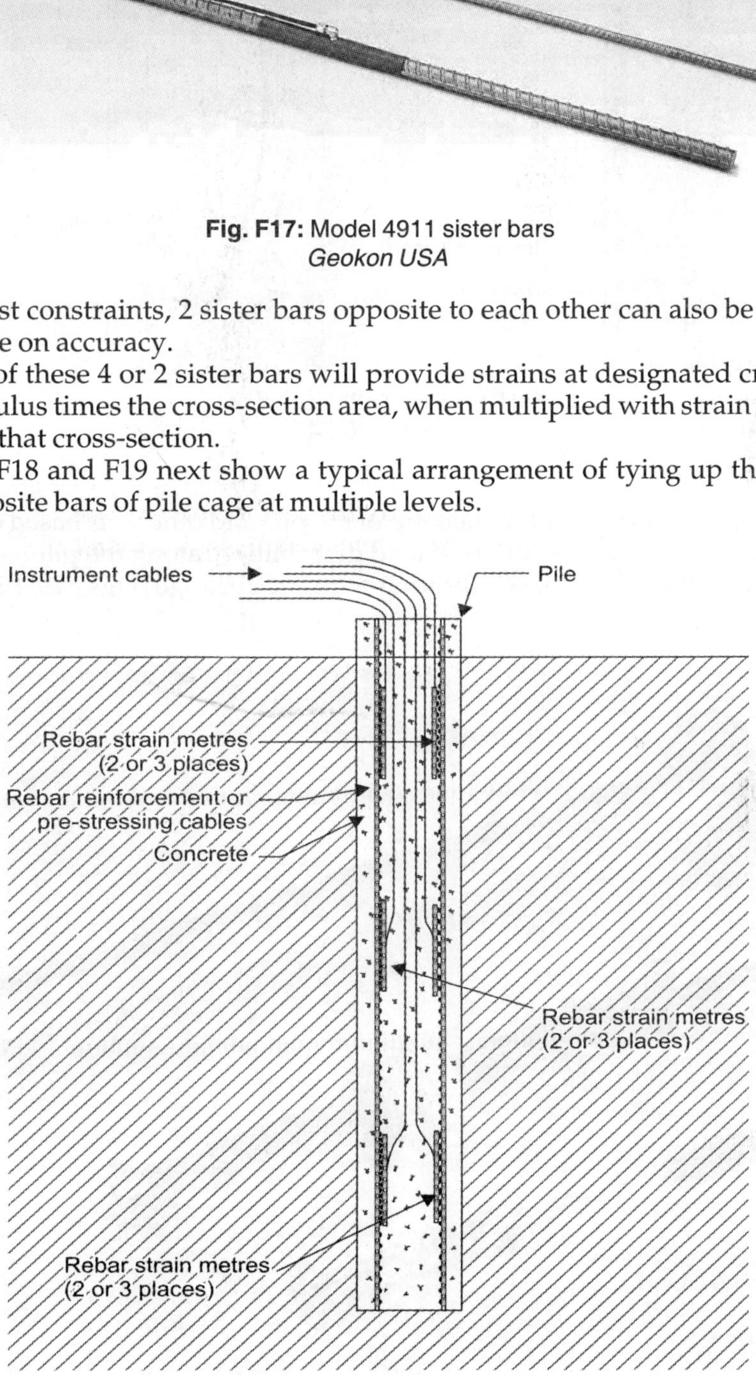

Fig. F18: Model 4911 Sister bar installation
Geokon USA

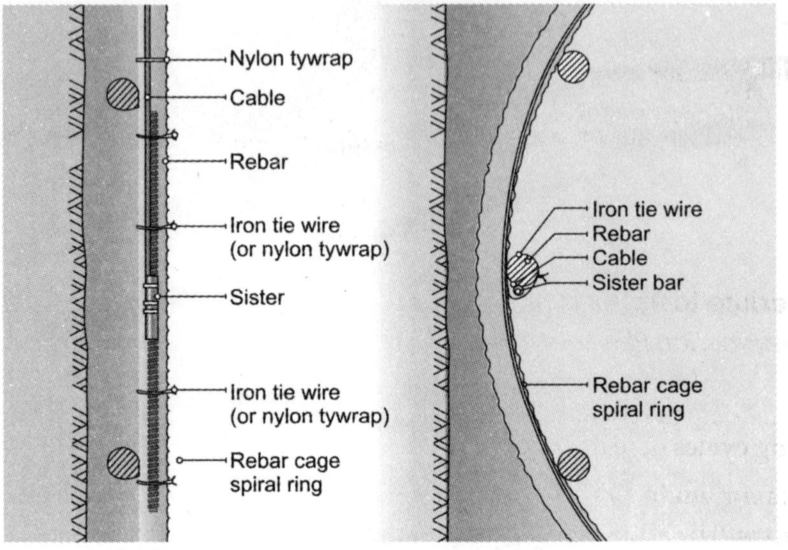

Fig F19: Model 4911 sister bar
Geokon, USA

One set near top and another set near tip of the pile and other sets based on the strata report will give fairly accurate value of load distribution, along the pile.

One set near top can be of 4200 vibrating wire strain gauge (Figs F20, F21 and F22).

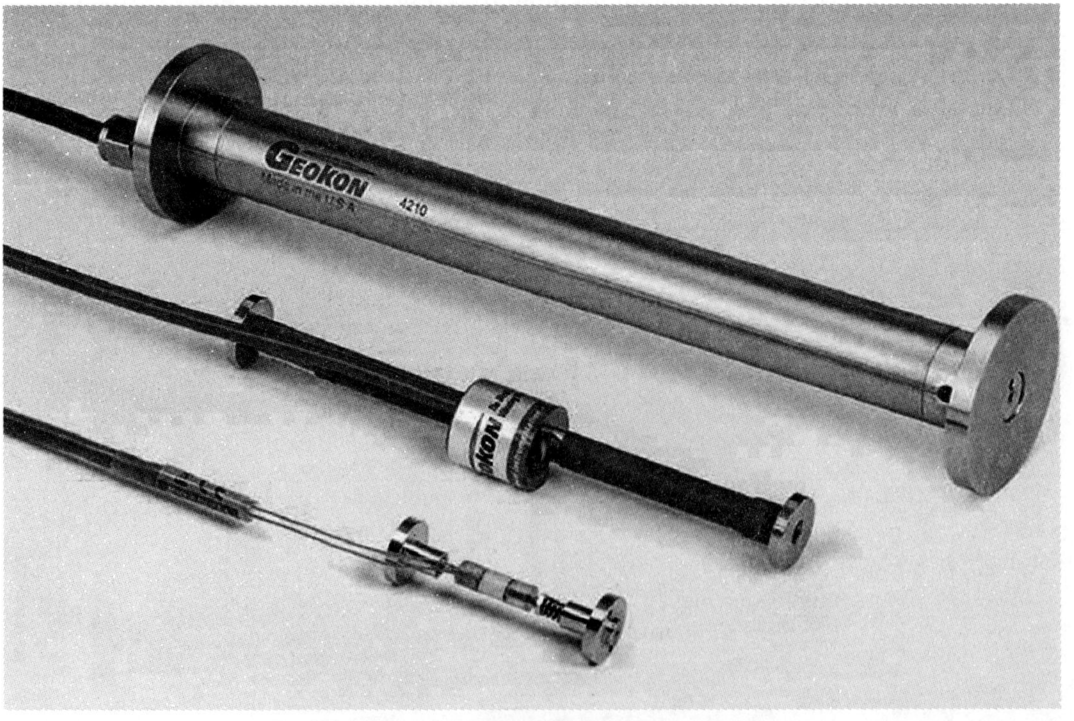

Fig. F20: Model 4200 vibrating wire strain gauge
Geokon USA

Fig. F21: Vibrating wire strain gauge 4200
Geokon, USA

F2.4 Procedure to Apply Load

F2.4.1 Compression Pile Load Test

The test shall conform to the modified *ASTM D 1143–81(Reapproved 1994) item 5.0 "Standard loading procedure"* with load sequence in percentage of design load. Following cycles of test will be performed:

Cycle 1 (maximum to 100% of the design load)

A. The load will be added gradually by increasing from initial 0 to 25%, 50%, 75%, and 100% of the design load.

B. When each load increment will be achieved, the next load increment will be added only when the settlement rate will be less than 0.25 mm per hour or after 2 hours, whichever shall occur first.

C. At each load increment, load, settlement and time will be recorded at 1, 2, 4, 8, 15, 30, 60, 90, 120, 240 minutes and every 2 hours with an accuracy of at least 0.01 mm.

D. The maximum load will be kept constant for at least 24 hours and then reduced to 75%, 50%, 25% and 0% of the design load, respectively. Each load will be maintained until the rate of settlement would not be greater than 0.25 mm per hour or after two hours, whichever shall occur first.

E. At "0" load, rebound movement will be recorded at 1, 2, 4, 8, 15, 30, 40, 60 minutes and every hour thereafter until a constant settlement will be reached.

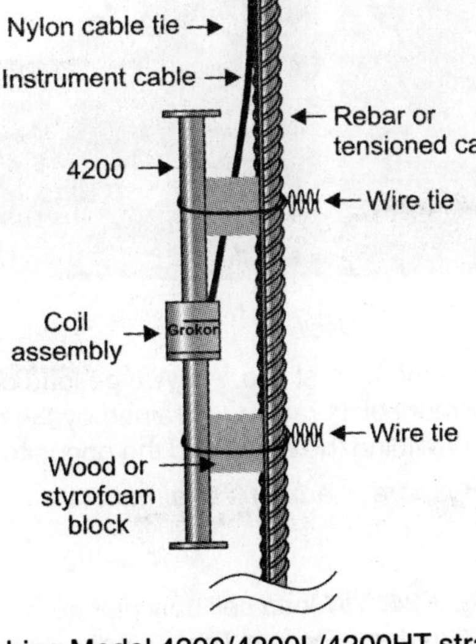

Fig. F22: Attaching Model 4200/4200L/4200HT strain gauges to rebar

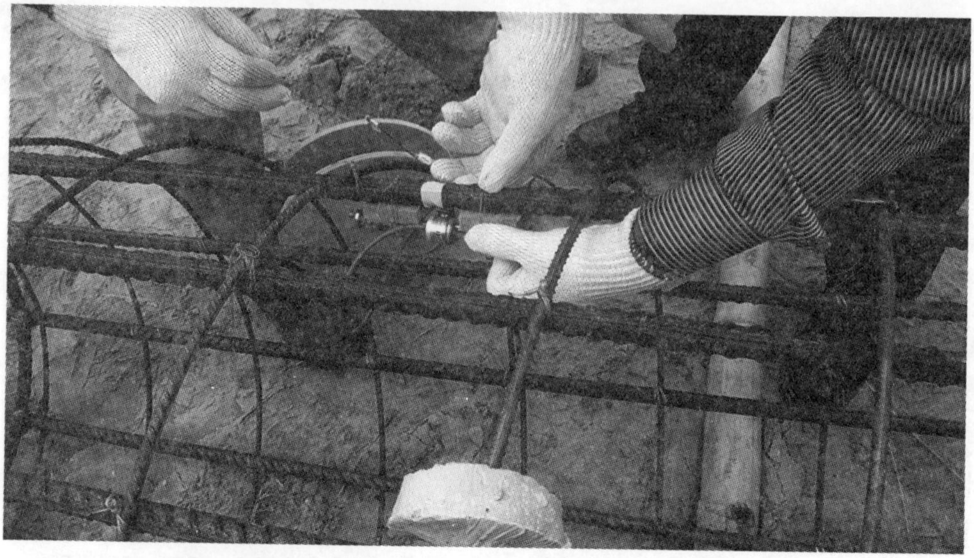

Fig. F23: VW strain gauge 4200 being installed on cage
Courtesy: Geokon USA www.geokon.com and Record Tech—India

A hydraulic jack is mounted on the top of the test pile. A VW type load cell is placed on its ram. When hydraulic jack is operated, the movement of its ram is restrained by the reaction frame or dead weight type load, as the case may be thus providing the equal and the opposite reaction force to the test pile

Courtesy: Strainology PTE Ltd, Singapore *admin@strainse.com*

Project location: Bangladesh

Fig. F24: VW load cell duly placed
Courtesy: Record Tech—India

Cycle 2 (Quick test, maximum to 200% of the design load)

A. The load will be added gradually by increasing from initial 0 to 25%, 50%, 75%, 100%, 125%, 150%, 175% and 200% of the design load.

B. When each load increment will be achieved, the next load increment will be added after every 5 minutes.

C. At each load increment, load, settlement and time will be recorded at 1and 5 minutes with an accuracy of at least 0.01 mm.

D. The load will then be reduced to 175%, 150%, 125%, 100%, 75%, 50%, 25% and 0% of the design load, respectively

E. At "0" load, rebound movement will be recorded at 1, 2, 4, 8, 15, 30, 40, 60 minutes and every hour thereafter until a no further settlement will be experienced.

F. The test pile will be considered failure when a rapid progressive movement of the pile in the direction of loading under a constant load or physical failure of the test pile is observed or a **settlement of more than 15% of the diagonal dimension of the pile.**

F2.4.2 Compression Testing Load Test For a Group of Piles

A group of 3 piles were to be tested. A real life scheme is being shown below:

Following is the plan view of "Test piles-S1–1, S1–2 and S1–3.

Anchor piles 10 in all, with id M1, M2 Through M10 were used to restrain the load. For testing S1–1, anchor piles M1, M2, M3, M6, M7 and M8 were used as above. Once the test was over the setup was moved to test S1–2, for which M2, M3, M4, M7, M8 and M9 were used.

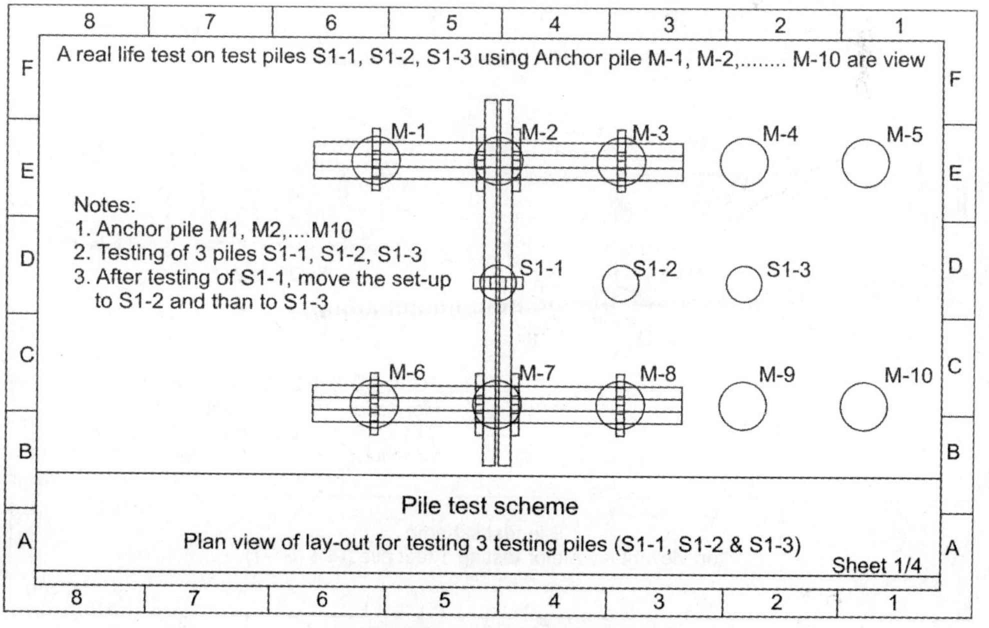

Fig. F25: Pile test plan

Courtesy: Record Tech electronics, Roorkee

Following the side view of the above when S1–1 was being tested.

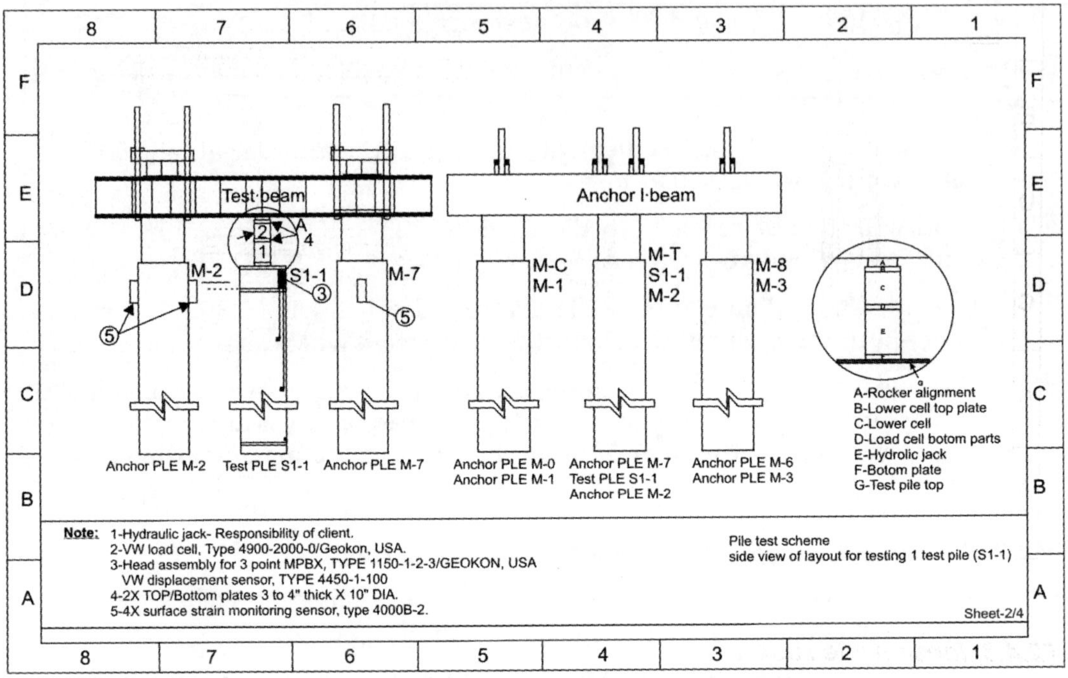

Fig. F26: Pile test view

Following is the top view of the pile under test and the 2 sides anchor piles.

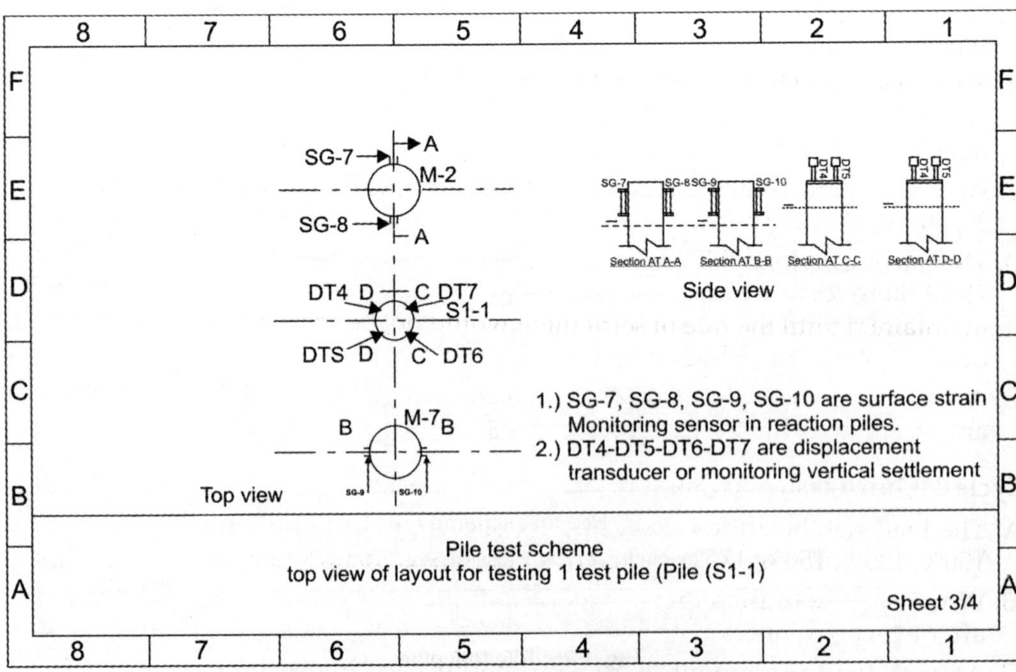

Fig. F27: Top view of layout for testing pile

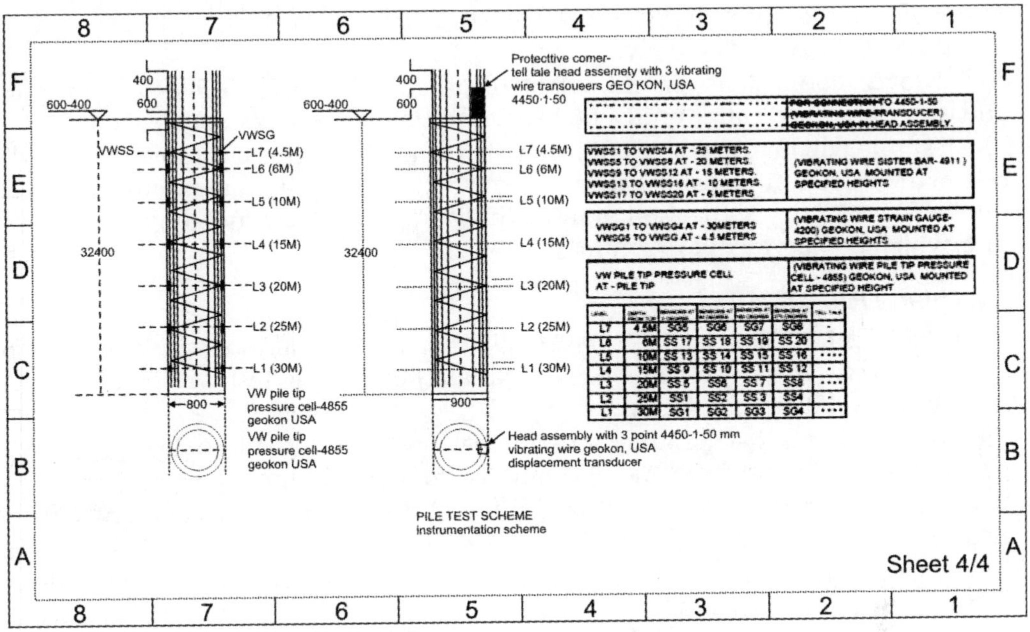

Fig. F28: Instrumentation scheme

F2.4.3 Tension Pile Load Test

The test shall conform to the *ASTM D 3689–90 (Reapproved 1995) "Standard test method for individual piles under static axial tensile load."* with load sequence as explained below. The test will be conducted until the pile exhibit signs of failure.

Cycle 1 (maximum to 100% of the design load)

A. The load will be added gradually by increasing from initial 0 to 25%, 50%, 75%, and 100% of the design load.

B. When each load increment will be achieved, the next load increment will be added only when the settlement rate will be less than 0.25 mm per hour or after 2 hours, whichever shall occur first.

C. At each load increment, load, settlement and time will be recorded at 1, 2, 4, 8, 15, 30, 60, 90, 120, 240 minutes and every 2 hours with an accuracy of at least 0.01 mm.

D. The maximum load will be kept constant for at least 24 hours and then reduced to 75%, 50%, 25% and 0% of the design load, respectively. Each load will be maintained until the rate of settlement would not be greater than 0.25 mm per hour or after two hours, whichever shall occur first.

E. At "0" load, rebound movement will be recorded at 1, 2, 4, 8, 15, 30, 40, 60 minutes and every hour thereafter until a constant settlement will be reached.

Cycle 2 (Quick test, maximum to 200% of the design load)

A. The load will be added gradually by increasing from initial 0 to 25%, 50%, 75%, 100%, 125%, 150%, 175% and 200% of the design load.

B. When each load increment is achieved, the next load increment will be added after every 5 minutes.

C. At each load increment, load, settlement and time will be recorded at 1 and 5 minutes with an accuracy of at least 0.01 mm.

D. The load will then be reduced to 175%, 150%, 125%, 100%, 75%, 50%, 25% and 0% of the design load, respectively.

I. At "0" load, rebound movement will be recorded at 1, 2, 4, 8, 15, 30, 40, 60 minutes and every hour thereafter until a no further settlement will be experienced.

J. The test pile will be considered failure when a rapid progressive movement of the pile in the direction of loading under a constant load or physical failure of the test pile is observed or a settlement of more than 15% of the diagonal dimension of the pile, which is equivalent to 90 mm.

F2.4.4 Lateral Pile Load Test

The test shall conform to the *ASTM D 3966-90 "Standard test method for individual piles under lateral load."* with load sequence as explain below. The test will be conducted until the pile exhibit signs of failure.

A. The load will be added gradually by increasing from initial 0 to 25%, 50%, 75%, 100%, 125%, 150%, 170%, 180%, 190% and 200% of the design load.

B. When each load increment will be achieved, the next load increment will be added.

C. At each load increment, load and time will be recorded at 1, 5, 10 minutes with an accuracy of at least 0.01 mm.

D. The test pile will be considered failure when a rapid progressive movement of the pile in the direction of loading under a constant load or physical failure of the test pile is observed.

During the load test, records including plots of load vs. time and load vs. settlement will be maintained progressively.

F2.4.5 Results of Test

The test results will then be reported in the form of:

- Time, load, pile head movements, settlements and reference beam movements
- Load-settlement curve
- Time-settlement curve
- Time-load curve
- Report and recommendations on the ultimate pile capacity

Inclinometers are used to profile tilt of the pile along the length when subjected to lateral loadings.

Inclinometer Theory

In the geotechnical field inclinometers are used primarily to measure ground movements such as might occur in unstable slopes (landslides) or in the lateral movement of ground around on-going excavations. They are also used to monitor the stability of embankments, slurry walls, the disposition and deviation of driven piles or drilled boreholes and the settlement of ground in fills, embankments, and beneath storage tanks.

In all these situations it is normal to either install a casing in a borehole drilled in the ground, to cast it inside a concrete structure, to bury it beneath an embankment, or the like. The inclinometer casing has four orthogonal grooves (Fig. F29) designed to fit the wheels of a portable inclinometer probe. This probe, suspended on the end of a cable connected to a readout device, is used to survey the inclination of the casing with respect to vertical (or horizontal) and in this way to detect any changes in inclination caused by ground movements.

In order to obtain a complete survey of the ground around the installed inclinometer casing it is necessary to take a series of tilt measurements along the casing.

Typically an inclinometer probe (Fig. F31) has 2 sets of wheels separated by a distance of 2 feet (English system) or 0.5 metre (Metric system). A casing survey would begin by lowering the probe to the bottom of the casing and taking a reading.

Subsequent surveys of the inclinometer casing, when compared with the original survey, will reveal any changes of inclination of the casing and locations at which these changes are taking place. Analysis of the change of inclination is best performed by computing the horizontal offset of the upper wheels relative to the lower wheels which has produced the tilting (·) over the reading interval (L) of the survey (usually the wheel base of the probe, 2 feet for English systems, 0.5 metre for Metric). At each position of the inclinometer the two readings taken on each axis (A+, A– and B+, B–) are subtracted from each other leaving a measure of sine Theta. This value is then multiplied by the reading interval (L) and the appropriate factor to output horizontal deflection in engineering units (inches for English, centimetres or millimetres for metric—Figs F33, F34, and F35).

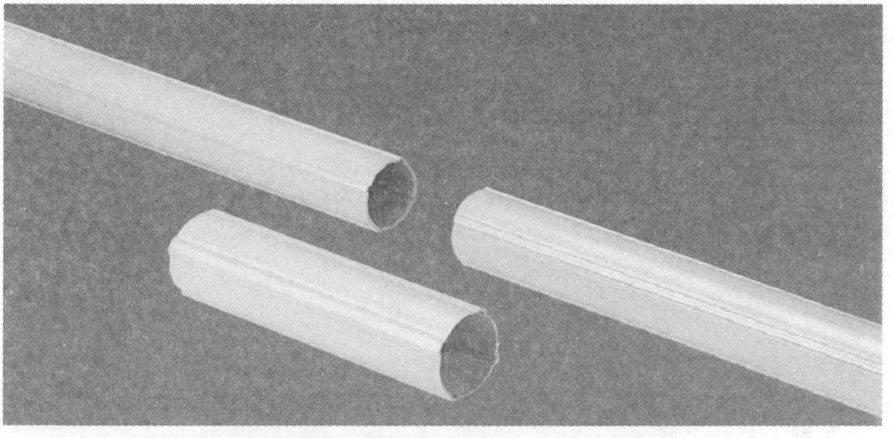

Fig. F29: Model 6500 Inclinometer casing
Geokon USA

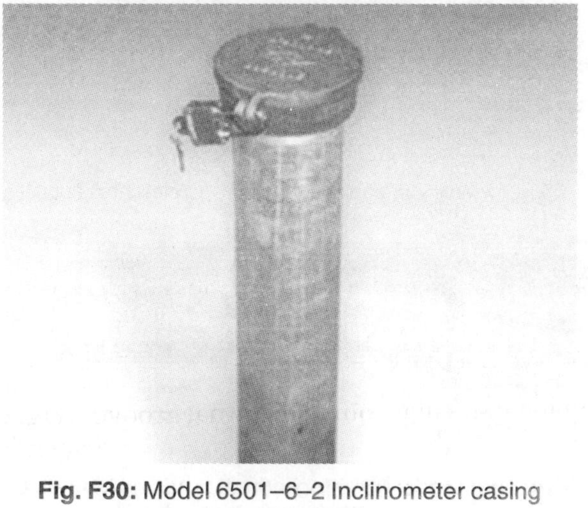

Fig. F30: Model 6501–6–2 Inclinometer casing
Geokon USA

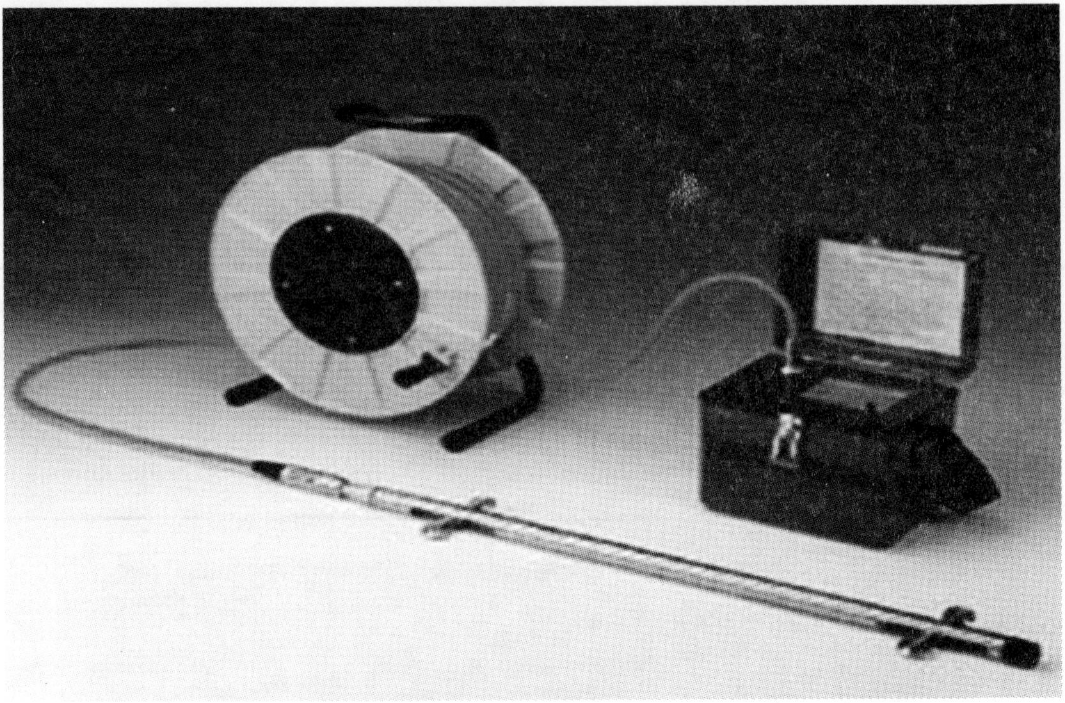

Fig. F31: Model 6000 inclinometer sensor with model GK-603
Courtesy: Geokon USA www.geokon.com

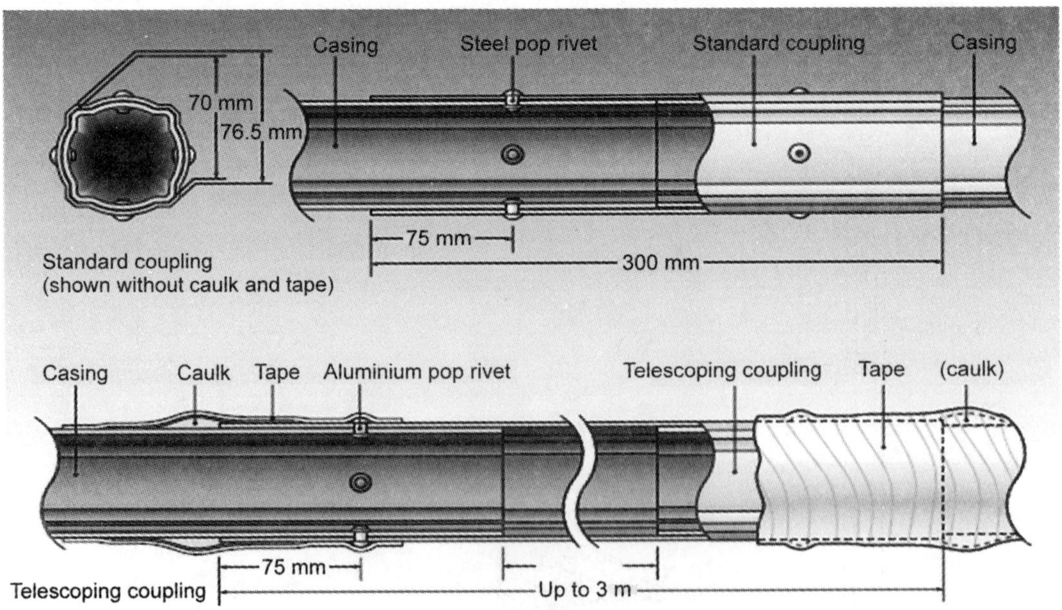

Fig. F32: Model 6500 Inclinometer casing dimension
Courtesy: Geokon USA www.geokon.com

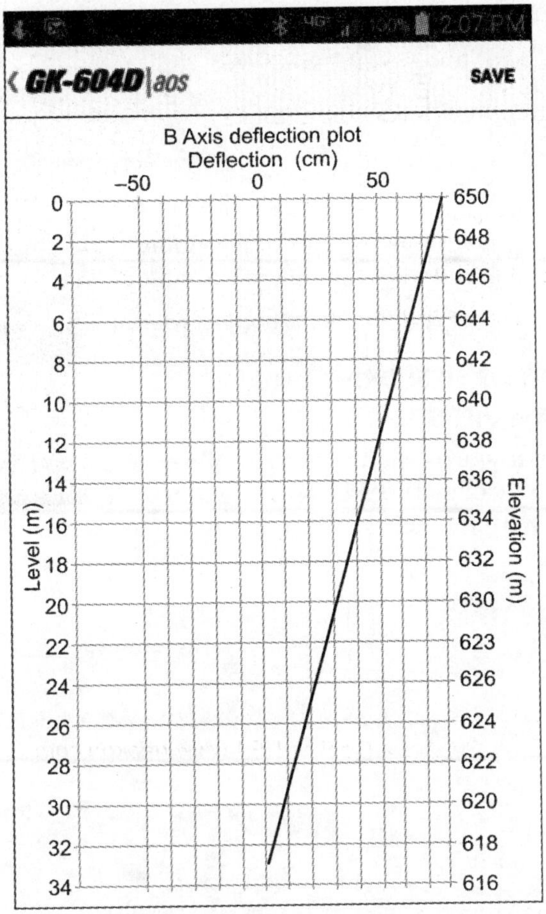

Fig. F33: Axis deflection

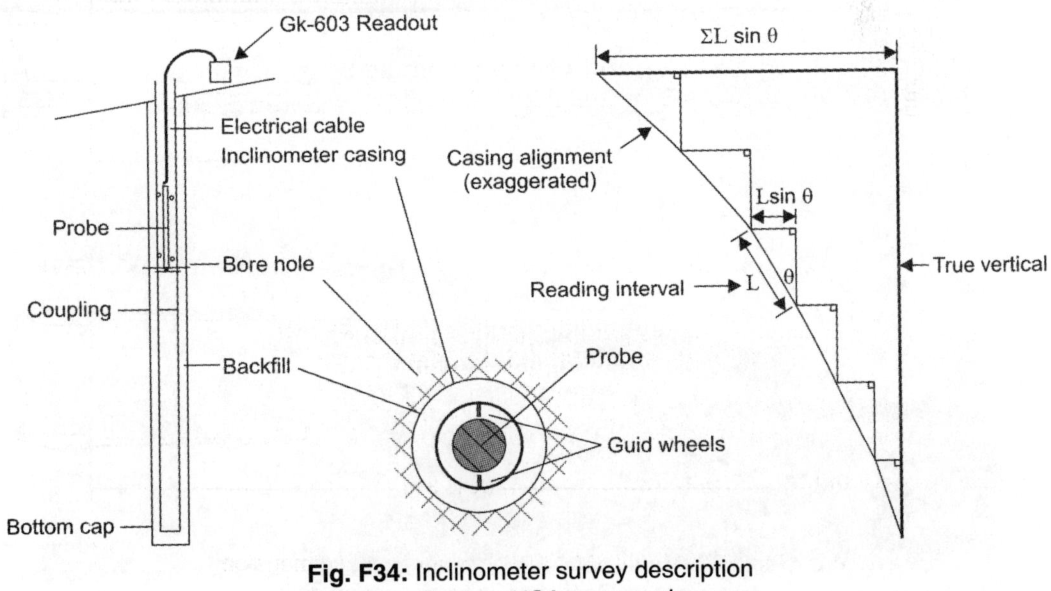

Fig. F34: Inclinometer survey description
Courtesy: Geokon USA www.geokon.com

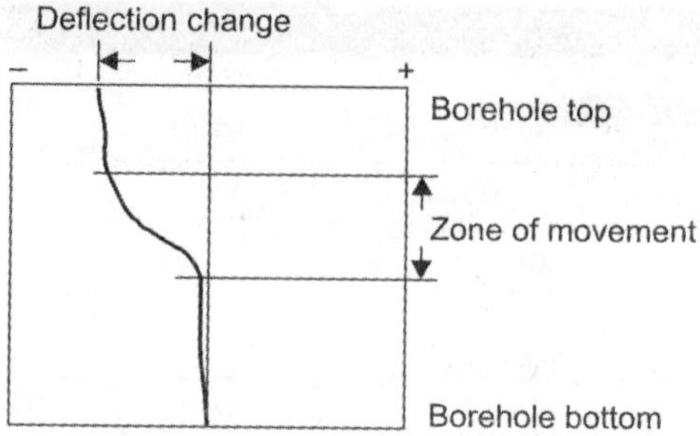

Fig. F35: Plot of borehole deflection

F3 INSTRUMENTATION OF STEEL DRIVEN PILES

To be Used as "Bridge PIERS"

At few cross-sections towards the tip of steel pile section, wedlable strain gauge type 4000, vibrating wire type of geokon, USA, were used. The mounting brackets (Fig. F36) were welded so that with the help of jig, Type 4000 gauge were used (Fig. F38).

Fig. F36: Mounting brackets duly welded on driven pile
Courtesy: Strainology Pte Ltd, Singapore admin@strainsg.com

To protect the gauges and wire, from damages whiles being driven, protective angle was welded all along the length of the pile.

Fig. F37: Driven pile

F3.1 Gauge Selection

Vibrating wire strain gauges are available for use on steel.

F3.2 Strain Gauge Protection

Strain gauge and cables need to be protected from being scraped off as the pile is driven into the ground (Figs F37 and F39).

Fig. F38: VW strain gauge 4000 duly mounted on driven pile
Courtesy: Strainology Pte Ltd, Singapore admin@strainsg.com

Fig. F39: Cable duly fixed with mounting brackets for protection
Courtesy: Strainology Pte Ltd, Singapore admin@strainsg.com

The test pile will be considered failure when a rapid progressive movement of the pile in the direction of loading under a constant load or physical failure of the test pile is observed or a settlement of more than 15% of the diagonal dimension of the pile.

F3.3 Conclusion

It is simple to introduce instrumentation. One has to only begin. Well begun is half done. While making a structure these instruments are useful in helping in validation of the designs. These will keep surveillance on the health of structure during its life. They become even more useful if there is any geotechnical activity later on in the neighbourhood, say a bridge is being made. For any guidance, contact author by mail *vijay.kumar@recordtek.com, or info@geokon.com, or admin@strainsg.com*

Appendix

G
Quality Assurance of Deep Foundation

By: Dr Ravikiran Vaidya, Principal Engineer, Geo Dynamics, 49, Atmajyoti Nagar, Ellora Park, Vadodara 390 023, email: ravikiran@geodynamics.net

Abstract: Low strain integrity, high strain dynamic pile testing (HSDPT) and cross hole sonic logging testing (CHL) has now been routinely conducted on various infrastructure and real estate projects across the country. The trend started with the "MSRDC 55 Flyovers Scheme" projects and is now extensively used all over the country including projects of metrorails at Delhi, Mumbai, Kochi, NHAI, railways, jetties, flyovers in cities and multistorey complexes across the country with most projects executed by the author. Integrity and dynamic pile testing have been valuable tools in ascertaining shaft quality and capacity for small and large diameter piles. The author has now conducted more than 150 co-relation studies in India and has more than 25000 integrity tests data generated annually. The present write-up tries to explain low strain integrity and dynamic testing, as well as cross hole sonic logging testing so that it is easier for the consultants/contractors to evaluate the data themselves and have more confidence in pile foundations and the test procedure. Low strain pile integrity testing procedure and data evaluation is described, followed by some case studies. Similarly, the procedure for dynamic pile testing is also described and a few co-relation studies are explained. Comparison is also made with some age old empirical methods like Hiley's formula for further understanding of the subject. Brief explanation regarding test method and application of cross hole sonic logging test is also presented.

INTRODUCTION

Low strain pile integrity testing is a quick, reliable and inexpensive tool that can be used to ascertain quality of pile shafts on site. Generally, the pulse echo method that uses only velocity measurement is widely used and recommended, although testing is also conducted by using an instrumented hammer to measure the intensity of impact. Today integrity testing is in code specifications of various countries in America, Europe, and Asia. It is also standardized as per ASTM D5882. In India, it is a part of the IS: 14893.

High strain dynamic test is quite often used to replace or supplement conventional static testing on construction and infrastructure projects. The test has been found to be reliable quick and inexpensive compared to static tests. It is standardized as per ASTM D4945 and various codes worldwide. Currently, it is one of the most widely used non-destructive test worldwide to ascertain pile static capacity.

The tests were developed in USA in 1975 by Dr. George Goble and his associates Frank Rausche and Garland Likins with active support from Federal Highway Administration. Simultaneously, it was also developed at the TNO Institute of Building and Construction Research, Netherlands.

Cross hole sonic logging generally applies to drilled shafts and requires that at least two tubes be installed in a drilled shaft prior to pouring the concrete, and that stress pulses are sent from one tube to the other at 20 to 50 mm vertical intervals. The arrival time of the pulse at the receiver tube indicates the quality and integrity of the concrete between the tubes.

LOW STRAIN PILE INTEGRITY TESTNG (PIT)

Application of Pile Integrity Testing

Low strain integrity testing using pulse echo method provides velocity data on structural elements (i.e. structural columns, driven concrete piles, cast in place concrete piles, concrete filled steel pipe piles, timber piles, etc.). This data assists evaluation of pile integrity and changes in pile cross-section qualitatively, continuity, and consistency of the pile material. However, this test method will not give information regarding the pile's load bearing capacity and should not be used as such.

It may be noted here that it is important that the equipment to be used or testing is also in accordance with ASTM D5882 or other relevant acceptable code. Generally, two equipment manufacturers, PDI of USA and TNO (now PROFOUND) of Netherlands are popularly used equipments and conform to various international standards although there may be several others in the recent times. It is also necessary that the person using these equipments be trained to operate and analyze the data.

Principle and Test Procedure

The procedure involves chipping of loose concrete from the pile top and cleaning the pile head so that it is flat, dry and accessible with sound concrete. Preferably, a disk grinder or carborundum stone may be used to grind the top of concrete to make it reasonably flat for testing. An accelerometer is fixed to the pile head somewhere near centre of the pile to collect the data. The test now involves generation of a low stress wave with the impact of a hand held hammer. This generated impacts in the 10 g to 100 g range and pile strains around 10^{-5}. The generated stress wave travels down the shaft to the pile bottom where it is reflected back to pile top. If this reflection occurs at the correct time and if no other earlier reflection waves are received at the pile top, then the pile shaft is probably free of major defects. The data is collected as an acceleration record and integrated to velocity form for further interpretation. Refer to Fig. G1 for

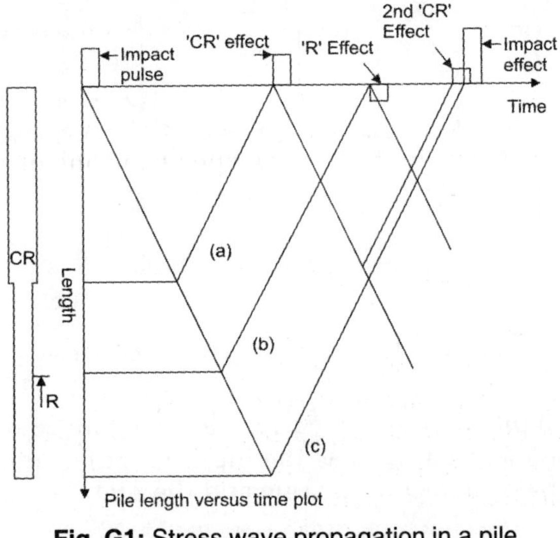

Fig. G1: Stress wave propagation in a pile

the test principle. Refer Fig. G2 for schematic of test set-up arrangements. A set of 3 to 5 datasets with each data set comprising of 3–6 blows is taken for analysis. As the impact is due to a light weight hammer, the corresponding strains are of very low magnitude, which would be measured only with great difficulty. Hence, the method is known as **low strain method.**

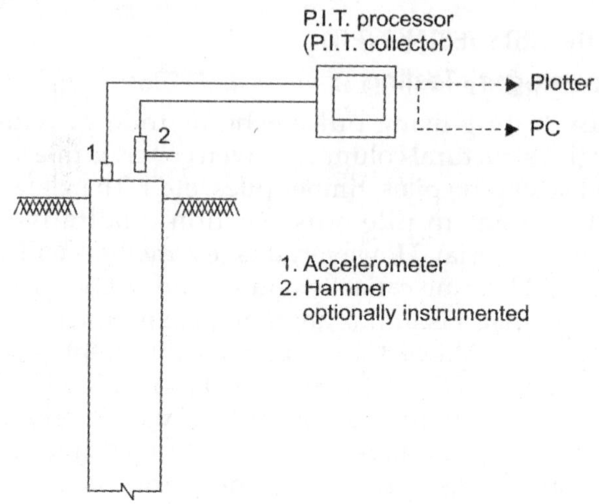

Fig. G2: Schematic of pile integrity testing

Interpretation of PIT Records and Case Studies

A typical good velocity curve is displayed in Fig. G3 and is a plot of velocity versus time for a pile of length 12.9 m. The first peak is the start of the pile. The second peak at 12.90 m indicates end of the pile. The wave speed through concrete is 4000 m/sec indicating good concrete quality. As apparent, the pile has no major anomalies and can be classified as good. Generally, in the Indian scenario a wave speed of 3500 m/sec to 4200 m/sec can be expected, unless otherwise due to factors like high performance concrete, heavier percentage reinforcement, use of permanent liners, etc.

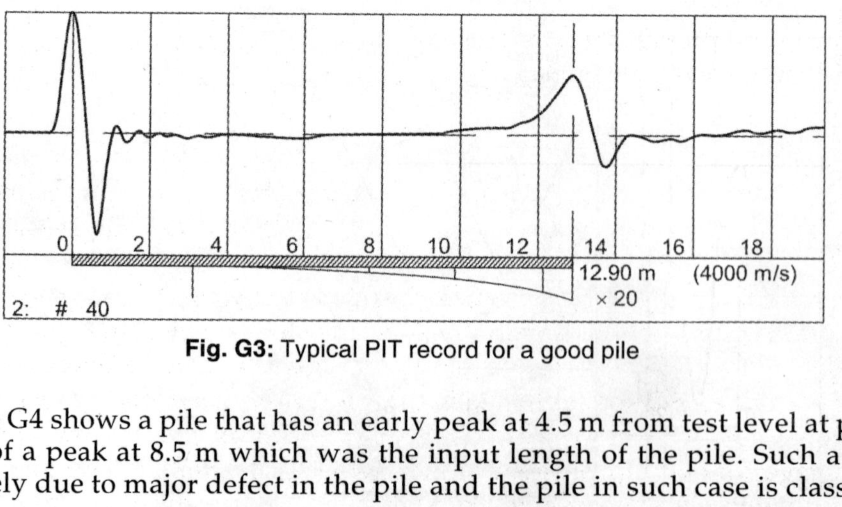

Fig. G3: Typical PIT record for a good pile

Figure G4 shows a pile that has an early peak at 4.5 m from test level at pile top, instead of a peak at 8.5 m which was the input length of the pile. Such a peak is most likely due to major defect in the pile and the pile in such case is classified as doubtful.

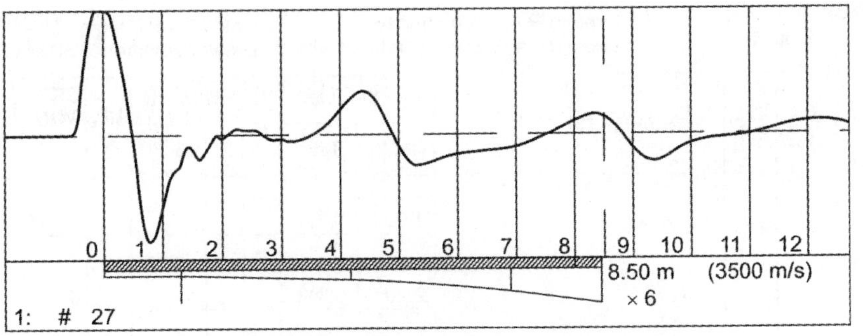

Fig. G4: Typical PIT record for a doubtful pile

Pile integrity tests were performed on 40 piles to determine their length. The tests were performed for multistoried residential building at Mumbai, since there were no previous records available. The results of such testing indicated most of the piles to be in the range of 10–11 m. Refer to Fig. G5 that shows the estimated pile length as 11.0 m. Actual coring in the pile concrete carried out on one pile to confirm the PIT results showed the pile length as 10.5 m. The core test results are shown in Fig. G6. Thus normally, it is possible to predict the pile length using PIT within a range of +5%–10%.

An integrity test was performed on a pile and concluded that there is a major defect in pile. The pile was then extracted and defect was confirmed as indicated by a PIT. Results of this test and excavated pile are presented in Fig. G7.

A static load test was performed on a pile and it failed; a PIT was performed to determine the potential cause of failure. PIT indicated a defect around 2 m from the pile top which was confirmed by excavating the pile. The findings are presented in Fig. G8.

It must be noted that inconclusive test results are also possible, particularly, when very large impedance increases (e.g. a large bulge or outgrowth in shaft diameters in a soft fill) near the pile top prevent a clear stress wave transmission.

Limitations

1. It requires experienced engineer trained in interpretation of results.
2. Adequate and proper site arrangements are necessary.
3. It is difficult to evaluate minor defects like localized loss of cover.

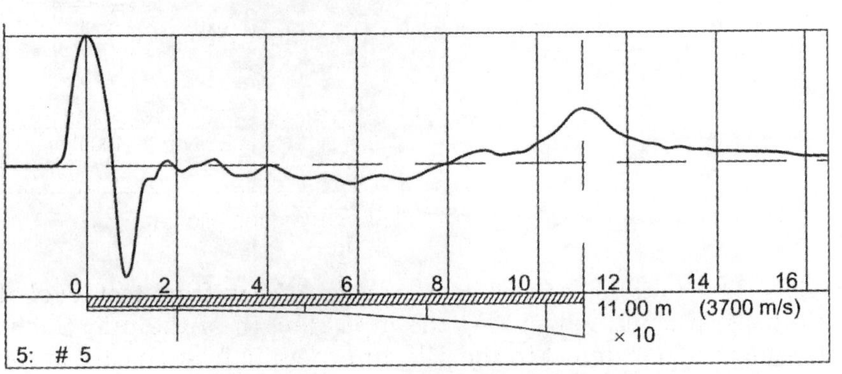

Fig. G5: PIT data to evaluate pile length

Bore hole Sine: 100 mm/Nx Method: Rotary	Ground R.L. : 100 mm/Nx Ground water depth : Not met with	Date of commence : 27 - 05 - 2001 Date of commence : 30 - 05 - 2001

Depth (m)	Bore Hole Dia (mm)	Thk. of Layer (mm)	Graphic Log	Stratum Description	Samples		Blows/15 cm				SPT N	CR / RQD %	Other tests / Remarks
					Depth	Type	15	30	45	60			
1					1.00	RUN1						97/97	
2					2.20	RUN2						88/88	
3					3.55	RUN3						97/70	
4					4.90	RUN4						97/74	
5	N x			Pile Concrete	5.90	RUN5						80/63	
6					6.65	RUN6						88/88	
7–8					8.25	RUN7						83/66	
9					9.10	RUN8						85/28	
10					10.00	RUN9						40/17	
		10.50			10.50	RUN10						44/20	
11–12					12.00	RUN11						07/nil	
13	N x			Highly weathered Reddish brown Volcanic Breccia Rock	13.00	RUN12						15/nil	
14					14.00	RUN13						34/nil	

Fig. G6: Coring data to confirm findings

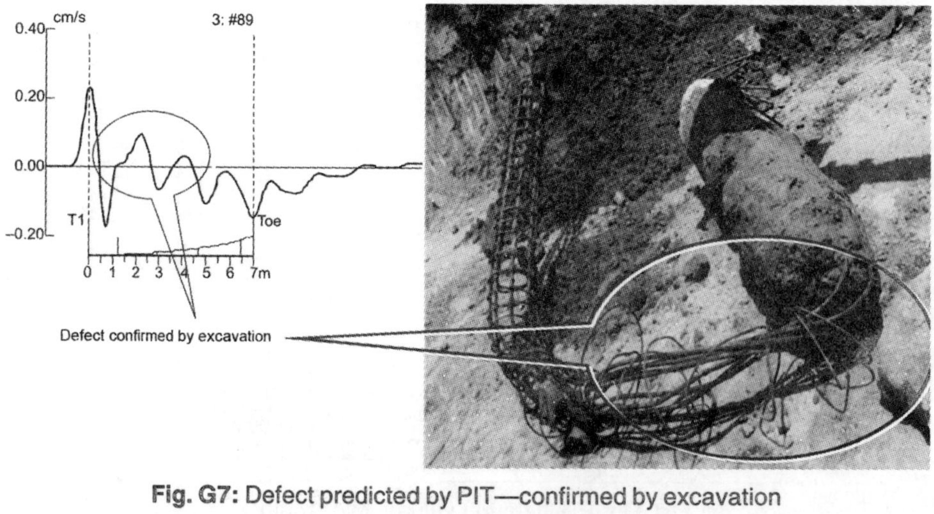

Fig. G7: Defect predicted by PIT—confirmed by excavation

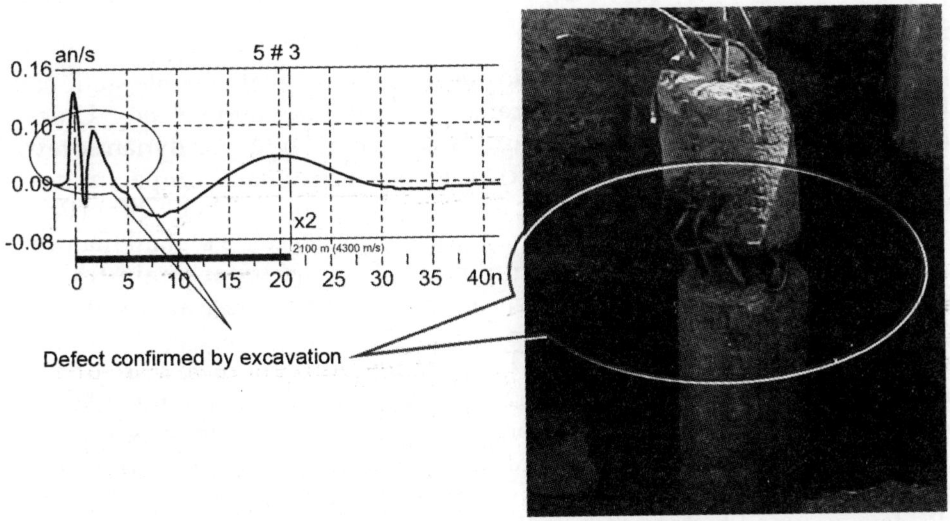

Fig. G8: PIT finds cause of static test failure

4. The method cannot generally tell integrity beyond the first major defect or a bulb.
5. It is difficult to identify if pile is tapered or if the pile is not installed plumb.
6. PIT is not applicable for micro piles, steel piles and driven jointed piles.
7. If piles are closely spaced with rebars extending above piles, ringing sound from rebars may cause distortion.

Benefits of Low Strain Integrity Testing

1. The test is simple, reliable if properly done and analyzed and can be quickly performed. The test determines defects in pile and with additional field information; it is possible to ascertain the nature of defect, viz. reduction in cross-section, cold joint, poor concrete quality, etc.
2. It can be used to identify piles for further static or dynamic testing instead of random selection thus ensuring better quality control.
3. The test can be used to confirm pile length with reasonable accuracy.
4. Cost of test is usually fraction of the entire cost of piling. Thus clients, contractors or consultants can use it on large or all numbers of piles at their project sites to quickly evaluate shaft integrity.

HIGH STRAIN DYNAMIC PILE TESTING

Application of Dynamic Pile Testing

The test can be used to evaluate various pile parameters, important of these are mentioned below.

- Static capacity of the pile at the time of testing.
- Static load test curve.
- Total skin friction and end bearing of the pile.
- Skin friction variation along the length of the pile.
- Compressive and tensile stresses developed in the pile during testing.
- Net and total displacement of the pile.
- Pile integrity and changes in cross-section if any.
- Hammer performance.

Principle of Testing

The basic purpose of high strain dynamic pile testing is to evaluate pile static capacity and its structural integrity using measurement of both force and velocity. Unless, any method uses an instrumentation to measure force it cannot be called **dynamic testing** as such, since the basic requirement of the test is to measure dynamic impact and convert it to static capacity. The use of dynamic formula as mentioned in IS: 2911 (Part: 3) is not dynamic load testing and should be strictly avoided as it is a blatant malpractice.

The method involves attaching a minimum of two pairs of strain transducers and accelerometers on diagonally opposite sides of the pile. These are fixed minimum 1.5 times the pile diameter below pile top.

Strains induced under the impact of a heavy falling hammer from a pre-determined height are measured with the help of strain transducers attached to the pile, whereas accelerometers record the accelerations generated in the pile. The pile driving analyzer converts strain to force, and acceleration records are converted to velocities. The resistance developed by the pile is then a function of force and velocity and includes few assumed factors such as the quake and damping parameters as inputs based on the soil type. The maximum pile top compression is obtained by integrating the pile top velocity. A more accurate value of these parameters is then obtained from CAPWAP analysis conducted on field data.

Thus briefly, it can be described follows the PDA uses the following basic equation to compute pile capacity, although a CAPWAP is generally required for more reliable capacity prediction.

$$R_s = (1 - J_c)\,(P_1 + Z.v_1)/2 + (1 + J_c)\,(P_2 - Z.v_2)/2$$

Here,

P_1, P_2 = Force at time t_1 and t_2 respectively
v_1, v_2 = Velocity at times t_1 and t_2 respectively
t_1 = Time of first peak of force/velocity
$t_2 = t_1 + 2L/c$
J_c = The damping factor

CAPWAP Analysis

The CAPWAP program is an analytical method that combines measured field data with pile wave equation type procedures, to predict the pile's static bearing capacity and soil resistance distribution. Measured force and velocity data is directly input from the PDA. Based on the measured velocity data, the program computes the force required to induce the imposed velocity. Both measured and computed forces are plotted as a function of time and the iterative analysis is continued till there is good agreement between both the curves. If the agreement is not satisfactory, the soil resistances at the pile point and along the pile are adjusted until a good match is obtained. This gives the frictional distribution along the sides, the end bearing component of the pile, as well as better estimates of the actual static pile capacity measured during field testing.

Method of Testing

The method involves attaching strain transducers and accelerometers to the sides of the pile 1.5 times pile diameter below the pile top. For large diameter piles of 1.5 m and more, it may be acceptable to fix the sensors at 2 m from top as it is not feasible to go substantially below the pile top in most cases. A pair of transducers is fixed onto opposite sides of the pile to detect bending in the pile if any during testing. These transducers are then connected through the main cable to a pile driving analyzer, which is a State of Art Data Acquisition System with ability to record strain and acceleration measurements and convert them from analog to digital form. The signals are triggered by the impact of a ram falling from a pre-determined height. The ram weight and fall height is determined in advance. As a thumb rule, the ram weight shall be 1–2% of the testing capacity of the pile and can even be 3–5% of the test load. It should also be a minimum 7–10% of the self-weight of the pile.

A typical output for a 1000 mm diameter pile tested with a 3 tonne hammer is attached in Fig. G9. The force and velocity curves are displayed and the project information/data input is in the lower left corner. The output is displayed in lower right corner along with sensor details.

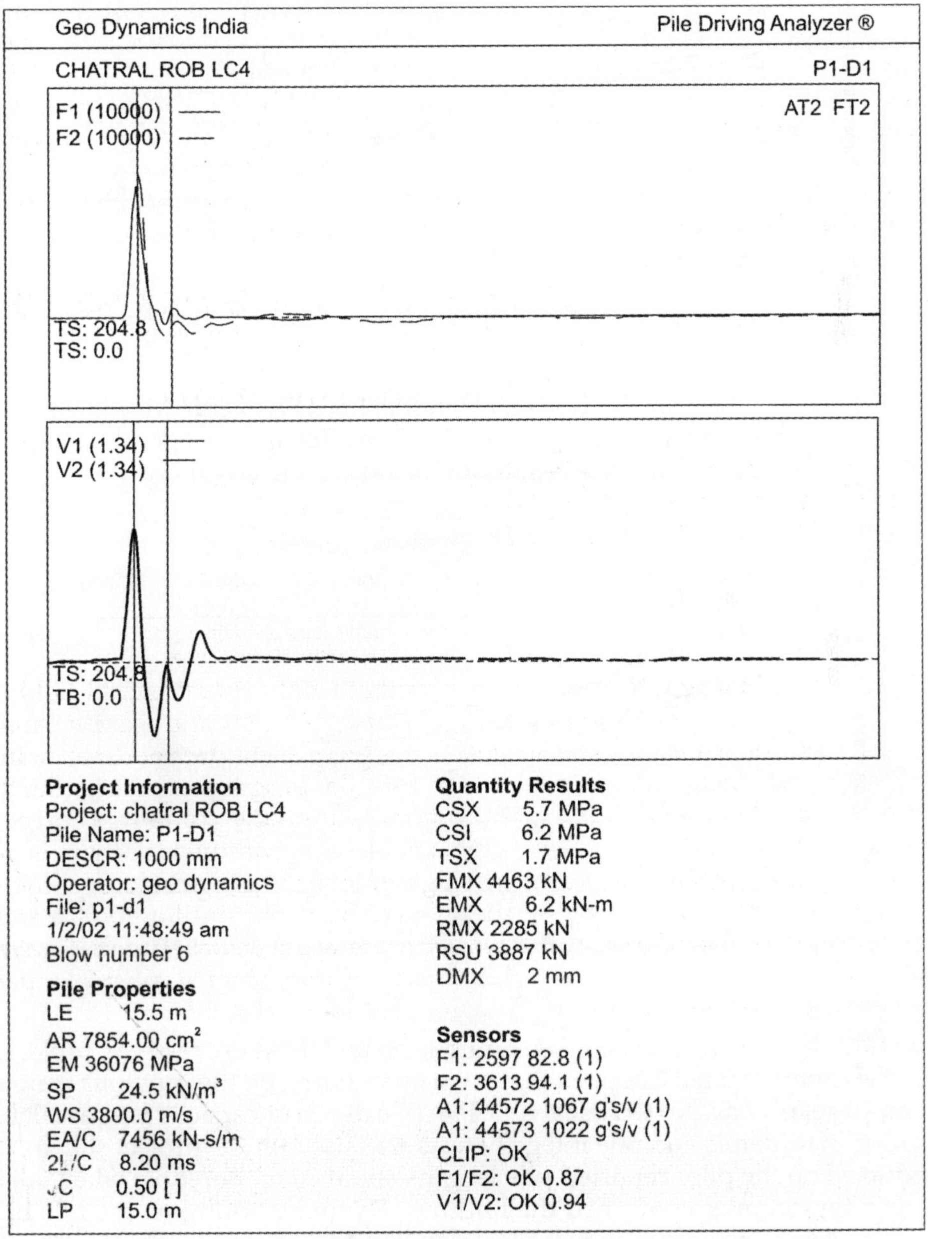

Fig. G9: PDA field output

Case Studies

For case histories described below, hammer weight equal to 1 to 1.5% of test capacity was used and the drop height varied from 1 to 3 m. Plywood sheets ranging from 25 to 50 mm and steel plates with thickness varying from 16 mm to 50 mm were placed on the pile top to avoid any damage to the pile during testing. The testing procedure as per ASTM D4945 was followed.

Figure G10 shows comparison between static and dynamic testing for JJ Flyover for a 1200 mm initial pile. The test load was 1375 tonnes and a 14 tonnes hammer was used. This is the heaviest hammer used for dynamic testing till date in India.

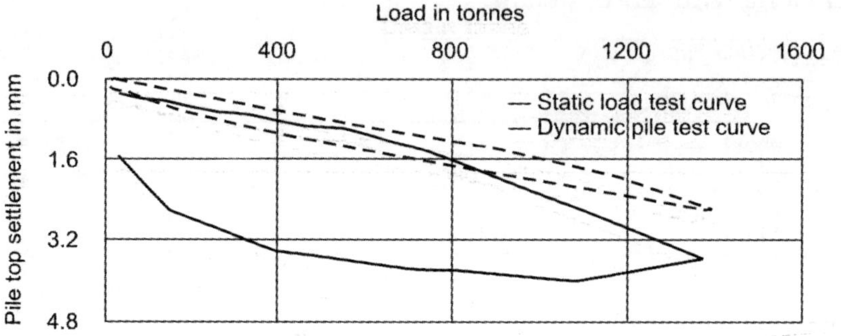

Fig. G10: Co-relation study between static and dynamic pile testing on initial pile at JJ Hospital Urban Viaduct, Mumbai

Figure G11 shows the same for Borivali National Park Flyover in Mumbai for a 1 m-diameter pile with a test load of 340 tonnes. Dynamic testing was performed prior to static testing in this case so that the results are not known in advance.

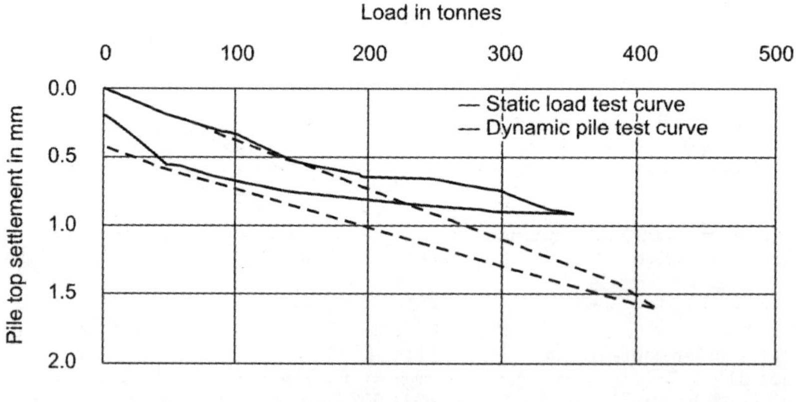

Fig. G11: Co-relation study between static and dynamic pile testing at Borivali National Park Flyover, Mumbai

Figure G12 shows case study for 500 mm pile on an NHAI expressway project near Baroda. The clients located 128 piles that were abandoned by the previous contractor more than 10 years back. No technical data like pile depth or capacity was available. A combination of dynamic and pile integrity tests was used on the project site. Initially, PIT conducted on the piles reported pile lengths about 20 m. Borehole taken adjacent to two such typical piles confirmed the findings. Dynamic testing on the pile helped establish its design load. Static test was also conducted by the client on the pile

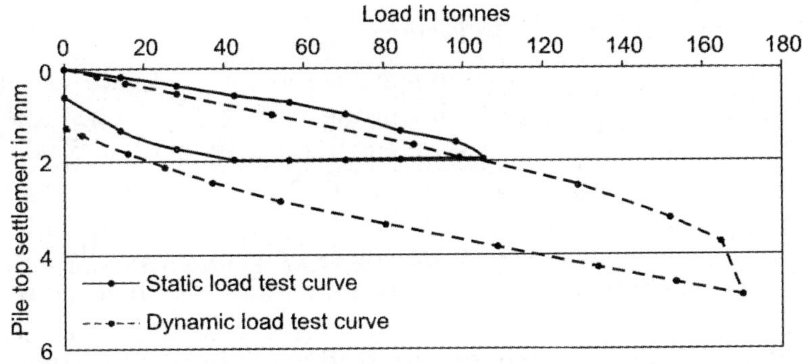

Fig. G12: Co-relation between static and dynamic pile testing on 500 mm diameter. Pile at Vadodara-Ahmedabad Expressway

dynamically tested for further confirmation. The static and dynamic test results matched well helping the client save lot of time and money, since most of the piles could be used in further construction at no additional cost.

Figure G13 shows one such co-relation study for a 1500 mm pile in Delhi and Fig. G14 shows such a study for a flyover project in Bangalore for a test load of 340 tonnes. A 5-tonne hammer equal to 1.5% of the test load was used in this case. As evident all the results match well.

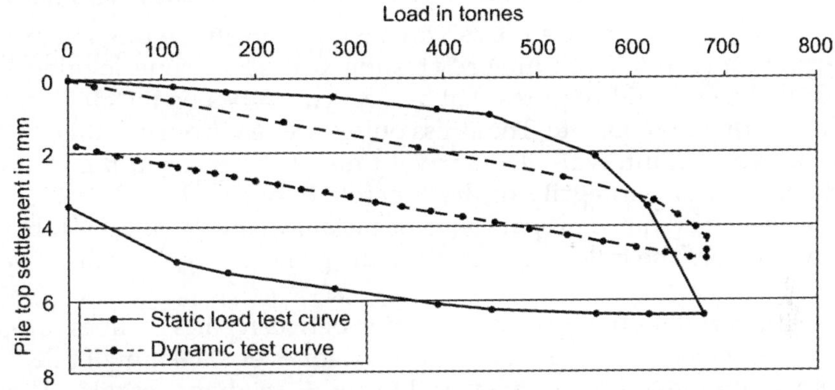

Fig. G13: Co-relation between static and dynamic pile testing on 1500 mm diameter pile for DMRC at New Delhi

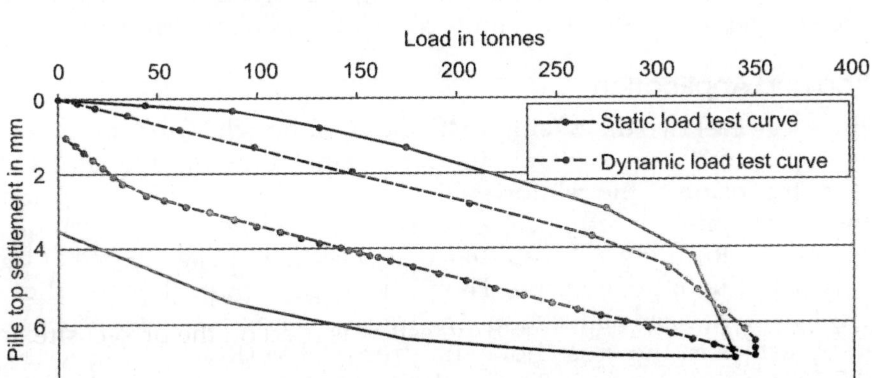

Fig. G14: Co-relation between static and dynamic pile testing at Bangalore

Benefits of High Strain Dynamic Pile Testing

1. The test is a part of various codes and specifications worldwide, as reliability has been proven in the last three decades including at least 1500 such studies in various parts of India. The test has been used for a load capacity of up to 5000 tonnes.
2. Requires minimum time and set-up arrangements compared to conventional static load testing.
3. Minimum hindrance to site during testing is helpful especially for flyovers or bridges in cities or where traffic density is high and it is difficult to conduct static testing.
4. Very useful for off-shore piles where static testing can be difficult, time-consuming and expensive.

Limitations of Dynamic Pile Testing

1. Highly experienced and trained personnel are required to conduct and interpret the test and hence co-relation study also helps to ascertain the reliability of test agency.
2. The reliability and the integrity of the testing company should be thoroughly ascertained by evaluating its past experience, technical papers published and presentations.

CROSS HOLE SONIC LOGGING TEST FOR PILES AND DIAPHRAGM WALLS TESTING

This is also one of the widely used tests worldwide to evaluate pile integrity and is easy to interpret compared to pile integrity testing. Cross hole sonic logging generally applies to drilled shafts and requires that at least two tubes be installed in a drilled shaft prior to pouring the concrete. The stress pulses are sent from one tube to the other at 20 to 50 mm vertical intervals. The arrival time of the pulse at the receiver tube indicates the quality and integrity of the concrete between the tubes. However, in general, such special advance preparations of installing the inspection tubes are not economically feasible. The following describes the principle, method and application of the test.

The CSL test determines the quality of concrete between pairs of tubes pre-installed in r. c. bored piles. Generally, one tube is required for every 300 mm of pile diameter. This implies a minimum of 4 tubes for every 1 m and 1.2 m diameter piles. The method is more recommended for pile diameters 800 mm and above as pipe installation is easier. The method also requires advance training and experience as with other tests, although defect evaluation is easier compared to low strain integrity tests.

Test Method and Application

For the CSL test at least two tubes (either MS or PVC although MS are preferred due to debonding issues) of 38 mm to 50 mm internal diameter are installed in advance by tying them to the inside of the reinforcement cage. For longer piles couplers may be used and welding is not preferred. The test is conducted normally seven days after concrete is done by lowering a transmitter with a certain voltage in one of the pipes and a receiver in the opposite pipe. The probes are then pulled simultaneously manually or by a motorized connection. The transmitter and receiver probes are also connected by cables to the cross hole analyzer (CHA) (Refer to Fig. G15) which records the first arrival time (FAT) and the energy attenuation as the probes are simultaneously raised to the top. As long as the FAT and the energy attenuation are roughly constant, one may deduce that the concrete quality is also uniform and the

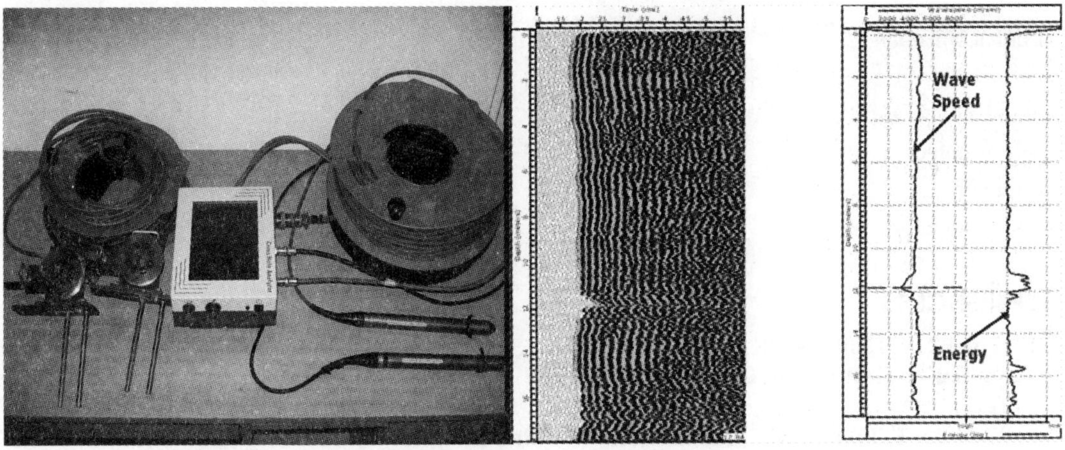

Fig. G15: Equipment for cross hole sonic test and typical output

pile is therefore acceptable. On the other hand, if at some level there is a noticeable increase in the FAT and/or in the energy attenuation, it means that the concrete at this level is inferior or defective. The test is then continued by changing the location of the probes to scan the cross-diagonals and the adjoining perimeters. A tomography analysis is also possible in case a minimum six scans of equal depth are obtained. The tomography analysis helps in analyzing both the extent and location of defect.

Single hole sonic logging is mostly used for smaller diameter piles where using CSL testing is difficult as multiple tubes cannot be installed inside such piles. Typically SSL is used for pile diameters up to 600 mm by installing a 38–50 mm PVC pipe inside the steel cage before concrete is cast. Although limited information is available due to only one tube, the method may determine major defects inside the pile and if defect is closer to the tube. Because of the character of the CSL method, it can detect defects, which may escape detection by other integrity testing methods. It is especially suitable for testing large-diameter piles and slurry-wall elements. If a defect is found, the steel tubes may be pierced at the corresponding depth and the pile repaired by grouting. The piles can be tested after the concrete has gained some strength, usually at an age of five days or more from casting.

For equidistant tubes, uniform concrete yields consistent first arrival times (FAT) with reasonable pulse wave speed and signal strengths. Non-uniformities such as contamination, soft concrete, and honeycombing, voids, or inclusions exhibit delayed arrival times (FAT) with reduced signal strength. Typically, the following classification explains interpretation of defects in the shaft. Here, the rating of the shaft integrity considers the increases in "first arrival time" (FAT) and the energy reduction relative to the arrival time or energy in a nearby zone of good concrete.

The wave speed obtained is a useful tool to evaluate concrete quality. However, because the tubes might not be perfectly straight or even parallel, a fixed absolute limit of a wave speed value cannot be used for evaluation for perimeter profiles. Wave speed is best determined from the test results from the major diagonals. The wave speed is also affected by age of concrete, localized bending of tubes, etc. and hence many times the energy is considered a more important parameter in evaluating the results. Typical output is presented in Fig. G15 along with the equipment.

Pile classification	*FAT increase relative to good concrete*	*Energy reduction relative to good concrete*
Satisfactory/good	0–10%	<6 db
Minor defect	11–20%	<9 db
Poor/flaw	21 to 30%	<9 to 12
Poor/defect	>31%	>12

REFERENCES

1. Nayak NV, Kanhere DK, Vaidya Ravikiran, "Static and High Strain Dynamic Test Co-Relation Studies on Cast-in-Situ Concrete Bored Piles" Proceedings of DFI 2000, September 2000, New York, USA.
2. Rausche, F, Likins, GE, and Hussain, M, "Pile Integrity by Low and High Strain Impacts", Proceedings of the Third International Conference on Stress Wave Theory on Piles, Ottawa, Canada, May 1988.
3. Vaidya Ravikiran and Shah DL, "Pile Diagnostics By Low Strain Integrity Testing", IGC 2001, "The New Millenium Conference", Indore, India, December 2001.
4. Berger JA and Cotton DM, "Low Strain Integrity Testing of Deep Foundations", Proceedings of Deep Foundation Institute Annual Meeting, Seattle, October 1990.
5. Rausche F, and Goble GG, "Determination of Pile Damage by Top Measurements." Behaviour of Deep Foundations, ASTM Symposium, Boston, 1978.
6. PIT and PDA Users Manual, May 1998, Pile Dynamics, Inc.
7. Prebharan, N, Brohms, B, Yu, R, and Li, S, 1990. Dynamic Testing of Bored Piles. Proceedings of the Tenth South East Asian Geotechnical Conference, Taipei, Taiwan.
8. Rausche, F, Hussein, M, Likins, G, and Thendean, G, 1994. Static pile load movement from dynamic measurements. ASCE, Geotechnical Publication No. 40, Proceedings of Settlement 94 Conference, College Station, Texas, Vol. 1.
9. Seidel, J and Rausche, F, 1984, Co-relation of static and dynamic pile tests on large diameter drilled shafts. Proceedings of the Second International Conference on Application of Stress Wave Theory to Piles, Stockholm, Sweden.

Bibliography

1. Alam Singh, "Soil Engineering in Theory and Practice". Vol. 18, 2. Asia Publishing House.
2. Bhandari, RK; Prakash, C, and Sharma, A. (1988): 'Failures in Cast-in-Place-Piles', Commorative Intemational Conference Maxican Soil Mech. Society, 25th Anniversary of its Foundation, Maxico August.
3. Das, BM, "Principles of Foundation Engg, Cengage Learning,
4. David, D. and Komornik, A. (1980). "Stable embedment depth of piles in swelling clays". 4th int. conf. on expansive soils, ASCE, vol. II p. 798.
5. Handbook of CBRI on under-reamed pile foundation. G.S. Jain & Associates, Roorkee.
6. IRC: 78–2014, "Standard Specification and Code of Practice for Road Bridges, Section VIII,
7. IS: 1888–1982, "Method of load test on soils". Indian Standards Institution, New Delhi-110002.
8. IS: 2911 (Part 4): 2013, "Design & Construction of Pile Foundations Code of Practice
9. IS: 2911 (Part-II)-1985. "Load test on piles". Indian Standards Institution, Manak Bhavan, New Delhi - 1
10. IS: 2911 (Part-III)-1980, Under-reamed piles" (first Revision), ISI, N. Delhi-110002.
11. IS: 2911 (Pt 1/sec. 3)-1979, "Code of practice for design and construction of pile foundation" Pt. I, concrete piles—Driven precast concrete piles.
12. IS: 2911 (pt. 1/sec. 1)—1979, "Code of practice for design and construction of pile foundations". Part I, concrete pile—Bored cast in situ piles.
13. IS: 456–1978, "Indian Standard code of practice for plain and reinforced concrete (3rd revision), ISI, N. Delhi-110002"
14. IS: 6426–1972, "Specifications for pile driving Hammer. ISI, N. Delhi-110002."
15. IS: 9716–1981, "Guide for lateral dynamic load test on piles". ISI, N. Delhi-110002.
16. IS: 2911 (Pt. 1/Sec. 2)-1979, "Code of practice for design and construction of pile foundations". Part I, concrete pile—Bored cast in situ piles.
17. Johnson, L.D. (1979), "Overview for design of foundations on expansive soils". Misc. paper GL-79–21, United States Army Engineer waterway experiment station, Vicksburg.
18. Leonards, G.A. "foundation eng". McGraw Hill Book Co. Inc. London.
19. Liu. Chang and Evett. J.B., "Soils & Foundations". Prentice Hall Inc. New Jersey.
20. Minikin, R.R., "Piling for Foundations". Crosby Lock wood & son, Ltd. London.
21. Mittal, Satyendra & Shukla, JP (2014), "Soil Testing for Engineers" Khanna Publishers, 2B Nath Market, Nai Sarak Delhi-110006.
22. Mohan, D. and Jain, G.S. (1955), "under-reamed pile foundations in black cotton soils". 25th Annual research committee meeting of CBIP (India).
23. Mohan, D. Jain, G.S. and Gupta, S.P. (1975), "Settlement characteristics of multi-underreamed piles", Proc. 5th ARC, Soil Mech. & found. Engg. Vol. 1.
24. Murthy, V.N.S. "Soil Mechanics and foundation engineering". Dhanpat Rai & Sons, Nai Sarak, Delhi-110006.
25. Poulous, H.G. and Davis, E.H. (1980), "pile foundation analysis and Design". Chapter 5 and 12, John Wiley & Sons.
26. Prakash, S., Ranjan G., Saran, S., "Analysis and design of foundations and retaining structures". Sarita Prakashan, Meerut.
27. Prakash, S., Ranjan, G. "Problems in soil engineering".
28. Proceedings, Symposium on Deep foundations, A.S.C.E.

29. Ramaiah, B.K. & Chickanagappa, L.S., "Soil Mechanics & foundation engineering". N.C. Jain & Bros. Roorkee.
30. Singh B. & Prakash, S., "Soil Mechanics & foundation engineering". N.C. Jain & Bros. Roorkee.
31. Skechy, K. & Varga, L. "Foundation engineering, soil exploration and spread foundation". Akademiali Kiado, Budapest.
32. Whitaker, T. "Design of pile foundations".

Index